A Guide to
Medicinal
Plants

An Illustrated, Scientific and
Medicinal Approach

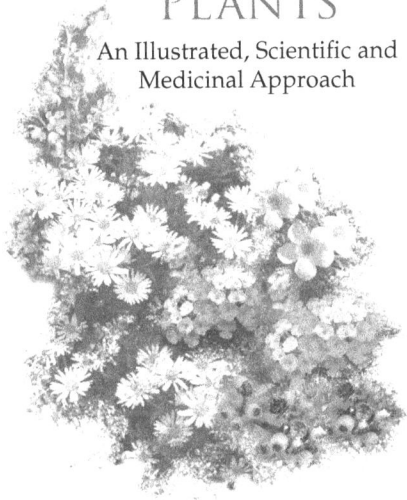

A GUIDE TO
MEDICINAL PLANTS

An Illustrated, Scientific and Medicinal Approach

Koh Hwee Ling
National University of Singapore, Singapore

Chua Tung Kian
Ministry of Education, Singapore

Tan Chay Hoon
National University of Singapore, Singapore

Other contributors:
Johannes Murti Jaya
Siah Kah Ying
Chin Kar Ling
Toh Ding Fung
Ching Jianhong
Li Lin

Photographers:
Chua Tung Kian
Koh Hwee Ling
Siah Kah Ying
Ching Jianhong
Johannes Murti Jaya

World Scientific
NEW JERSEY · LONDON · SINGAPORE · BEIJING · SHANGHAI · HONG KONG · TAIPEI · CHENNAI

Published by

World Scientific Publishing Co. Pte. Ltd.

5 Toh Tuck Link, Singapore 596224

USA office: 27 Warren Street, Suite 401-402, Hackensack, NJ 07601

UK office: 57 Shelton Street, Covent Garden, London WC2H 9HE

British Library Cataloguing-in-Publication Data
A catalogue record for this book is available from the British Library.

First published 2009 (Hardcover)
Reprinted 2016 (in paperback edition)
ISBN 978-981-3203-53-2

A GUIDE TO MEDICINAL PLANTS
An Illustrated, Scientific and Medicinal Approach

ISBN-13 978-981-283-709-7
ISBN-10 981-283-709-4

Typeset by Stallion Press
Email: enquiries@stallionpress.com

Disclaimer

The contents of the book serve to provide both general and scientific information about medicinal plants and their uses and are not intended as a guide to self-medication by consumers or to treatment by health care professionals. The general public is advised to discuss the information contained herein with a physician, pharmacist, nurse or other authorised health care professionals. Neither the authors nor the publisher can be held responsible for the accuracy of the information itself or the consequences from the use or misuse of the information in this book.

The resources are not vetted and it is the reader's responsibility to ensure the accuracy of the information cited. Readers are reminded that the information presented is subject to change as research is on-going and there may be interindividual variations. While every effort is made to minimise errors, there may be inadvertant omissions or human errors in compiling these monographs.

About the Authors

Koh Hwee Ling is an Associate Professor at the Department of Pharmacy, Faculty of Science of the National University of Singapore. She obtained her BSc (Pharmacy) and MSc (Pharmacy) degrees at the National University of Singapore and a PhD degree at the University of Cambridge (UK), with support from the Economic Development Board of Singapore — Glaxo Human Resource Development Scholarship and the Herchel Smith Endowment Fund. Her research areas include the quality control and safety of Traditional Chinese Medicine and herbal medicine, drug discovery from medicinal plants, and the design and development of novel therapeutics. She is a registered pharmacist and a technical/expert assessor with the Singapore Accreditation Council-Singapore Laboratory Accreditation Scheme (SAC-SINGLAS). She reviews for various international journals and grant awarding bodies, and publishes in various international peer-reviewed journals including *Drug Discovery Today, Journal of Chromatography A, Drug Safety, Journal of Pharmaceutical and Biomedical Analysis, Journal of Agricultural and Food Chemistry* and *Food Additives & Contaminants*.

Chua Tung Kian is an Education Officer at River Valley High School, Singapore. He obtained his BSc (Applied Biology-Chemistry) degree at the University of Leeds, UK, with support from the Public Service Commission, and a MSc (Chemistry) degree at the National University of Singapore, with support from the Ministry of Education, Singapore. He also holds a Post-Graduate Diploma in Education with distinction from the National Institute of Education, Singapore, and was awarded a gold medal and a book prize for his achievements. He is currently teaching Chemistry at the H2 level and is an author of six textbooks and assessment books published both overseas and locally.

Tan Chay Hoon is an Associate Professor at the Department of Pharmacology, Yong Loo Lin School of Medicine of the National University of Singapore and consultant psychiatrist at the National University Hospital. She obtained her MBBS in 1980, Master of Medicine (Psychiatry) in 1986 and PhD (Pharmacology) in 1997. She represented Singapore in the "Technical Review of Antidepressant Medications" Task Force to review the international use of antidepressants. She is appointed by the Ministry of Health as member of the Chinese Proprietary Medicine Advisory Committee and National Medication Network Committee. Her main research area is in the field of clinical and experimental studies of psychoactive drugs. She has edited and contributed chapters to psychiatric textbooks both in Singapore and in Asia.

This book is dedicated to our families for their support and to our students who continue to inspire us!

Foreword

At first glance, *A Guide to Medicinal Plants: An Illustrated, Scientific and Medicinal Approach* appears to be a medical compendium of plants intended as a guide and reference resource for professionals in the field. To my delight and I am sure of anyone who picks up this book, I discovered it contains nuggets of information that would interest a great many readers, from schoolchildren to teachers, from undergraduates to researchers, from homemakers to business people and of course, the healthcare professionals.

It is an authoritative and well-researched work on seventy-five, mostly familiar plants that have medicinal value. These grow well here and in the tropics. Although there have been books on medicinal plants published locally, none can match this comprehensive work which, as the authors state, is the first of its kind.

It contains information in well laid-out sections that would interest the healthcare professionals and at the same time, provide the general public valuable insights into the traditional use of plants as medicines. Today such plants are being studied intensively to elucidate their bioactive composition in the hope of discovering novel therapeutics and potential cures for major diseases such as cancer and AIDS.

This guidebook is user friendly. It provides the reader the ability to identify the seventy-five plants through their scientific, vernacular and common names as well as through the descriptions and high quality photographs. Readers are also free to go straight to the section that interests them most. A strength of this book is the detailed references provided for each plant and these are provided towards the end of the book for the serious reader or researcher.

One word of caution that the authors themselves have provided in the text but is worth repeating. Do not mistake this scientific book for a do-it-yourself medication guide. As the authors aptly state, "the information collated is not meant to be a guide for self medication by consumers or for treatment by healthcare professionals."

I commend the authors for this labour of love and have no doubt that this book will be much sought after by both the healthcare professionals and the lay public.

Professor Leo Tan Wee-Hin
President
Singapore National Academy of Science

Preface

"All things are poison and nothing is without poison, only the dose permits something not to be poisonous" i.e., "the dose makes the poison".

Paracelsus

This book presents up-to-date information on a total of 75 medicinal plants. It is a single, comprehensive yet easy to read book on various important information on medicinal plants for both the general public and health professionals (clinicians, pharmacists, nurses and Complementary and Alternative Medicine practitioners).

This is the first publication of its kind on medicinal plants growing in Singapore. Information collated includes plant description, origin of the plants, traditional medicinal uses, phytoconstituents, pharmacological activities, adverse reactions and reported drug-herb interactions. In this era of evidence-based medicine, scientists are increasingly looking towards the traditional uses of medicinal plants for clues to the discovery of potential lead compounds and novel therapeutics.

With the growing interest in drug discovery, this book is useful and timely as many of the plants found growing in Singapore are still understudied. Besides native medicinal plants, some of the plants featured in this book also include those that originated from other parts of the world. It will appeal to both local and overseas readers. Colourful photographs of each plant are also included for ease of reference and aesthetic appeal. There is no minimum level of knowledge required to read this book yet it is useful for academics, scientists and professionals as it provides a comprehensive reference list at the end of the book. This book will also appeal to working professionals, clinicians, pharmacists, nurses, educators and researchers. It serves as a quick reference to the medicinal uses and properties of medicinal plants. Educators and students in complementary medicine and health, pharmacognosy, medicinal chemistry, natural products, pharmacology, toxicology, pharmacovigilance, medicine, pharmacy, nursing, botany, biology, chemistry and life sciences will find the information useful. Greater understanding of such plants will enhance their appreciation of nature and their various fields of study.

The authors hope that this book will inspire and stimulate further research and greater interest in nature, biodiversity, bioconservation, drug discovery and our natural resources, the medicinal plants.

Guide to Using This Book

This book is intended for both the general public as well as health professionals. Hence, it is quite a challenge to present the vast amount of information collated from published literature over the years, in a simple way that will benefit and interest different groups of readers. In general, the information presented is fully referenced. Wherever possible, the original terms in the references are used. Difficult botanical and medical terms are explained in simple terms in the glossaries provided.

Monographs of the plants are arranged in alphabetical order of the scientific name. Each monograph consists of information which includes the scientific name with the family name in parenthesis, common name(s), coloured photographs, description, origin, phytoconstituents, traditional medicinal uses, pharmacological activities, dosage (if available), adverse reactions, toxicity and reported drug-herb interactions. Authors' notes are added in some cases as well. Cross listings of the scientific names and common names, and vice versa, are provided in the appendix for easy reference.

Traditional medicinal uses refer to those uses that have been reported and may not have been studied scientifically. Pharmacological activities refer to the biological activities that have been reported in scientific publications involving mainly *in vitro* or *in vivo* tests using animals, and very rarely, in clinical trials. Activities due to extracts as well as pure components are reported. Due to space constraints, only the keywords are given. Interested readers can refer to the original publications for more details. The full list of references is provided at the end of the book according to the plant names in alphabetical order.

"No information as yet" means that no such information is available, the authors have not found the relevant information, or the information found is not cited due to lack of clarity or completeness. A word of caution: although some dosages have been reported and cited, these have largely not been verified in clinical trials. For phytoconstituents, it is not possible to list down all reported constituents. Hence only selected constituents, especially those peculiar to the plant are shown, typically with the most important ones listed first. Interested readers can refer to the references cited for further details.

Although medicinal plants are generally safe when used appropriately according to the traditional methods, some are inherently toxic. In addition, inappropriate use (wrong plant parts, dose, frequency, route of administration, preparation, etc.) or abuse may lead to undesirable consequences. The information collated is not meant to be a guide for self medication by consumers or for treatment by health care professionals. There is a fine line between a poison and a useful drug afterall.

Acknowledgement

The authors would like to express our heartfelt gratitude to:

Professor Wee Yeow Chin for his professional advice and constructive criticisms

Professor Ben-Erik van Wyk, Professor Zhao Zhong Zhen and Professor Hugh Tan for their professional advice

Staff and students who have contributed to this book

The National University of Singapore for financial support of an Academic Research Fund (R-148-000-079-112)

Staff of Nparks (Singapore) and Singapore Botanic Gardens, especially Professor Benito Tan, Mr Lua Hock Keong, Ms Serena Lee, Ms Patricia Yap and the park rangers

Staff of Singapore Science Centre

Many like-minded friends and colleagues for their interesting and inspiring discussions on medicinal plants

Staff of World Scientific Publishing for their professionalism and support

Contents

Plant Monographs

1. *Abrus precatorius* L. (Leguminosae)

Rosary Pea, Indian Licorice, Precatory Bean

Seeds of *Abrus precatorius*

Description: *Abrus precatorius* L. is a perennial climber with a slender stem. Leaves are pinnate and 5–8 cm long. Leaflets are rhomboid, numbering 20–24 or more, opposite and are 1.2–1.8 cm long. Leaf margin is entire. It bears pink flowers arranged in dense axillary racemes. Pods are oblong, cylindrical, inflated, 5–6 cm by 1 cm and contains 3–6 round, glossy, black and red seeds.[1–3]

Origin: Native to Pakistan, India, Ceylon and tropical Africa; and introduced widely in the New and Old World.[4]

Phytoconstituents: Abrectorin, abricin, abridin, abrins A–D, (+)-abrine, abruslactone A, abrusgenic acid, abrusogenin, abrusoside A–D, precatorine, abruquinones, abraline, abrusic acid, abruquinone G and others.[2,5–18]

Traditional Medicinal Uses: A decoction of the leaves has been prescribed for scurvy, cough, bronchitis, sprue and hepatitis and as a refrigerant. They are also

2

applied on painful swellings, eye inflammation, cancer, syphilis and on leucodermic spots.[19] The leaves are also effective in the treatment of coryza, cough, fever, and jaundice resulting from viral hepatitis and intoxications.[6] The seeds have been used to treat fever, malaria, headache, dropsy and to expel worms.[3] A decoction of the seeds is applied for abdominal complaints, conjunctivitis, trachoma and malarial fever.[2] Central Africans use powdered seed as an oral contraceptive.[5] It is also used to lower high blood pressure and relieve severe headache.[5] The seeds are very toxic and can be applied externally to treat bacterial infection and accelerate the bursting of boils and to cure mastitis and galactophoritis.[6] The seed has purgative properties and is used as an emetic, tonic, aphrodisiac, and for nervous disorders. The poultice can be used as suppository, abortifacient, or tonic for pregnant women and children and to treat severe headaches.[19] Water from the boiled roots is used to cure cough, bronchitis, sore throat and also applied as an emetic agent.[2,5]

Pharmacological Activities: Antibacterial,[20] Anthelmintic,[21] Antiviral,[17] Anti-inflammatory,[10,22,23] Anticancer/Antitumour,[11,24–30] Antiplatelet,[23] Antiprotozoal,[31] Immunomodulatory,[27,32–35] Antioxidant,[23] Antiplasmodial,[17] Antitubercular[17] and Molluscicidal.[36]

Dosage: In Central Africa, 200 mg powdered seeds are used as an oral contraceptive which can last 13 menstrual cycles.[5] Approximately 14 g of powdered seeds are used as tonic for pregnant women and children.[19] About 5–7 of seed grains are prescribed for pertussis.[37]

Adverse Reactions: Ingestion of *Abrus* seeds resulted in pulmonary oedema and hypertension.[38] Abrine can cause coma, confusion, convulsions, dehydration, gastroenterosis and hypotension.[37]

Toxicity: The LD_{50} of abrin in mice is 0.02 mg/kg body weight.[2] Ingestion of seeds causes severe stomach cramping accompanied by nausea, severe diarrhoea, cold sweat, fast pulse, organ failure, coma and circulatory collapse.[2,5,39] Oral administration of a 50% ethanol extract of *A. precatorius* seeds (250 mg/kg) in albino rats for 30 and 60 days induced infertility in males which was reversible.[40] Dose-dependent degenerative changes in the testicular weights, sperm count, spermatogenesis and Leydig cells were observed in testes of rats treated with steroidal fraction of the seeds.[41] The methanol extract of the seeds caused a concentration-related impairment of percentage human sperm motility with an EC_{50} of 2.29 mg/ml.[42]

Contraindications: No information as yet.

Drug-Herb Interactions: No information as yet.

2. *Adiantum capillus-veneris* L. (Polypodiaceae)

Black Maidenhair Fern, Southern Maidenhair Fern, Venus Hair Fern

Adiantum capillus-veneris fern

Description: *Adiantum capillus-veneris* L. is a perennial fern with short creeping stems. Leaf blades are lanceolate, pinnate, 10–45 cm by 4–15 cm and glabrous. Ultimate segments are various but generally cuneate or fan-shaped to irregularly rhombic, about as long as well as broad with its base broadly to narrowly cuneate. Plant is delicate, brittle and has dark stalks.[1–6]

Origin: Native to America, Mexico, West Indies and South America. It can also be found in temperate regions of Eurasia and Africa.[6]

Phytoconstituents: Adiantoxide, adiantone, isoadiantone, isoadiantol, hydroxyadiantone, capesterol and others.[1,4,7–11]

Traditional Medicinal Uses: The fern is considered an astringent, demulcent, depurative, diaphoretic, diuretic, emmenagogue, emollient, expectorant, laxative, stimulant, sudorific and a tonic.[1] It is also used for alopecia, asthma, bladder ailments, catarrh, chest ailments, chills, cold, dropsy, dysmenorrhoeal, fever, head ailments, hepatosis, labour, lung ailments, respiratory problems, rheumatism, sclerosis, snake bite, sores, sore throat, splenosis, stones and other urinary calcification.[1,4] The plant expels worm, induces vomitting and relieves fever. Used externally, it is poulticed on snakebites and as a treatment for impetigo.[1,3] The Russians use the herb for rhinitis while the French, with orange flowers and honey, uses it for pulmonary catarrh.[1] A handful of leaves is made into a tea and drunk as an expectorant, astringent, tonic for coughs, throat afflictions and bronchitis.[1] The plant is also used as a hair wash for dandruff and to promote hair growth in Latin America.[1,3] The ashes mixed with vinegar and olive oil is rubbed into the scalp to cure alopecia.[5] In Traditional Chinese Medicine, the leaves are used for bronchial diseases and as an expectorant;[3] and in Africa, the leaves are smoked to prevent head and chest colds.[1]

Pharmacological Activities: Antibacterial.[12]

Dosage: Taken internally as a tea prepared from powdered dried fronds. The standard single dose is 1.5 g of drug to 1 cup of fluid per dose.[13]

Adverse Reactions: No known side effects with therapeutic dosages.[13]

Toxicity: No information as yet.

Contraindications: Should not be used during pregnancy.[13]

Drug-Herb Interactions: No information as yet.

3. *Allamanda cathartica* L. (Apocynaceae)

Allamanda, Common Allamanda, Golden Trumpet

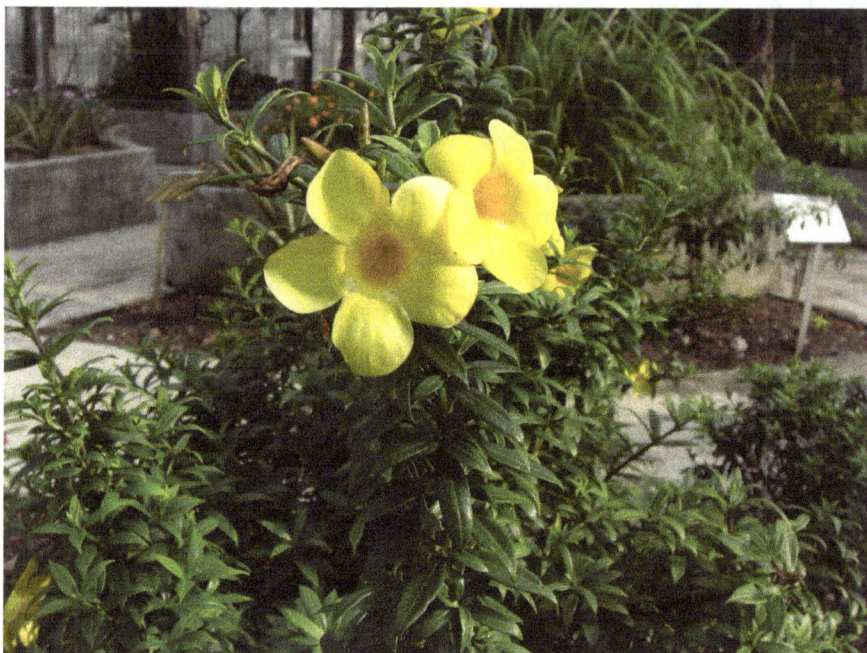

Allamanda cathartica flowers and leaves

Description: *Allamanda cathartica* L. is a woody shrub that can grow up to 4 m tall. The stems exude milky white sap when incised. Leaves are simple, exstipulate, glossy, leathery and glabrous. Leaf blade is oblong-lanceolate, 8–15 cm by 4–5 cm and arranged in opposites of 3–5 sessiles. Flowers are large, tubular, bright yellow and 4–5 cm long.[1–3]

Origin: Native to South America; cultivated in China for medicine.[4]

Phytoconstituents: Allamandin, plumericin, plumieride, ursolic acid and others.[3,5–10]

Traditional Medicinal Uses: The plant has been used as a purgative to induce vomiting at low dosage.[3] Its leaves are cathartic and the bark is used as a hydragogue for ascites.[11] In Surinam's traditional medicine, its roots are used against jaundice, for complications with malaria and enlarged spleen.[11]

Pharmacological Activities: Anthelmintic,[12] Antifungal,[7,13–15] Antineoplastic,[16] Antivenom[17] and Wound healing.[18]

Dosage: No information as yet.

Adverse Reactions: No information as yet.

Toxicity: Every part of the plant was reported to be poisonous.[19] The sap of the plant was reported to cause mild and occasional oral irritation and slight nausea when sucking cut stems. Rash or dermatitis were also reported when sap was in contact with sensitive skin.[19]

Contraindications: No information as yet.

Drug-Herb Interactions: No information as yet.

4. *Aloe vera* Mill. (Aloaceae)

Aloe, Lidah Buaya

Aloe vera   Young and old plants

Description: *Aloe vera* Mill. is a short-stemmed, up to 50 cm, succulent herb with thick green leaves that have a sharp and pointed apex, 15–50 cm by 4–7 cm, arranged in a rosette around the short stem. Blade is green to variegate with small white or glaucous dots, irregular bands, lanceolate, tapering from base to apex, glabrous with green and spiny-toothed margins. The leaves contain a thick colourless juice. Flowers are yellow, orange or red, crowded into a rosette and in panicles.[1–3]

Origin: Native to North Africa; cultivated in China for medicine and widely used as indoor ornamental plants.[3,4]

Phytoconstituents: Aloin (barbaloin), arabinose, aloe-emodin, aloetinic acid, emodin, aloeresin A–C, aloesone, aloeride and others.[2,5–12]

Traditional Medicinal Uses: The plant has been used in cosmetic preparations for the treatment of pimples, acne and mouth ulcers.[2] It has also been used to control bleeding, itching of piles, and relief from arthritic pains.[2] The Chinese uses the plant juice as a mild laxative, wash for piles, abscesses and scabies. In the Philippines, it is used to treat dysentery and pain in the kidneys.[2] The plant has been found to treat bacterial infection, as a cathartic, emmenagogue, purgative and vermifuge. It can be used in the treatment of burns, oedema, pain, swellings and wounds; treatment of leukemia, lung cancer; treatment of constipation, eczema, piles and pertussis.[6,13–15] The whole

8

plant has also been used for the treatment of rectal fissures and piles while the root is used to treat colic.[16] The juice from the leaves is used to increase menstrual flow.[13] The jelly is used as aperient, for wounds, applied on the abdomen in fever, after confinement, on swelling and is especially useful in correcting constipation due to intake of iron medication.[17,18] Fresh juice of the leaves is cathartic and cooling and used for various eye diseases.[16] The dried juice is applied with lime juice for reducing swellings and promoting granulation in ulcers. In Malaysia, it is used for treating wounds, fever, swellings and put on the abdomen of women after confinement. The mixture of sugar with sap obtained from heated leaves is taken for asthma. The mucilaginous flesh and the sap are used for burns. The watery extract is used as a hair tonic. It is also used in cosmetics for decreasing wrinkles and other skin problems. It is mixed with milk and given for dysentery and pains in the kidney.[15] It is used in Ayurveda to alleviate pain and is also mentioned in folk medicine of Arabian Peninsula for the management of diabetes.[19]

Pharmacological Activities: Angiogenic,[20] Antifungal,[21–23] Antidiabetic,[24,25] Anti-inflammatory,[12,26–30] Anticancer,[31–38] Antimicrobial,[39] Antioxidant,[11,40–45] Antiproliferative,[12,38] Chemopreventive,[35,46] Gastric mucosal protection,[47,48] Hepatoprotective,[19,49,50] Neuroprotective,[51] Hypolipidaemic,[52] Immunomodulatory,[11,33,53–57] Immunostimulatory,[58] Antimutagenic,[31] Alloantigenic,[59] Antileishmanial,[60,61] Prevention of kidney stones,[62,63] Radioprotective[64,65] and Wound healing.[66–69]

Dosage: Single dose of powdered Aloe, 50–200 mg at bedtime; tincture BPC 1949 (1:40, 45% ethanol), 2–8 ml. Aloes should only be taken for short periods, maximum 8–10 days.[70] Doses of 10–30 mg act as a bitter stomachic; 60–200 mg as a laxative and 300–1000 mg as a purgative.[18,71] A dose of 1 teaspoon after meals, or otherwise advised by manufacturers and practitioners has been reported.[72] To prevent kidney stones, a dose of 2 to 3 tablespoon daily is reported. As a laxative, the recommended dose is 500 to 1000 mg daily. For burns or wound healing, fresh gel from plant may be applied topically and liberally. For haemorrhoids, as a stool softener, 0.05 to 2 g of dry aloe extract is administered. In the treatment of HIV, 800 to 1600 mg of acemannan daily (equivalent to 0.5 to 1 L of *Aloe vera* juice) is administered. To relieve constipation, 20 to 30 mg hydroxyanthracene derivatives daily, calculated as anhydrous aloin is prescribed.[73]

Adverse Reactions: Barbaloin was shown to have a laxative effect.[70] Ingestion of *A. vera* is associated with diarrhoea, electrolyte imbalance, kidney dysfunction, and conventional drug interactions; episodes of contact dermatitis, erythema, and phototoxicity have been reported from topical applications.[74] *A. vera* could also induce acute liver damage.[75]

Toxicity: Severe gastrointestinal cramping can occur if the latex (which is just below the leaf surface) is taken internally.[72] Toxic doses cause severe haemorrhagic diarrhoea and kidney damage, and sometimes death. The lethal dose of the dried plant extract is stated to be 1 g/day taken for a period of several days.[76]

Contraindications: Contraindicated in intestinal obstruction, acute inflammatory intestinal diseases (e.g., Crohn's disease, ulcerative colitis), appendicitis and idiopathic abdominal pain. Should not be used during pregnancy. Should not be given to children below 12 years of age.[77] Should avoid application of *Aloe* topically on deep, vertical wounds. Contraindicated in menstruation and if the person has kidney complaints.[72]

Drug-Herb Interactions: Increase the actions of cardiac glycosides and antiarrhythmic drugs (chronic use of aloe causes potassium loss), thiazide diuretics, loop diuretics, licorice and corticosteroids.[77] Aloe gel, when taken orally, can reduce the absorption of many medications. Thus, it should be taken two hours apart from all medications.[72] A study reported that *Aloe vera* preparations improved the absorption of both vitamins C and E.[78]

5. *Andrographis paniculata*
(Barm.f.) Nees (Acanthaceae)

Hempedu Bumi, Sambiloto, Chuan Xin Lian

Andrographis paniculata plants

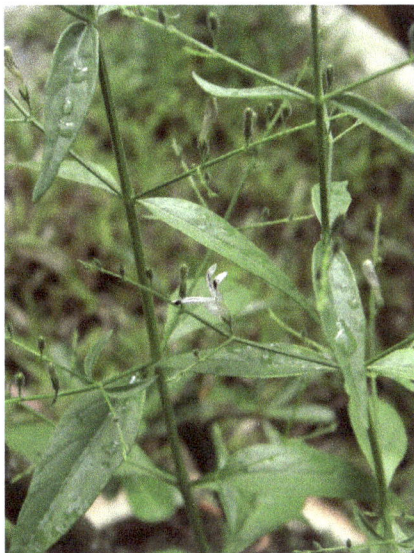

Andrographis paniculata flowers

Description: *Andrographis paniculata* (Barm.f.) Nees is an annual herb that grows up to 1 m in height. Stems are glabrous and articulated. Leaves are simple, opposite and exstipulate. Blade is dark green, bitter, glossy, simple, lanceolate, opposite and 4–8 cm by 1.3–2.5 cm. Its small and white flowers grow in terminal or axillary panicles. Both the bracts and the 5-lobed calyx are small. Fruits are upright, fusiform, capsular and contains numerous seeds.[1]

Origin: Native to Indian subcontinent and cultivated elsewhere.[2]

Phytoconstituents: Andrographolide, andropanolide, andrographic acid and andrographidine A, andrographatoside, andropaniculosin A and andropaniculoside A and others.[3–14]

Traditional Medicinal Uses: The plant is used orally to prevent and treat common cold, influenza, pharyngotonsilitis, allergies and sinusitis. Traditionally, it is used for many conditions including anorexia, atherosclerosis, insect and snake bites, bronchitis, prevention of cardiovascular disease, diabetes, hypertension, cholera and as a tonic.[1,15,16]

Pharmacological Activities: Antiapoptotic,[17] Antibacterial,[18–20] Antifungal,[18] Anticancer,[21–30] Antidiabetic/Hypoglycaemic,[31–36] Antifertility,[37] Anti-inflammatory,[38–48] Antioxidant,[45,49–51] Antiplatelet,[52,53] Antiprotozoal,[54–56] Antiviral,[57,58] Cardioprotective,[59] Chemopreventive,[60] Hepatoprotective,[61–63] Hypotensive,[64] Immunomodulatory,[35,42,65,66] Psychopharmacological activities,[67] Vasorelaxant[68] and Cytotoxic.[12]

Dosage: For decreasing symptoms of common cold, doses of 400 mg of standardised andrographolide are required three times daily; for preventing common cold, a dose of 200 mg daily for 5 days in a week; for relieving fever and sore throat in pharyngotonsilitis, doses of 3 g and 6 g daily were used.[16] Use 6–9 g for influenza with fever, sore throat, ulcers in the mouth, acute or chronic cough, colitis, dysentery, urinary tract infection, carbuncles, sores and venomous snake bite.[69]

Adverse Reactions: Orally, large doses of Andrographis may cause gastrointestinal distress, anorexia, emesis and urticaria. Androgapholide taken orally at 5 mg/kg three times a day may cause headache, fatigue, rash, abnormal taste, diarrhoea, itching, lymphadenopathy, anaphylactic reactions, etc.[16]

Contraindications: Contraindicated in pregnancy, likely to be unsafe due to abortifacient effect.[16]

Toxicity: No toxic effect was observed after administration of a decoction of *Andrographis paniculata* leaves to rabbits.[70] LD_{50} of andrographolide in mice through oral route is > 40 g/kg body weight, which indicates low toxicity.[71]

Drug-Herb Interactions: Simultaneous application of *A. paniculata* and warfarin did not produce significant effects on the pharmacokinetics of warfarin, and practically no effect on its pharmacodynamics.[72]

6. *Ardisia elliptica* Thunb. (Myrsinaceae)

Mata Pelanduk/Ayam, Sea-Shore Ardisia, Shoebutton Ardisia

Flowers of *Ardisia elliptica*

Ardisia elliptica trees

Ardisia elliptica fruits

Description: *Ardisia elliptica* Thunb. is a small shrub that can grow up to 10 m tall. The leaves are obovate, 6–9 cm long with smooth margins. They have an acute apex and a cuneate leaf base. The leaves have a leathery texture. The plant bears whitish pink axillary inflorescences and the drupes are globular, 1–1.2 cm in diameter and grow in clumps, pale red when immature, and turning dark purplish upon maturity.[1–3]

Origin: Native to tropical and temperate Asia.[4]

Phytoconstituents: Rapanone,[5] bauerenol, α- and β-amyrin,[6,7] syringic acid, isorhamnetin and quercetin,[8] bergenin,[9] 5-(Z-Heptadec-4'-enyl)resorcinol and 5-pentadecylresorcinol.[10]

Traditional Medicinal Uses: The decoction of the leaves is used by the Malays to treat pain in the region of the heart.[10] The *Kadazan Dusun* tribes in Malaysia used paste made from the leaves of *Ardisia elliptica* to treat herpes and measles.[1,11] The fruits are used in Thai Traditional Medicine to cure diarrhoea with fever.[8]

Pharmacological Activities: Antiplatelet[10,11] and Antibacterial.[8]

Dosage: No information as yet.

Adverse Reactions: No information as yet.

Toxicity: No information as yet.

Contraindications: No information as yet.

Drug-Herb Interactions: No information as yet.

7. *Areca catechu* L. (Palmae)

Betel Nut Palm, Areca Nut, Pinang

Fruit of *Areca catechu*

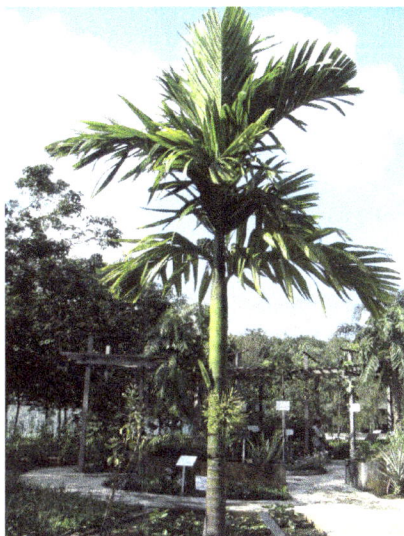

Areca catechu tree

Description: *Areca catechu* L. is a tall, slender palm that can grow up to 10 m. Leaves are dark green, pinnate and up to 1.2–2 m long. Inflorescence is branched and male flowers grow in one row surrounding the female flower at the base of branch. Fruit is a one seeded ovoid berry about 5 cm long.[1-4]

Origin: Originate from the Philippines.[5]

Phytoconstituents: Arecoline, arecaidin, arecaine, catechin, glucides, guvacine, guvacoline, arecolidine, isoguvacine, nicotine and others.[1,6–8]

Traditional Medicinal Uses: In Irian Jaya, parts of this tree are used on wounds, swellings and other skin afflictions. The pericarp is effective in the treatment of flatulence, oedema, dysuria and hyperemesis of pregnancy. On the Finschhafen coast, Papua New Guinea, the inner seed is chopped, heated over fire and pressed on sores caused by sea urchins. Chewing the betel nut with lime and the leaves of catkins or *Piper betel* gives a stimulant effect as well as an attributed sedative effect.[9] This may also be used to soothe a mad person. The red mixture is applied to ulcers in New Britain and to treat sores caused by venereal disease in Northern Province; whereas in New Ireland, the scraped bark is mixed with sea water with a leaf of *Indocarpus fagiferus* and drunk to treat asthma.[9] The Malays use a decoction of the leaves to treat diarrhoea in children.[1] The dried ripe fruits have been used by the Chinese to expel tapeworms and roundworms, treat diarrhoea, indigestion, lumbago, urinary problems and increase menstrual flow.[1,10] The kernel of the fruit is chewed as a narcotic, fresh or cured with slacked lime and betel leaves.[2] The fruits are also used for beriberi, dysentery, dyspepsia, dysuria, oedema and malaria.[6] The fruit can also be applied on venereal sores (ground fresh nut with betel leaf, *Nigella sativa* and roots of *Gymnema hirsutus*, cooked in mustard oil or butter and applied).[11]

Pharmacological Activities: Analgesic,[12] Anthelmintic,[6,13] Antibacterial,[14] Anticancer/Antineoplastic,[15,16] Anticonvulsant,[17] Antidepressant,[18,19] Antihypertensive,[20,21] Antimitotic,[22] Antioxidant,[23] Apoptotic,[24] Hypocholesterolaemic,[25] Immunomodulatory,[26,27] Immunostimulatory,[28–31] Antihyaluronidase,[32] Antivenom,[33] Cell growth inhibitor,[34] Molluscicidal[35] and causes periodontitis.[36]

Dosage: For the treatment of diarrhoea, 30 g areca powder in 200 cm^3 water, simmered for 1 hour is taken before breakfast. If expulsion does not take place within 9 hours, 50 cm^3 of 50% magnesium sulphate solution may be taken.[11] A decoction of the pericarp has been prescribed in a daily dose of 6 to 12 g to treat flatulence, oedema, dysuria and hyperamesis during pregnancy. To treat diarrhoea and dysentery, a daily dose of 0.5 to 4 g of the kernel has been used.[7] For sore throat, the pressed juice is used as a gargle. 2 g of fresh nut can be chewed for 15 min or more before spitting it out. Another reported usage is rolling the leaves and placing them between teeth and gums/lips.[37]

Adverse Reactions: Excessive chewing can cause dizziness, nausea, vomiting, diarrhoea, and seizures.[38]

Toxicity: A dose of 8–10 g is toxic to humans.[6,39] Heavy consumption may cause the development of cancer in the upper and middle third of the oesophagus respectively[40] and chronic kidney disease.[41] Betel quid (a mixture of areca nut and flavouring ingredients with or without processed tobacco leaves) chewing resulted in a statistically significant increase in the risk of total and cerebrovascular deaths in the elderly population.[42] At higher doses, bradycardia, reflex excitability, tremor, spasms and eventual paralysis may occur. Long term effect as stimulant causes malignant tumours of oral cavity through formation of nitrosamines.[39] Betel quid chewing during pregnancy has a substantial effect on a number of birth outcomes, including sex ratio at birth, lower birth weight and reduced birth length.[43] It is toxic during pregnancy[44,45] as it also possesses cytotoxic and genotoxic activities.[46,47]

Contraindications: Should not be used during pregnancy and lactation. Should not be given to children. Patients with oral or oesophageal cancers, ulcers, oesophagitis, or renal disease should avoid its use.[37]

Drug-Herb Interactions: Decrease action of antiglaucoma agents. Increase action of beta-blockers, calcium channel blockers, cardiac glycosides (digoxin, digitoxin). For neuroleptics, extrapyramidal symptoms can occur.[37] Avoid taking alcohol and atropine.[38]

8. *Asplenium nidus* L. (Aspleniaceae)

Bird's Nest Fern

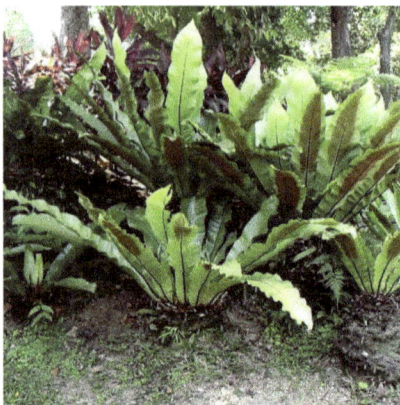

Asplenium nidus growing on the ground

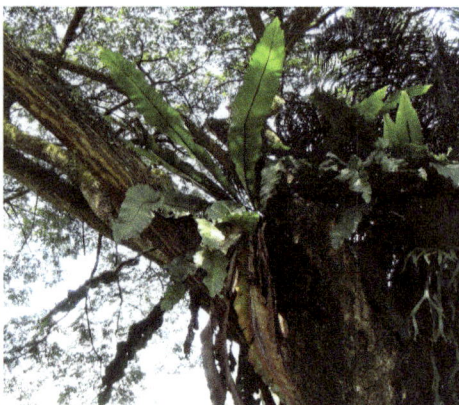

Asplenium nidus on a tree

Description: *Asplenium nidus* L. is a common epiphytic fern found growing on trees. Fronds are long, simple, green and grow from a central rhizome attached to the tree branch in the shape of a nest. Parallel lines of spores are found on the undersides of the fronds and radiate away from the midrib towards the leaf margin.[1]

Origin: Native to tropical Africa, temperate and tropical Asia and Australasia.[2]

Phytoconstituents: Kaempferol-3-*O*-gentiobiosie-7,4′-bisglucoside, kaempferol-3-*O*-diglucoside, kaempferol-3,7-diglycoside and kaempferol-3-*O*-vicianoside.[3]

Traditional Medicinal Uses: *A. nidus* is regarded as depurative. Infusion of the fronds is used to ease labour pains by Malaysia native tribes. The Malays pound the leaves in water and apply the resulting lotion to feverish head.[4] Two young fronds are eaten when they are still coiled, just after menstruation, in the morning as a contraceptive. Tea made from the fronds is recommended for general weakness.[5,6]

Pharmacological Activities: Oxytocic activity.[6]

Dosage: No information as yet.

Adverse Reactions: No information as yet.

Toxicity: No information as yet.

Contraindications: No information as yet.

Drug-Herb Interactions: No information as yet.

9. *Aster tataricus* L.f. (Compositae)

Tatarian Aster, Tatarian Daisy

Aster tataricus flowers

Aster tataricus shrub

Description: *Aster tataricus* L.f. is a small shrub with abundant fibrous roots. Leaf blades are oblanceolate to lanceolate, margins serrate or entire, 4–18 by 1–5 cm and acute. Flower heads are in large bunches with white petals and yellow centre.[1-3]

Origin: Native to Siberia.[3]

Phytoconstituents: Shinone, friedelin, epifriedelinol, shinoside A–C, asterinin A–F, astins A–E, astertarone A&B and others.[4-11]

Traditional Medicinal Uses: The underground rootstock is used as a purgative, treats colds, coughs with excessive sputum or with blood and painful menstruation.[2] It is also used as a bechic-expectorant.[12] Used as an aromatic tonic in chronic gastroenteritis.[13]

Pharmacological Activities: Anticancer[14,15] and Antioxidant.[4]

Dosage: No information as yet.

Adverse Reactions: No information as yet.

Toxicity: No information as yet.

Contraindications: No information as yet.

Drug-Herb Interactions: No information as yet.

10. *Azadirachta indica* A. Juss. (Meliaceae)

Neem

Fruits of *Azadirachta indica*

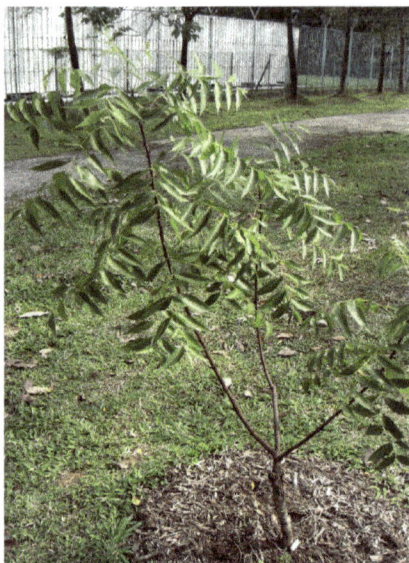

Azadirachta indica tree

Description: *Azadirachta indica* A. Juss. is a tall evergreen tree, growing up to 30 m in height. Leaves are pinnate with opposite or alternate, lanceolate, serrated and glabrous 8–16 leaflets, 20–32 cm long. Flowers are yellowish white. Fruits are small, ellipsoid, about 5 cm long and green.[1-3]

Origin: A native of India and China, cultivated and naturalised throughout India, Malay Peninsula, Indonesia and Pakistan.[4]

Phytoconstituents: Azadirachtin O–Q, nimbin, deacylnimbin, salanin, nimbidin, nimbinin, nimbidol, azadirone, melianol, meliacinol, nimbothalin, nimonol, azharone and others.[3,5-28]

Traditional Medicinal Uses: It is used for the treatment of a variety of human and veterinary ailments including head lice, mange, fleas, fever, convulsions, leprosy, scrofula, rheumatism, asthma, worm infestations, treat bacterial infection, insecticide, local application for indolent ulcer and consumed as tonic after childbirth.[8] It is used for boils, heart disease, fever,

22

tuberculosis, diarrhoea, jaundice, dysentery, to promote healing, measles, smallpox, sores, inflamed gums, syphilis, leprosy, piles, urinary diseases, to expel worms, purgative, emollient, local stimulant, treat fever (crushed leaves added to lemon), disinfectant (oil from nuts), astringent, contraceptive and tonic.[29,30] It is also used for dermatological problems in Nigeria.[31] Neem has also been used to protect crops, stored grains and library books from insects. Neem leaves buried in grain bins are used to keep the stored crops insect-free. Crushed seeds soaked in water produce a potent pesticide that does not harm mammals, birds, earthworms and bees.[3]

Pharmacological Activities: General review.[32] Antibacterial,[33–37] Anticancer,[38–46] Anticarcinogenic,[47–55] Anticonvulsant,[56] Antifertility,[57,58] Antifungal,[35,59–61] Anti-inflammatory,[62] Antimalarial,[63–68] Antimicrobial,[69] Antioxidant,[70–74] Antiproliferative,[75,76] Antipyretic,[77] Antiviral,[78,79] Gastroprotective,[80–82] Hepatoprotective,[83–85] Hypoglycaemic,[86,87] Hypotensive,[88–90] Immunostimulatory,[38,91–93] Neuroprotective,[94] Anthelmintic,[95] Antihaemorrhagic,[37] Antileishmanial,[96] Antimutagenic,[97–99] Molluscicidal,[100] Insecticidal and Insect repellent.[101–110]

Dosage: Approximately 100 g of bark is soaked in 1 L of water and approximately 3 ml of this infusion is consumed daily for one month as a male contraceptive.[30]

Adverse Reactions: Nausea, vomiting, anorexia, hypersensitivity, Reye's syndrome (infants).[111]

Toxicity: Mice injected with the tetranortriterpenoid fractions (>86 mg/feeding) into the tail vein died within 24 hours.[8] Various acute, subacute and chronic toxicity tests of extracts of neem have been reported, especially on human and animal fertility.[112] Non-aqueous extracts appeared to be the most toxic neem-based pesticide products.[112]

Contraindications: Should not be used during pregnancy and lactation, and in children. Should not be used in persons with hypersensitivity to neem.[111]

Drug-Herb Interactions: No information as yet.

[**Authors' Note:** An extensive review[112] on the safety of neem derived pesticides concluded that the use of neem derived pesticides as an insecticide should not be discouraged.]

11. *Barringtonia asiatica* L. (Lecythidaceae)

Beach Barringtonia, Fish-killer tree, Putat Laut

Barringtonia asiatica fruit

Developing fruits of *B. asiatica*

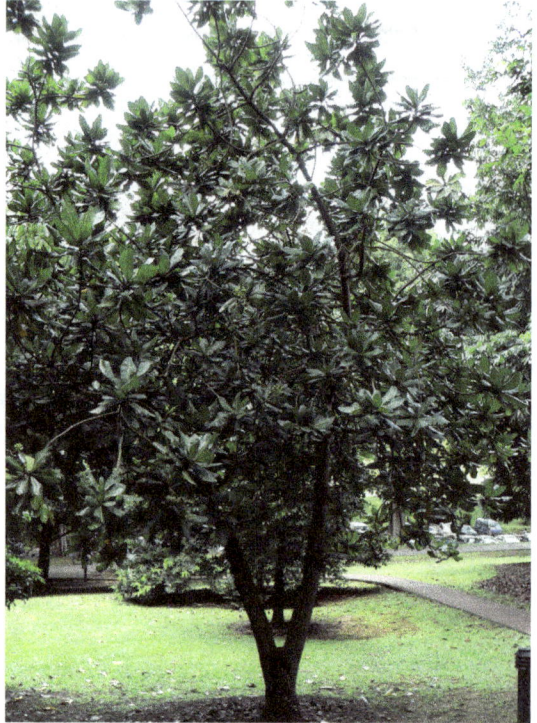

Barringtonia asiatica tree

Description: *Barringtonia asiatica* L. is a large tree bearing large simple leaves, 20–30 cm long which taper to the leaf base. Flowers are large and white with several white stamens. The flowers are actinomorphic and have four petals. The fruit is oblong, green, large, 8–10 cm across and contains one seed.[1–3]

Origin: Native to Africa, temperate and tropical Asia and Australasia.[4]

Phytoconstituents: A_1-barrinin, ranuncoside VIII, A_1-barrigenin and others.[4,5–7]

Traditional Medicinal Uses: The plant is used to treat fungal infections,[8] burns and wounds.[9] The leaves are heated and used to treat stomachache and rheumatism in the Philippines. Its fruits are used as a fish poison and the fruit juice for controlling scabies while the seeds are used for the expulsion of intestinal worms and also as a fish poison.[10] They are also used to treat sores, cough, influenza, sore throat, diarrhoea, swollen spleen after malaria.[11] In other provinces of Vietnam, the fresh nut is scraped and applied to sores; dried nut is ground into a powder, mixed with water and drunk to cure coughs, influenza, sore throat, bronchitis, diarrhoea and swollen spleen.[12] The bark is used in the treatment of tuberculosis.[11] In Yambio (Sudan), the inner bark is crushed and mixed with water and drunk to ease the aching associated with malaria. It is also used in combination with other plants as a medicine to treat tuberculosis in New Ireland and the Solomon Islands.[12]

Pharmacological Activities: Insect repellent,[5] Antibacterial and Antifungal.[13]

Dosage: To relieve the aching associated with malaria, inner bark is crushed and mixed with water and drunk, 2 cups per day for 2 days.[12]

Adverse Reactions: No information as yet.

Toxicity: No information as yet.

Contraindications: No information as yet.

Drug-Herb Interactions: No information as yet.

12. *Barringtonia racemosa* (L.) K. Spreng
(Lecythidaceae)

Putat Kampong, Samundrapandu

Barringtonia racemosa flowers

Barringtonia racemosa fruits

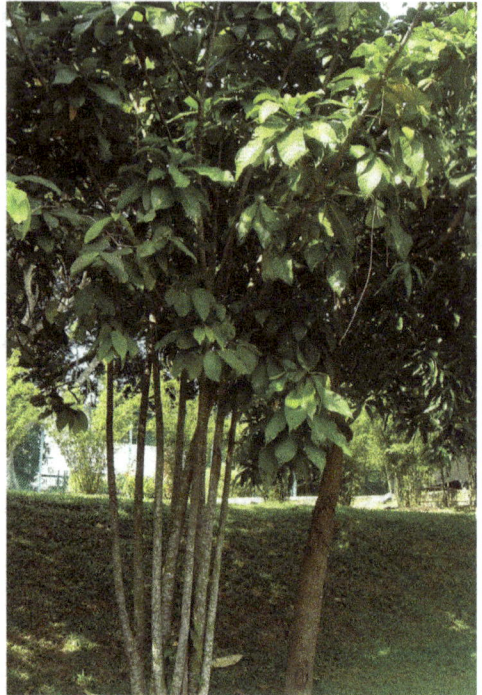
Barringtonia racemosa tree

Description: *Barringtonia racemosa* (L.) K. Spreng is a small tree with large, simple leaves about 20 cm long. Flowers are large with numerous pinkish stamens. The fruit is ellipsoid, green, 8 cm long, and turns red upon maturity.[1,2]

Origin: Native to Africa, temperate and tropical Asia, Australasia and the Pacific.[3]

Phytoconstituents: Nasimalun A and B, barringtonin, R_1-barrigenol, R_2-barrigenol, barringtogenol, barringtogenic acid and others.[4–9]

Traditional Medicinal Uses: The plant is widely used in the form of a decoction in Sri Lankan traditional medicine.[10] The leaves and bark are used for rat and snake bites, rat poisoning and on boils.[10,11] The fruits are used for cough, asthma and diarrhoea.[11,12] Kernels of the drupe are mixed with milk to treat bilious diseases and jaundice whereas the seed has been used as an insecticide and tonic.[11] The seeds along with other ingredients are employed in preparations for the treatment of itch, piles and typhoid fever, while the bark is also used for gastric ulcers.[10] Its roots act as a coolant and deobstruent.[11]

Pharmacological Activities: Antinociceptive,[10] Antibacterial,[12] Glucosidase and Amylase Inhibition,[13] Anticancer [14] and Cytotoxic.[15]

Dosage: No information as yet.

Adverse Reactions: No information as yet.

Toxicity: Aqueous extracts (500, 750, 1000 or 1500 mg/kg) of *B. racemosa* bark in male rats did not produce any unwanted side effects or toxicity or alter fertility, gestational length, peri- and neonatal development and appeared to be non-teratogenic.[10] *B. racemosa* seed extract was found to be devoid of acute and short-term toxicity to mice, when administered daily, intraperitoneally for a fortnight up to a dose of 12 mg/kg. The treated mice showed conspicuous toxic symptoms only at 24 mg/kg. The LD_{50} of male mice for a single i.p. dose is 36 mg/kg.[14]

Contraindications: No information as yet.

Drug-Herb Interactions: No information as yet.

13. *Bauhinia purpurea* L. (Leguminosae)

Butterfly Tree

Leaves of *Bauhinia purpurea*

Flower of *Bauhinia purpurea*

Bauhinia purpurea tree

Description: *Bauhinia purpurea* L. is a deciduous tree. Leaves are simple and stipulate. Blade is butterfly shaped, coriaceous with 9–11 pairs of secondary veins. Flowers are showy, pink and arranged in axillary. The fruits are darkish pods, 1.5 by 15 cm and woody.[1,2]

Origin: Native to India and grown in the Asia-Pacific as ornamental plants.[2]

Phytoconstituents: Bauhiniastatins 1–4, bauhinoxepin C–J, bauhibenzofurin A, bauhispirorin A, bauhinol E and others.[3–9]

Traditional Medicinal Uses: The root is grated with water and is drunk to treat common fever. The flowers are used as laxative and leaves applied to sores and boils, and for cough treatment.[10]

Pharmacological Activities: Antibacterial,[9,11] Antifungal, Antimalarial, Cytotoxic,[9] Anticancer,[8] Anti-inflammatory[9,12] Antinociceptive, Antipyretic[12] and Thyroid hormone regulating.[13]

Dosage: No information as yet.

Adverse Reactions: No information as yet.

Toxicity: No information as yet.

Contraindications: No information as yet.

Drug-Herb Interactions: No information as yet.

14. *Bixa orellana* L. (Bixaceae)

Annatto, Lipstick Tree

Fruits of *Bixa orellana*

Bixa orellana flower

Bixa orellana tree

Description: *Bixa orellana* L. is a small tree with simple and spiral leaves, 10–20 cm by 6.3–12.5 cm, dark green, ovate, acuminate, truncate at the base and glabrous. Flowers are large, 5 cm in diameter, pinkish or white, arranged in terminal panicles. Fruits are dehiscent, ovoid capsules containing 15–20 trigonous seeds in bright red pulp.[1–3]

Origin: Originate from tropical America. Cultivated Pantropically.[4]

Phytoconstituents: Bixin, valencene, β-elemene, β-selinene, copaene, δ-cadinene, spathulenol, γ-cadinene, δ-elemene, ledol, α-muurolene, α-cadinol and others.[5–10]

Traditional Medicinal Uses: The leaves have been used to treat snakebites and jaundice and the seed is considered a good cure for gonorrhoea.[11] The bark of the root is used to treat fever and as an aperient.[11] In Cambodia, the leaves are a popular febrifuge while in Indonesia, water in which the leaves are rubbed is poured over the head of children with fever. In Malaysia, the leaves are used in a postpartum medicine and in the Philippines the leaves are pounded in coconut oil and heated, then applied to the abdomen to relieve tympanites. Pastes of the fresh leaves are rubefacient and used in dysentery. In Vietnam, lotions or baths of leaves are used during fever. Its unripe fruits are emollient in leprosy.[3] Alcoholic extracts of seed coat are taenifuge and laxative.[3] Decoctions of barks are used for catarrh. Infusions of seeds are used to treat asthma and excessive nasopharynx mucus production.[3] Traditionally, it is also used as a gargle for sore throats and oral hygiene.[12] In Trinidad and Tobago, the leaves and roots are used for hypertension, diabetes and jaundice.[13] Leaves and seed pods are used as a female aphrodisiac.[14]

Pharmacological Activities: Antibacterial,[12,15–18] Anticancer,[19] Anticonvulsant,[16] Antidiarrhoeal,[16] Anti-inflammatory,[19,20] Antioxidant,[3,16] Antiplatelet,[21] Hypoglycaemic,[3,22] Immunostimulatory,[3] Sedative,[16] Antigenotoxic and Antimutagenic,[23] Antifungal, Antileishmanial[24] and Radioprotective.[3]

Dosage: Approximately 9 seed pods are boiled in 3 cups of water for 10 mins and drunk as a diuretic before each meal. 10 g powdered seed/40 ml oil for topical pastes.[25] As a female aphrodisiac, 3 leaves in 0.5 L of water and red paste of seed pods.[14]

Adverse Reactions: Urticaria and angiooedema are possible adverse reactions with annatto dye. A patient developed these symptoms and hypotension within 20 min of ingestion of annatto containing fibres.[3]

Toxicity: Toxic to dogs dosed with 60 mg/kg trans-bixin.[26] However, annatto containing 5% bixin was non-genotoxic and non-carcinogenic to rat livers even at the highest concentration tested at 1000 ppm (4.23 bixin/kg body weight/day).[27] Annatto given through gavage to Wistar rats on days 6–15 of pregnancy showed no adverse effect on the mothers and foetus.[28]

Contraindications: Trans-bixin is hyperglycaemic and should not be ingested by patients with diabetes mellitus.[25]

Drug-Herb Interactions: No information as yet.

15. *Calophyllum inophyllum* L. (Guttiferae)

Indian Laurel, Penaga Laut, Borneo Mahogany

Flowers of *Calophyllum inophyllum*

Calophyllum inophyllum tree

Description: *Calophyllum inophyllum* L. is a large tree with broad, glossy, leathery, elliptic-oblong leaf blades, 8–16 cm by 4–8 cm, and with numerous parallel side veins. The tree bears sweetly scented white flowers in erect racemes. Fruits are globose, 2 cm across and are green in colour.[1,2]

Origin: It is found in Africa, tropical and temperate Asia, Australasia and the Pacific.[3]

Phytoconstituents: Inophynone, canophyllol, canophyllic acid, calophyllolide, inophyllolide, inophyllum B, C, P, and E, jacareubin, (+)-calanolide A, inocalophyllins A and B, calophinone, calophyllumin C, inophyllin A and others.[4–28]

Traditional Medicinal Uses: The whole plant is used as a crude drug for curing rheumatism and skin affections in South India.[14] Its juice is a purgative and the seed oil is specific for rheumatism and various skin diseases (i.e., scabies, ringworm and dermatosis). The bark is used for internal haemorrhage and as an astringent.[29] In Buso, Papua New Guinea, the milky latex from the leaves is diluted with water and the solution is applied to irritated eyes.[30] The gum is emetic and purgative.[29]

Pharmacological Activities: Antibacterial,[31,32] Anticancer/Antineoplastic,[19] Anti-inflammatory,[33] Antiplatelet,[34,35] Antipsychotic,[33] Antiviral,[15.18.36–38] Photoprotective,[39] Molluscicidal[40] and Piscicidal.[5]

Dosage: No information as yet.

Adverse Reactions: No information as yet.

Toxicity: The unrefined oil is toxic.[41] There was a significant difference in the plasma cholesterol levels of the rats fed with *C. inophyllum* oil when compared with the control. Mild, focal to severe and widespread lesions were found in the kidneys, hearts and livers of rats fed with *C. inophyllum* seed oil.[42]

Contraindications: No information as yet.

Drug-Herb Interactions: No information as yet.

16.   *Cananga odorata* (Lam.) Hook. f. & Th.
(Annonaceae)

Kenanga, Ylang-Ylang

Flowers of *Cananga odorata*

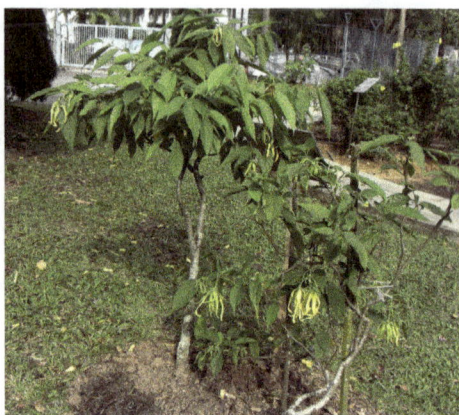

Cananga odorata shrub

Description: *Cananga odorata* (Lam.) Hook. f. & Th. is a shrub which grows to a height of 4 m. Its bark is grey and smooth. Leaves are simple, alternate, exstipulate, oblong to broadly elliptic, large, 3–6.5 cm across, 8.5–29 cm long, with distinct venation pattern. Flowers are fragrant, in clusters on older branches. Fruit turns black on maturity and has many seeds in two rows.[1–3]

Origin: Originates from Indochina, Malesia and tropical Australia.[4]

Phytoconstituents: Acetogenin, aporphine, liriodenine, canangone, α-humulene, β-cubebene, germacrene D, cananodine, γ-eudesmol and others.[3,5–7]

Traditional Medicinal Uses: It is used for asthma, malaria, fever, cholera, typhoid, scabies, dermatitis, ulcer and wounds.[8] The seeds are used for stomach complaints with fever and in Indonesia, the bark is used for scabies.[3] In Malaysia, a paste of fresh flowers is applied to the chest for asthma and to treat malaria. In Solomon islands, a paste of fresh flowers is applied to boils while in India, the essential oil from the flowers makes an external remedy for cephalgia, ophthalmia and gout.[3,9]

Pharmacological Activities: Antibacterial,[10,11] Antifungal,[11,12] Antihypertensive,[3,13] Antioxidant,[10] Antineoplastic[7,12,14,15] and Antiprotozoal.[12,16]

Dosage: No information as yet.

Adverse Reactions: No information as yet.

Toxicity: A 50% ethanolic root bark extract administered orally to male albino rats at the dose of 1 g/kg body weight/day for 60 days resulted in decreased epididymal sperm motility and sperm count, and morphological abnormalities in the sperms.[17] However, it is non-toxic at the current level of intake as a food ingredient (0.0001 mg/kg/day). Although sometimes Ylang-Ylang oil has been reported to cause dermal sensitisation reactions in animals and humans, it is unclear what constituent(s) within the essential oil comprise the offending agent.[18]

Contraindications: No information as yet.

Drug-Herb Interactions: No information as yet.

17. *Capsicum annuum* L. (Solanaceae)

Chilli, Red Pepper

Capsicum annuum fruits

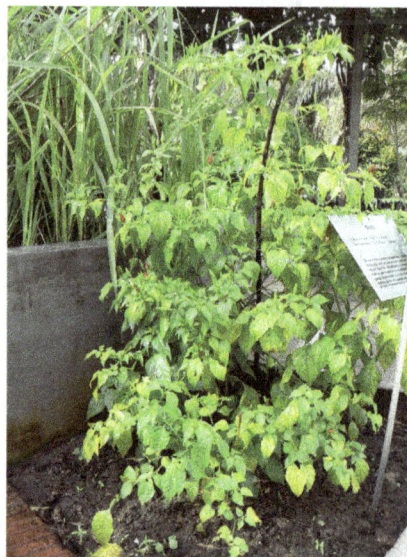

Capsicum annuum plant

Description: *Capsicum annuum* L. is a small herb that can grow up to 1 m tall. Leaves are oblong-ovate, ovate, or ovate-lanceolate, 4–13 cm by 1.5–4 cm with entire margin. Flowers are small, white or tinged purple. Fruits are mostly red, but can be green, orange, yellow and can grow up to 15 cm. Seeds are pale yellow, discoid or reniform and 3–5 mm.[1–5]

Origin: Native to Mexico and South America, widely cultivated throughout the world.[5]

Phytoconstituents: Capsaicin, capsicosides E–G, capsianosides 1–4, capsianosides VIII, IX, X, XIII, XV and XVI, solanidine, solanine, solasdine, scopoletin and others.[6–15]

Traditional Medicinal Uses: The leaves are used to treat toothache. The fruits are used to stimulate gastric activities and increase blood circulation.[4] It is also a stimulant, carminative, and used locally for neuralgia and for rheumatism.[15] Uterine pain associated with childbirth is treated with soup containing the fruit.[16] The Commision E approved *Capsicum annuum* for painful muscle spasms in areas of shoulder, arm and spines. Preparations are used to treat arthritis, neuralgia, lumbago and chilblains.[17]

Pharmacological Activities: Antibacterial,[18,19] Antifungal,[20,21] Anticancer,[22,23] Antioxidant,[10,24–26] Antiprotozoal,[27] Hypocholesterolaemic/ Hypolipidemic,[19,28] Immunomodulatory,[29] Antimutagenic[30,31] and Pesticidal.[32]

Dosage: Liquid extract is prepared by percolating 100 gm of the plant extract with 60 mg of ethanol, to be used as an antirheumatic. External daily dose of semi solid preparations containing maximum of 50 mg of capsaicin in 100 gm neutral base is also used as an antirheumatic and applied to the affected area not more than 3 or 4 times daily.[33]

Adverse Reactions: Internally, it may cause gastrointestinal cramping, pain, and diarrhoea. Topically, it may cause painful irritation of mucous membrane.[34]

Toxicity: Oral LD_{50} values for capsaicin are 161.2 mg/kg (rats) and 118.8 mg/kg (mice), with haemorrhage of the gastric fundus observed in some of the animals that died. However, capsaicin is considered to be safe and effective as an external analgesic counterirritant.[35] Rabbits fed with *C. annuum* powder at 5 mg/kg per day in the diet daily for 12 months showed damaged liver and spleen. A rabbit skin irritation test of *C. annuum* fruit extract at 0.1% to 1.0% produced no irritation but caused neoplastic changes in the liver and intestinal tumours were observed in rats fed red chili powder at 80 mg/kg per day for 30 days.[35] High doses administered over extended period of time can cause chronic gastritis, kidney damage, liver damage and neurotoxic effects.[33]

Contraindications: Should not be used during pregnancy and lactation, in people with hypersensitivity and in children. Should not be used on open wounds or abrasions, or near the eyes.[34]

Drug-Herb Interactions: Reported with concomitant administration with aspirin and salicylic compounds.[33,36] Decrease the actions of α-adrenergic blockers, clonidine (anti-hypertensive) and methyldopa (antihypertensive). Hypertensive crisis with monoamine oxidase inhibitors.[34]

18.   *Cassia fistula* L. (Leguminosae)

Golden Shower Tree, Indian Laburnum, Purging Cassia

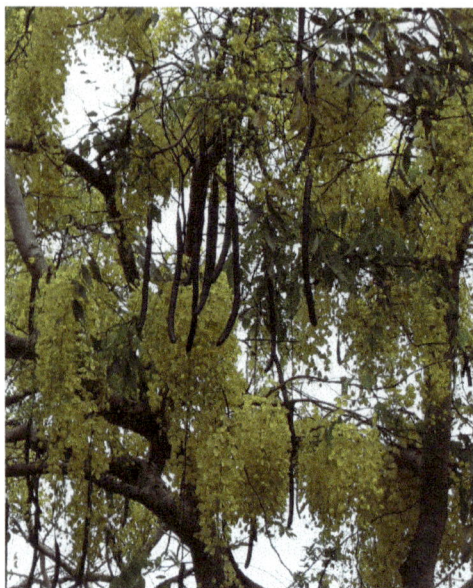

Cassia fistula fruits and flowers

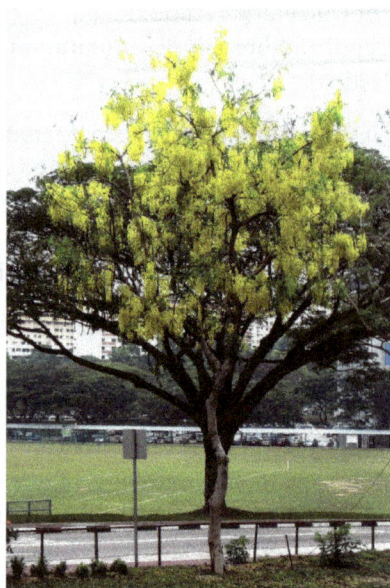

Cassia fistula tree

Description: *Cassia fistula* L. is a large tree, which grows to 10 m tall. Leaves are alternate, pinnate, 3–8 pairs of leaflets, broadly ovate and pointed. Flowers are dense, bright yellow and about 4–5 cm across. Fruit pods are long, 30–60 cm, cylindrical, brown in colour and contains many seed.[1–3]

Origin: Native to India and Sri Lanka.[2,4]

Phytoconstituents: Fistucacidin, chrysophanic acid, chrysophanol, clitorin, sennosides A and B, chrysophanein and others.[2,4–9]

Traditional Medicinal Uses: The whole plant is used for anthrax, burns, cancer, constipation, convulsions, delirium, diarrhoea, dysentery, dysuria, epilepsy, fever, influenza, gravel, haematuria, pimples, syphilis, tumours and worms.[4,10] The leaf is used for skin diseases (juice), healing ulcers, for ringworm and irritation of skin (juice of young leaves), facial paralysis and rheumatism (paste).[10] The raw black pulp found between the seeds is a popular remedy for constipation.[3,4] It is also used as a cathartic, for rheumatism and snakebite (pulp), treats bacterial infections (pulp mixed with leaves of *Cassia angustifolia*), liver complaints, heart disease, reduce fever, as abortifacient, demulcent and is useful in liver, throat, eye diseases, convulsions and sores. The seed is an emetic, carminative, appetiser, and is used for constipation, jaundice, cancer on face and syphilis.[10] The roots act as a purgative while the rootbark is used for cleansing wounds.[4] The root is also used as an astringent, tonic, febrifuge, for skin diseases, leprosy, tuberculous glands, syphilis and epilepsy.[10]

Pharmacological Activities: Antimicrobial,[11–14] Anticholinergic,[15] Antifertility,[16] Anti-inflammatory,[17] Antineoplastic,[18] Antioxidant,[19–22] Depressant,[23] Hepatoprotective,[24–26] Hypocholesterolaemic,[27] Antileishmanial,[28] Larvicidal[29] and Wound healing.[30,31]

Dosage: No information as yet.

Adverse Reactions: No information as yet.

Toxicity: In cases of overdose or prolonged administration, loss of electrolytes, especially potassium ions, aldosteronism, albuminuria, haematuria, inhibition of intestinal motility and muscle weakness may occur. Rarely, cardiac arrhythmia, nephropathy, oedema, and accelerated osteoclasis may arise.[32]

Contraindications: Contraindicated with acute inflammatory diseases of intestine and appendicitis. Should not be used during pregnancy and while nursing. Should not be used in children under 12 years of age.[32]

Drug-Herb Interactions: Interaction with anthranoid laxatives.[33] Enhancement of effects of cardioactive steroids may occur. Effects of antiarrhythmics may also be affected.[32]

19. *Catharanthus roseus* (L.) G. Don
(Apocynaceae)

Madagascar Periwinkle, Rose Periwinkle

Catharanthus roseus flower

Catharanthus roseus shrub

Description: *Catharanthus roseus* (L.) G. Don is a herb, up to 80 cm tall. Stem is woody, slightly branched and all parts contain white milky latex. Leaves are simple, dark green, glossy, obovate-elliptic, 4–5 cm by 2–3 cm with prominent lateral veins on the abaxial surface. Flowers are bisexual, white, purple, pink, red or white with a red or pink centre. Fruits consist of pairs of greenish succulent follicles, 2–3 cm long and contain small oblong seeds.[1,2]

Origin: Native to Madagascar. Cultivated or naturalised in all tropical countries.[2,3]

Phytoconstituents: Vinblastine, vincristine, leurosine, akuammicine, carosine, catharanthine, catharicine, catharine, catharosine, cathovaline, catharanthiole, vindoline, vindolinine, vincaleucoblastine, secologanin, mauritianin, rosicine and others.[4–11]

Traditional Medicinal Uses: The plant is used as a remedy for diarrhoea, malaria, diabetes, astringent, diaphoretic, bechic, emmenagogue, menstrual pain, hypertension, insomnia and depurative after parturition in Indochina, the Philippines, Jamaica, West Indies, South Africa, Southeast Asia, India and Queensland.[2,5–7,12,13] The plant is also used for cold, cough, fever and bronchitis.[14] In Malaysia, the crushed leaves are applied to scalds, burns, sores, mumps, swollen neck, tonsillitis and insect bites.[13] In Puerto Rico and Cuba, the flowers are decocted and used as an eyewash.[12]

Pharmacological Activities: Anticancer/Antineoplastic,[5,12,15–19] Antioxidant,[20] Antiangiogenic,[21] Chemopreventive,[22] Hypoglycaemic[23–28] and Wound healing.[29,30]

Dosage: Leaves are useful in treating oliguria, haematuria, diabetes mellitus, and menstrual disorders in a daily dose of 4 to 8 g as a decoction or liquid extract.[7] A decoction of 30 g of the plant is taken for diabetes, dysentery, enteritis, menstrual pains, hypertension, insomnia and cancer in Malaysia.[13] For the treatment of cold and sore throat, tea is made from 9 pink flowers in 1 pint of water and sipped throughout the day.[31] Dilute infusions of roots are used to treat diabetes.[32]

Adverse Reactions: *Catharanthus roseus* pollen can trigger IgE-mediated respiratory allergy in the people living in close proximity.[33]

Toxicity: No information as yet.

Contraindications: No information as yet.

Drug-Herb Interactions: No information as yet.

[**Authors' Note:** Clinically, vinblastine and vincristine are anticancer drugs administered by intravenous injection or infusion as solutions of their sulphate salts.[34]]

20. *Celosia argentea* L. (Amaranthaceae)

Feather Cockscomb, Red Spinach

Flowers of *Celosia argentea*

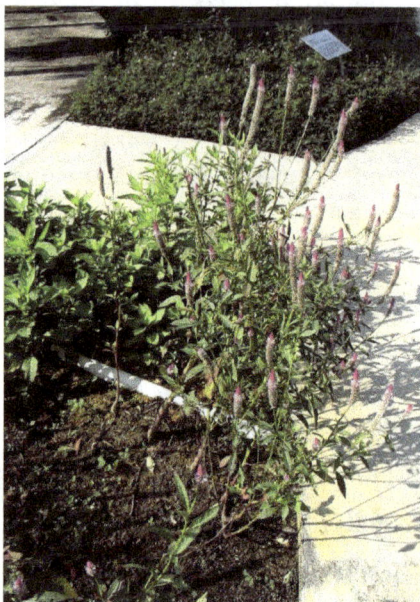

Celosia argentea plant

Description: *Celosia argentea* L. is an annual tropical herb, up to 1 m tall. Stems are cylindrical and the aerial part is branched. Leaves are simple, small, spirally arranged, about 5–8 cm by 1–3 cm, alternate and exstipulate. The blade is lanceolate and ovate. The apex is acuminate. It bears several pinkish or white flowers which are minute. The fruits are globose and seeds are black.[1–5]

Origin: Native to India.[6]

Phytoconstituents: Celosian, nicotinic acid, celogenamide A, celogentin A–D, H, J and K, moroidin and others.[3,5,7–11]

Traditional Medicinal Uses: *C. argentea* is used internally for haematological and gynaecologic disorders and externally to treat inflammation and as a disinfectant. The whole plant is used to treat dysentery and dysuria, and used externally as poultices for broken bones.[5] The plant is used for eye

42

and liver ailments in Yunnan, China[12] and also for the treatment of mouth sores and blood diseases and used as an aphrodisiac.[13] The petioles are used to treat sores, wounds, boils and swellings.[4,12] The seeds are used for the treatment of conjunctivitis and hypertension. In China, the seeds are used for haemorrhage, menorrhagia and opthalmia.[5,12,14] In Indonesia, the flowering tops are used for bleeding lungs whereas in Malaysia, the red flowering tops are prepared as decoctions which are given in cases of white discharges, excessive menstruations, haematuria, dysentery, proteinuria, bleeding piles and bleeding nose.[5]

Pharmacological Activities: Antibacterial,[15] Antimitotic,[16] Antineoplastic,[17] Diuretic,[5] Hypoglycaemic,[18] Hepatoprotective,[11,19] Immunomodulatory,[5,17,20] Cytoprotective[21] and Wound healing.[22]

Dosage: No information as yet.

Adverse Reactions: No information as yet.

Toxicity: No information as yet.

Contraindications: Leaves should not be eaten by menstruating women.[23]

Drug-Herb Interactions: No information as yet.

21. *Centella asiatica* (L.) Urban (Umbelliferae)

Indian Pennywort, Asiatic Pennywort

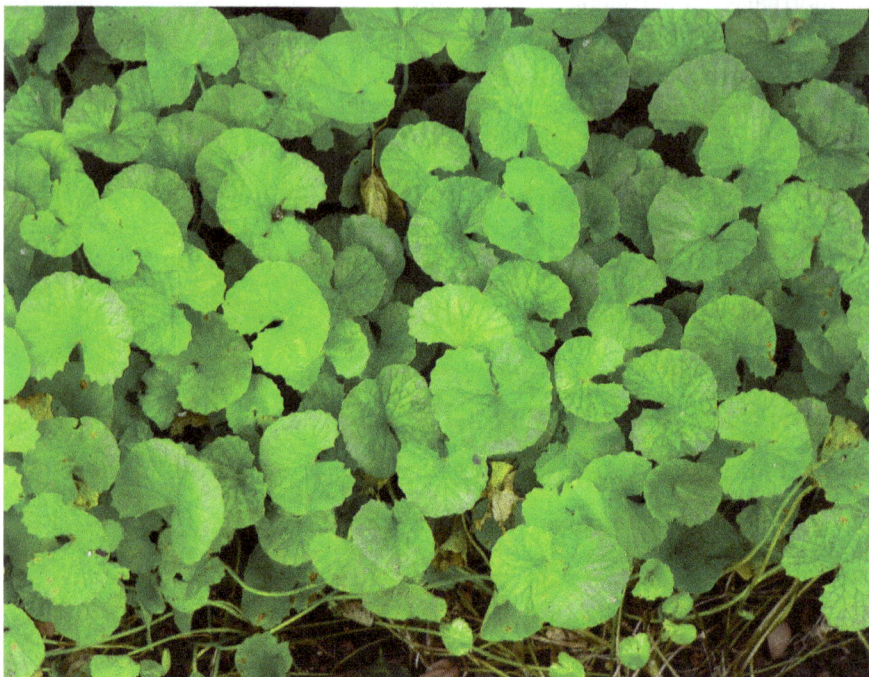

Top view of *Centella asiatica* herb

Description: *Centella asiatica* (L.) Urban is a creeping, perennial herb with long slender horizontal stolons, characterised by long internodes. Leaves are green, fan-shaped or round-reniform, 1–4 cm by 1–7 cm with a crenate or dentate margin. Flowers are umbels with white or light purple-to-pink petals and bear small oval fruit.[1,2]

Origin: Native to India, found in tropical America, Africa, West Pakistan, China, Japan and the Pacific.[3]

Phytoconstituents: Asiaticoside, asiatic acid, brahmic acid, brahmoside, centellic acid, centellose, indocentelloside, madecassic acid, madecassoside, thankuniside, vellarin, bayogenin, centellin, asiaticin, and centellicin.[4–15]

Traditional Medicinal Uses: The plant is used in cooling drinks when boiled, for diarrhoea, diuretic, gravel, leprosy treatment, stones, wound healing and as a tonic.[2,4] It is also part of a mixture to treat colic in Indonesia.[16] The Chinese uses it to improve appetite, aid digestion, treat sores and ulcers. In India, it is used to treat skin, nervous system and blood diseases.[2,7,17] The plant is used to treat fever and rheumatism, as a detoxicant, sedative and peripheral vasodilator.[7,18] The plant is reported as treatment for cancer, circulatory stimulant, hypotensive, stimulant, tonic, cicatrizant (leaf), used for the treatment of haematuria, gonorrhoea, peptic ulcer and sore throat.[5] It is also used for tuberculosis, headache (decoction), dysentery (decoction with other ingredients), boils and tumours (paste applied), leucoderma, anaemia, urinary discharges, bronchitis, insanity, leprosy (decoction), mental deficiency, dysentery (juice or paste on empty stomach for 2–3 days), cough (decoction with ginger and black pepper), cooling (paste with pepper and salt), tonic (juice with palm jaggery given to women after childbirth), elephantiasis, enlarged spleen (ointment from leaves), cures stuttering or stammering, diuretic, small pox and as a local stimulant.[19] External application in the form of poultices is prescribed for contusions, closed fractures, sprain and furunculosis.[18]

Pharmacological Activities: Antibacterial,[20] Antidepressant,[21] Antiemetic,[22] Antineoplastic,[8,23–26] Antioxidant,[27–32] Antithrombotic,[33] Anxiolytic,[34,35] Gastroprotective,[9,36] Immunomodulatory,[37,38] Antigenotoxic,[39] Nerve-regenerative,[40] Radioprotective[41,42] and Wound healing.[43,44]

Dosage: Approximately 0.6 g dry weight of whole plant taken three times a day (condition not indicated).[45] Larger amounts of fresh leaves are sometimes eaten as vegetable and dietary supplement.[45] 60 mg of *C. asiatica* extract given daily for varicose veins.[46] 30 or 60 mg of Total Triterpenic fraction of *Centella asiatica* (TTFCA) three times a day for improving venous hypertension.[47] Titrated extract of *C. asiatica* given 60 or 120 mg daily for chronic venous insufficiency.[48] For the treatment of fever, measles, haematemesis, epistaxis, diarrhoea, dysentery, constipation, leucorrhoea, jaundice, dysuria, furunculosis, dysmenorrhoea, varices, daily dose of 30 to 40 g of fresh plant in the form of extracted juice or decoction is taken.[18]

Adverse Reactions: Allergic contact dermatitis.[49] Sedation, increased blood glucose and cholesterol levels.[50]

Toxicity: Hepatotoxic.[51]

Contraindications: Should not be used during pregnancy and lactation and not to be given to children.[50]

Drug-Herb Interactions: Reported with ephedrine, theophylline, atropine and codeine.[52] Decrease effectiveness of antidiabetic and antilipidemic drugs.[50]

[**Authors' Note:** Topical creams containing the active component asiaticoside is available commercially for wound healing.]

22. *Cerbera odollam* Gaertn. (Apocynaceae)

Pong Pong Tree, Indian Suicide Tree, Sea Mango

Fruit of *Cerbera odollam*

Cerbera odollam leaves

Cerbera odollam tree

Description: *Cerbera odollam* Gaertn. is a medium-sized tree with smooth and grey bark. Leaves are simple, few, without stipules, 12–16 cm by 3–5 cm, arranged in a spiral. The blade is succulent, dark green, glossy and lanceolate. Flowers are white, large and bisexual with a yellow eye in the throat of the corolla tube and arranged in terminal. Fruits are round, waxy surfaced, large, 5–10 cm across, with fibrous husk covering the single seed. The fruit turns from green to reddish brown upon maturity.[1–4]

Origin: Native to Indian subcontinent, Indochina, Malesia and the Pacific.[5]

Phytoconstituents: Cerberin, cerleaside A, 17α-neriifolin, 17β-neriifolin, thevetin B, acetyl-thevetin B, diacetylneriifolin, cerberoside, odollin and others.[3,6–12]

Traditional Medicinal Uses: In Malaysia, rheumatism is treated with embrocations of the fruits.[11] The seeds are poisonous and have been used to poison rats and dogs.[3,11] The seeds are also narcotic. In Indonesia, oil obtained from the seeds is rubbed on the body as remedy for colds, scabies, and rheumatism. In the Philippines, oil of the seeds is also used to treat rheumatism.[11] It is used in Burma as an insecticide or insect repellent when mixed with other oils.[13] The bark, latex and roots are used as purgatives and emetics in India.[3,11,13]

Pharmacological Activities: Antineoplastic[8,10] and Antipsychotic.[14]

Dosage: No information as yet.

Adverse Reactions: No information as yet.

Toxicity: Seeds are poisonous and have been used to poison rats and dogs.[3,11] The seeds have a long history as an ordeal poison in Madagascar, due to the highly toxic cardiac glycosides they contain. The kernel contains cerberin, cerberoside and odollin which are toxins.[15] Humans poisoned by oral consumption of half to one seed kernel would result in sinus bradycardia, wandering pacemaker and second-degree sino-atrial block and nodal rhythm.[15,16] Other symptoms include nausea, retching and vomiting.[16] The poisoning may result in hyperkalemia, which can cause death.[17]

Contraindications: No information as yet.

Drug-Herb Interactions: No information as yet.

23. *Cissus quadrangularis* L. (Vitaceae)

Grape Leaf, Veld Grape

Flowers of *Cissus quadrangularis*   *Cissus quadrangularis* plant

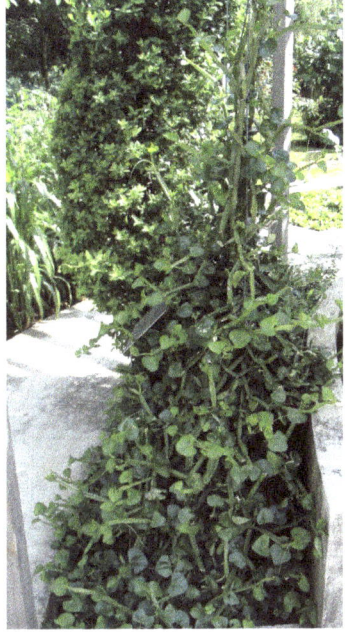

Description: *Cissus quadrangularis* L. is a herbaceous plant with a thick, quadrangular, succulent stem and is constricted at regular intervals. It has long, slender and simple tendrils. Leaves are ovate, entire, crenate-serrate, 3–5 cm by 5–13 cm and glabrous on both sides. Flowers are pink and white, 2 mm long, while the berries contain one or two seeds.[1,2]

Origin: India, Pakistan, Indonesia, Malaysia, East Africa, Sri Lanka and Arabia.[2]

Phytoconstituents: Quadrangularins A–C, δ-amyrin, δ-amyrone, resveratrol, piceatannol, pallidol, parthenocissine A and others.[3–7]

Traditional Medicinal Uses: Whole plant is used for urinary *schistosomiasis* in Mali.[8]

Pharmacological Activities: Antibacterial and Antioxidant,[9] Analgesic and Anti-inflammatory,[10] Antimalarial,[11] Gastroprotective[12–16] and Antiosteoporotic.[17]

Dosage: No information as yet.

Adverse Reactions: No information as yet.

Toxicity: No information as yet.

Contraindications: No information as yet.

Drug-Herb Interactions: No information as yet.

24. *Cocos nucifera* L. (Palmae)

Coconut Palm, Kelapa

Cocos nucifera leaves

Fruits of *Cocos nucifera*

Cocos nucifera trees

Description: *Cocos nucifera* L. is a tall palm with a ringed stem that can grow up to 30 m tall. Leaves are pinnate, 2–6 m long with numerous pairs of narrow leaflets. Flowers are arranged in large panicles among the leaves with female flowers near the base of the inflorescence. The fruit is symmetrical, ovoid and about 20–30 cm across. Thick fibrous husk encloses the hard shell (endocarp) in which the fleshy pericarp adheres.[1,2]

Origin: Native to the Pacific.[3]

Phytoconstituents: Trans-zeatin, dihydrozeatin, dihydrozeatin-O-glucoside, *meta*-topolin riboside, N^6-isopentenyladenine, N^6-benzylaminopurine and others.[4,5]

Traditional Medicinal Uses: Its fruit juice is used to treat poisoning, cholera and is a diuretic. Fresh coconut juice with rice flour is poulticed onto carbuncles, gangrenous sores and indolent ulcers.[6] Coconut water is also used for fever, urinary complaints and to stop vomiting.[7] Juice of the green, unripe coconut is boiled and drunk to relieve diarrhoea. In Somoa, coconut oil is used as a laxative and to relieve stomach ailments.[8] The decoction of husk fibre has been used in northeastern Brazil traditional medicine for the treatment of diarrhoea and arthritis.[9] Coconut water has also been used as short-term intravenous hydration and resuscitation fluid in emergencies.[10] The rootbark is used as an astringent, styptic and in haemorrhages.[6] It is also used for uterine diseases, bronchitis, liver complaints and dysentery.[7]

Pharmacological Activities: Analgesic,[11] Antibacterial,[9,12] Antifungal,[13] Antineoplastic,[14,15] Antioxidant,[11,16] Antiprotozoal,[17] Antiviral,[9] Hypoglycaemic,[18] Hypolipidaemic,[19,20] Hypotensive,[21] Immunomodulatory[22] and Antitrichomonal.[23]

Dosage: No information as yet.

Adverse Reactions: A case report of occupational allergic conjunctivitis due to coconut fibre dust has been reported.[24]

Toxicity: No information as yet.

Contraindications: No information as yet.

Drug-Herb Interactions: No information as yet.

25. *Coix lacryma-jobi* L. (Gramineae)

Job's Tears, Adlay

Fruits of *Coix lacryma-jobi*

Coix lacryma-jobi plants

Description: *Coix lacryma-jobi* L. is a small herbaceous plant which grows up to 2.5 m. Leaf sheaths are glabrous and the leaf blades are narrowly lanceolate, 20–50 cm by 1.5–4 cm. The midrib is prominent. Fruit is tear-shaped, 8 mm by 1.1 cm, with glossy berries which turn black upon maturity.[1–3]

Origin: Native to tropical and temperate Asia.[4]

Phytoconstituents: Coixol, coixenolide, α-coixins, γ-coixins, syringaresinol, mayuenolide, coixan A–C, coixic acid and others.[5–9]

Traditional Medicinal Uses: A decoction is believed to benefit the blood and breath and used to wash newborns to prevent diseases.[10] The kernels are used to treat lung and chest complaints, rheumatism, dropsy and gonorrhoea.[3] Fruit is used for intestinal or lung cancers and warts. The fruit is also used as a vermifuge and for hypertension. As the seed is diuretic and refrigerant, a decoction is used for appendicitis, arthritis, beriberi, bronchitis, cancer, diarrhoea, dryskin, dysuria, oedema, hydrothorax, inflammation, pleurisy, pneumonia, pulmonary abscesses, rheumatism and tuberculosis.[10,11] Seeds are also used for the treatment of enteritis, persistent diarrhoea in children, urinary lithiasis, rheumatism and acrodynia.[12] An infusion of the seeds is prescribed for bronchitis, pulmonary abscess, pleurisy and hydrothorax.[11] The root is given along with roots of long pepper and other herbs for fever with drying of saliva and intense thirst, for dysentery, diarrhoea and puerperal fever.[13]

Pharmacological   Activities: Anti-inflammatory,[14,15]   Anticancer/
Antineoplastic,[16–21] Antioxidant,[8,18] Hypoglycaemic,[9,22] Hypolipidaemic[22–24]
and Hypotensive.[25]

Dosage: For the treatment of enteritis, persistent diarrhoea in children, oedema, urinary lithiasis, rheumatism and acrodynia, doses of 10 to 30 g daily in the form of powder or decoction are given.[12]

Adverse Reactions: No information as yet.

Toxicity: Embryotoxicity in pregnant rats was observed. Oral administration of 1 g/kg body weight of water extracts caused an increase in foetal resorptions and postimplantation mortality.[26]

Contraindications: No information as yet.

Drug-Herb Interactions: No information as yet.

26. *Crinum asiaticum* L. (Amaryllidaceae)

Crinum Lily, Spider Lily, Bawang Tanah

Crinum asiaticum flowers

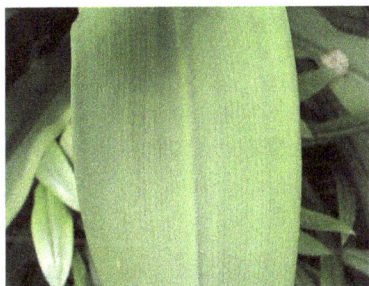
Leaf blade of *Crinum asiaticum*

Crinum asiaticum plant

Description: *Crinum asiaticum* L. is a bulbous herb with a flowering stalk in the centre of the plant. The leaves are narrowly lanceolate, acuminate, 0.5–1.5 m long, greenish and have a hairy texture. The flowering stem is about 1–1.2 m. Flowers are white, 6–12 in an umbel. Filaments are reddish. The fruit is irregularly globose, 4–5 cm across, with one or a few large green seeds.[1–4]

Origin: Native to Tropical Asia, introduced in Northern America.[5]

Phytoconstituents: Crinamine, lycoricidine, hamayne, isocraugsodine, palmilycorine, lycoriside, ambelline, crinasiatin, hippadine, bakonine, pratorimine, crinine, powelline, ungeremine, criasbetaine, crinasiatine, phenanthridone II and others.[6–17]

Traditional Medicinal Uses: In Southeast Asian countries, *C. asiaticum* has a considerable medicinal reputation as a potent folk medicine in the treatment of injury and inflamed joints.[18] The plant has been used for carbuncles and cancer.[7] In Indonesia, the oiled and heated leaves are useful to treat wounds by poisoned arrows, bites and stings. In Malaysia, poultices of the leaves are applied to swellings, swollen joints, lumbago, pains and in cases of headache and fever. The leaves are also an emollient. In Northwest Solomon Islands, the leaves make a topical treatment for inflammation.[19] In Malaysia, the leaves are used as a rheumatic remedy and to relieve local pain.[18] On Karkar Island and in Simbu, Papua New Guinea, the latex from the leaves is applied to cuts. In India, the leaves are applied to skin diseases and inflammation.[20] The crushed leaves are used to wash piles or mixed with honey and applied to wounds and abscesses.[3] Its seeds are considered purgative and emmenagogic.[7] In the Trobriands, Papua New Guinea, the stem fibres are used to stop bleeding and in New Ireland, the milky sap from the stem is used for stonefish wounds.[20] The bulb is an emetic and counter-irritant. In Papua New Guinea, juice obtained from bulb is ingested regularly for 2 months to treat gonorrhoea. In the Philippines, the bulbs are crushed and applied as an ointment.[19] Juice from the fresh bulbs, taken several times per month induces vomiting. It is also instilled in the ear to treat otitis.[21] The root is also an emetic, diaphoretic and nauseant when fresh.[22] In a Finschhafen area village, Papua New Guinea, the cut root is cooked in a banana leaf, then cooled and placed on an aching tooth. Roots are used in New Caledonia, Indonesia, and Malaysia in a poultice for wounds, ulcers and swellings.[20]

Pharmacological Activities: Analgesic and Antibacterial,[19] Anti-inflammatory,[18] Antiviral,[18,23] Anticancer,[13,24] Antitumour,[15] Mast cells degranulation,[10] Antimitotic and Membrane stabilising.[8]

Dosage: No information as yet.

Adverse Reactions: No information as yet.

Toxicity: No information as yet.

Contraindications: No information as yet.

Drug-Herb Interactions: No information as yet.

27. *Cymbopogon citratus* (DC.) Stapf. (Gramineae)

Lemon grass

Cymbopogon citratus plants

Description: *Cymbopogon citratus* (DC.) Stapf. is a grass composed of dense leafy clumps that grows up to 1.8 m tall. Leaf sheath is tubular and acts as a pseudostem, 12–25 cm long, thickening towards the base and 1–2 cm in diameter. Leaf blades are sessile, simple, green, linear, glabrous, 60–90 cm by 1–2 cm, and possess parallel venation. Inflorescence is a raceme.[1–3]

Origin: Native to South Asia, Southeast Asia and Australia.[3]

Phytoconstituents: Citral, citronellal, cymbogonol, α-terpineol, citronellic acid, α-camphorene, geranial, isoorientin, isoscoparin and others.[1,4–8]

Traditional Medicinal Uses: The plant is used to treat digestive problems and relieve cramping pains.[4] A decoction of the plant is used by the Chinese to treat coughs, colds and blood in sputum.[2,9] The entire plant is used to treat bacterial infection and possesses fever-reducing and stomachic properties.[5] In many Asian countries, leaves in water provide a bath to reduce swelling, to remove body odour, improve blood circulation, treat cuts, wounds, bladder problems and leprosy.[9] Its oil is used as an insect repellent, for aerosols, deodorants, floor polishes and household detergents.[4,5] The oil is carminative for cholera, and is prescribed for dyspepsia, vomiting, fever, and headache,[10] and used externally to treat eczema.[5] Its roots are taken to induce sweating, increase flow of urine,[9] treat coryza and influenza.[5]

Pharmacological Activities: Analgesic,[11,12] Anthelmintic,[13] Antibacterial,[14–18] Antifungal,[16,19–25] Anticancer/Antineoplastic,[26,27] Antimalarial,[28] Antioxidant,[8,29,30] Antiplatelet,[31] Hepatoprotective,[32] Hypoglycaemic,[33] Sedative,[34] Vasorelaxant,[35] Antimutagenic,[36] Insecticidal[37] and Radioprotective.[38]

Dosage: A dose of 10 to 20 g of roots is used for treating coryza, influenza and fever. It is also prescribed for dyspepsia and vomiting and as a carminative, by using 3 to 4 drops of the essential oil diluted in water.[5]

Adverse Reactions: Application of thick ointments with the volatile oil on the skin has led to rare incidence of allergy.[39]

Toxicity: Alveolitis occurred as a result of inhalation of the volatile oil.[39] It is also reported to be cytotoxic and genotoxic.[40]

Contraindications: Should not be used during pregnancy.[41]

Drug-Herb interactions: No information as yet.

28. *Dolichos lablab* L. (Leguminosae)

Lablab Bean, Hyacinth Bean

Dolichos lablab leaves

Dolichos lablab flowers and beans

Description: *Dolichos lablab* L. is a woody climbing herb which can reach a length of 5 m. Leaves are pinnate and generally 3-foliolate. Leaflets are acute, entire, 6–12 cm by 5–9 cm. Flowers are white or purplish pink. Fruits are green pods, 6 cm long by 2 cm wide, flattened, contain 4–5 seeds and turn light brown when mature.[1–4]

Origin: Native to Africa.[5]

Phytoconstituents: Dolichin, arabinogalactan 1 & 2, lablabosides A–F, phytin, pantothenic acid, saponin I, putrescine, spermidine, spermine and others.[3,6–14]

Traditional Medicinal Uses: The plant is decocted for alcoholic intoxication, cholera, diarrhoea, globefish poisoning, gonorrhoea, leucorrhoea and nausea.[15] Its seeds are used to stimulate gastric activities, as antidote against poisoning, to treat colic, cholera, diarrhoea, rheumatism and sunstroke.[4,16] The juice from the fruit pods are used for inflamed ears and throats. The fruit is also astringent, digestive, stomachic and used to expel worms. The seeds

59

are reportedly alexiteric, aphrodisiac, febrifuge, stomachic and used for menopause and spasms.[15] The flowers are used to treat dysentery, inflammation of uterus and to increase menstrual flow.[4,15] They are also used for leucorrhoea, menorrhagia, and summer heat disorders, as they have alexiteric and carminative properties.[15]

Pharmacological Activities: Antifungal,[6,17] Antiviral[6] and Haemagglutinating activities.[18–20]

Dosage: The reported dose of the seeds is 8 to 16 g daily in the form of powder or a decoction for the treatment of nausea, vomiting, diarrhoea, enteritis, abdominal pains and alcoholism.[21]

Adverse Reactions: No information as yet.

Toxicity: No information as yet.

Contraindications: No information as yet.

Drug-Herb Interactions: No information as yet.

29. *Elephantopus scaber* L. (Compositae)

Elephant's Foot, Tutup Bumi, Tapak Sulaiman

Elephantopus scaber flowers

Elephantopus scaber herbs

Description: *Elephantopus scaber* L. is a tropical herb that can grow up to 30–60 cm tall. Leaves are simple, without stipules and when fully developed, form a rosette on the ground. Blades are obovate or oblong obtuse, hairy, large, 5–10 cm by 1.2–3 cm. The petioles are short, hairy, often crowded at the base of stem. Flowers are small and whitish pink.[1–3]

Origin: Native to Africa, tropical and temperate Asia and Australasia.[4]

Phytoconstituents: Molephantin, crepiside E, deoxyelephantopin, stigmasterol, stigmasteryl, scabertopin, lupeol and others.[5–9]

Traditional Medicinal Uses: In Vietnam, the plant is considered diuretic and administered at parturition.[10] A decoction of whole plant is bechic, used to treat pulmonary disease and scabies. In India, it is used as a tonic, laxative, analgesic, aphrodisiac and to treat inflammation. In the Philippines, the plant is febrifuge, diuretic and emollient.[10] In Malaysia, it is used as a preventive medicine after childbirth, to expel intestinal worms, for coughs and venereal diseases. In Chinese medicine, it is used to cure "dumpheat", which includes indigestion, swollen legs and loss of appetite.[1] In Indonesia, the plant is prescribed when there is a yellowish discharge from the vagina.[1] The plant is also considered as diuretic, emollient, used for relief of anuria and blennorrhoea, remedy for leucorrhoea and anaemia, to treat fever, inflammation and as a disinfectant.[6] The whole plant is decocted for abscesses, cold, dysentery, oedema, gastroenteritis, gonorrhoea, influenza, pharyngitis and snakebite.[11] In Burma, a decoction of the stem and leaves is used in cases of menstrual disorders.[10] The roots are used for fever in children, on pimples, wounds of cattle, as an abortifacient, for urinary complaints, amoebic dysentery and other digestive problems.[12]

Pharmacological Activities: Antibacterial,[13] Anticancer/Antineoplastic,[14,15] Anti-inflammatory,[16] Antiviral[17] and Hepatoprotective.[18,19]

Dosage: No information as yet.

Adverse Reactions: No information as yet.

Toxicity: Molephantin isolated from plant demonstrated cytotoxic activity.[5]

Contraindications: No information as yet.

Drug-Herb Interactions: No information as yet.

30. *Euphorbia hirta* L. (Euphorbiaceae)

Asthma Weed

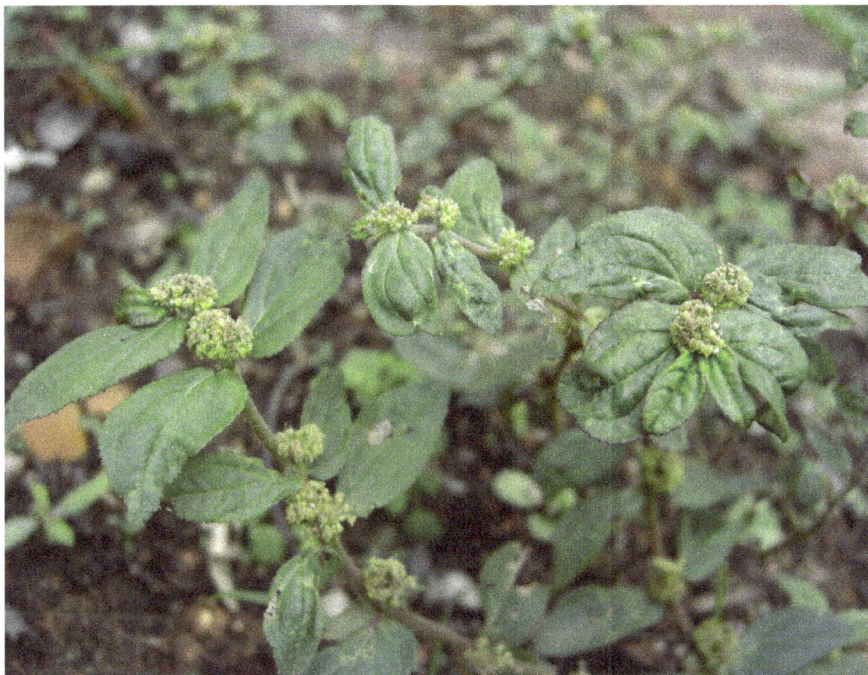

Euphorbia hirta herb

Description: *Euphorbia hirta* L. is an annual herb, which can grow up to 30 cm tall, and branched near the base. Leaves are elliptic-oblong, 2–3 cm by 0.8–1.5 cm. Inflorescence is small, in axillary dense clusters, each with one female flower and 4–5 male flowers inside.[1]

Origin: Native to the Pantropic.[2]

Phytoconstituents: Euphorbon, euphosterol, camphol, leucocyanidol, xanthorhamnin, taraxerol, taraxerone, myricitrin, euphorbianin and others.[3–6]

Traditional Medicinal Uses: The whole plant is decocted for athlete's foot, dysentery, enteritis, fever, gas, itch, and skin conditions.[3] It is also regarded as anodyne, depurative, diuretic, lactogogue, purgative, and vermifuge. The plant is used for asthma, bronchitis, calculus, colic, cough, dyspnoea, eruptions, excrescences, influenza, fractures, gonorrhoea, headache, hypertension, measles, nausea, ophthalmia, sores, splinters, stomachache, tumours, urogenital ailments, warts and wounds.[3,7] In Central Province of Papua New Guinea, the plant is boiled and the solution is taken by patients who pass blood in the urine. The Chinese use the plant to treat fever, dysentery and skin conditions. In the Philippines and Indonesia, the plant is used to treat bowel problems.[8] The latex is used on warts and abscesses.[7]

Pharmacological Activities: Analgesic,[9] Antibacterial,[10–14] Anti-diarrhoeal,[6,15] Anti-inflammatory,[9,16,17] Antiplatelet,[18] Antiprotozoal,[19,20] Antipyretic,[9] Anxiolytic,[21] Diuretic,[22] Sedative,[23] Antianaphylactic[24] and Molluscicidal.[25,26]

Dosage: No information as yet.

Adverse Reactions: No information as yet.

Toxicity: Toxic to brine shrimp.[11]

Contraindications: No information as yet.

Drug-Herb Interactions: No information as yet.

31. *Eurycoma longifolia* Jack (Simaroubaceae)

Tongkat Ali, Ali's Umbrella, Pasak Bumi

Eurycoma longifolia leaves

Eurycoma longifolia fruits

Eurycoma longifolia tree

Description: *Eurycoma longifolia* Jack is a small tree with compound leaves on branches that can grow up to 1 m long. The numerous leaflets are opposite or subopposite, lanceolate to ovate-lanceolate, 5–20 cm by 1.5–6 cm, with smooth margins. Flowers are tiny, reddish, unisexual and are densely arranged. The drupes are ovoid with a distinct ridge, 1–2 cm by 0.5–1.2 cm and they turn dark reddish brown when ripe.[1–3]

Origin: Native to Malesia and Indochina.[4]

Phytoconstituents: Eurycomalactone, eurycomanol, eurycomanone, eurylactone, eurylene, laurycolactone A and B, longilactone, pasakbumins A to D, eurycomalide A and B, piscidinol A and others.[3,5–19]

Traditional Medicinal Uses: The plant is used to cure indigestion and lumbago. It is used as a tonic after childbirth, to relieve pains in the bone and for treatment of jaundice, dropsy, cachexia and fever.[3,20] Tongkat Ali is one of the most well known folk medicines for intermittent fever (malaria) in Southeast Asia.[16] Decoction of the leaves is used for washing itches, while the fruits are used in curing dysentery.[3] Its bark is used as a vermifuge.[3] The taproots are used to lower high blood pressure, while the root bark is used for the treatment of fever and diarrhoea.[18] The roots of this plant are used as folk medicine for the treatment of sexual insufficiency, aches, persistent fever, malaria, dysentery, glandular swelling and also as health supplements.[18]

Pharmacological Activities: Antianxiety,[21] Antibacterial,[22] Anticancer,[6,8,17,18,24–26] Antitumour,[23,27] Antimalarial/Antiplasmodial,[5,13,18,23,28–32] Antischistosomal,[23] Antiulcer,[33] Aphrodisiac[15,18,34–38] and Plant growth inhibitor.[39]

Dosage: 1 g daily is recommended to be the maximum dose for supplemental use.[39]

Adverse Reactions: No information as yet.

Toxicity: One animal study found that the LD_{50} in mice was 1500–2000 mg/kg of the alcohol extract and 3000 mg/kg of the water extract. A subacute toxicity study with the alcohol extract indicated that 600 mg/kg daily was associated with signs of toxicity while 200 mg/kg daily was not, and another study found no toxic effects at 270–350 mg/kg daily but toxic effects were observed at 430 mg/kg daily.[39] Eurycomanone was identified as the most toxic component from its butanol extract.[40]

Contradindications: Should be used with caution and preferably not for extended periods without taking periodic breaks from use when it is used as a supplement. Tongkat Ali should not be taken by methods other than oral administration.[39]

Drug-Herb Interactions: No information as yet.

32. *Hibiscus mutabilis* L. (Malvaceae)

Cotton Rose, Chinese Rose

Hibiscus mutabilis flower

Hibiscus mutabilis tree

Description: *Hibiscus mutabilis* L. is a small tree that can grow up to 5 m tall. Leaf blades are heart-shaped, broadly ovate to round-ovate or cordate, 5–7-lobed, 10–15 cm in diameter, and papery. Abaxially, they are densely stellate and minutely tomentose; adaxially they are, sparsely stellate and minutely hairytoothed, 8–15 cm wide. Flowers are solitary and with multi-petals, white colour in the morning, changing to pink in the afternoon.[1–3]

Origin: Native to China.[4]

Phytoconstituents: Isoquercitrin, hyperoside, rutin, quercetin, naringenin, tetracosanoic acid, daucosterol, salicylic acid, quercimeritrin, meratrin and others.[5–10]

Traditional Medicinal Uses: The plant is used for fistulae, pustules and tumours.[11] The leaves and flowers are used as an analgesic, to expel phlegm, treat excessive bleeding during menstruation, painful urination, inflammation and snake bites.[3] The leaves and flowers are also used as demulcent, diuretic and treat bacterial infection. They are used to treat boils, particularly on the chin, in the form of a poultice made of powdered dried leaves and flowers mixed with concentrated tea infusion which makes the boils burst earlier and less painfully. They are also used for treating impetigo, prurigo, metritis, leucorrhoea, mastitis, nephritis, cystitis, dysuria and infections.[12] The leaf is applied to swellings, crushed and compressed and applied onto abscesses, burns, and ulcers. It is also used as anodyne, alexipharmic, demulcent, expectorant, and refrigerant.[11] The flowers are used for lung ailments, with leaves for burns, inflammation, and snake bite. They are also prescribed for cough, dysuria and menorrhagia.[11]

Pharmacological Activities: Anti-inflammatory.[13]

Dosage: For the treatment of impetigo, prurigo, metritis, leucorrhoea, mastitis, nephritis, cystitis, dysuria and infections, a dose of 5 to 20 g of leaves and flowers is taken daily in the form of a decoction.[12]

Adverse Reactions: No information as yet.

Toxicity: No information as yet.

Contraindications: No information as yet.

Drug-Herb Interactions: No information as yet.

33. *Hibiscus rosa-sinensis* L. (Malvaceae)

Hawaiian Hibiscus, China Rose, Bunga Raya

Leaves of *Hibiscus rosa-sinensis*

Hibiscus rosa-sinensis flower

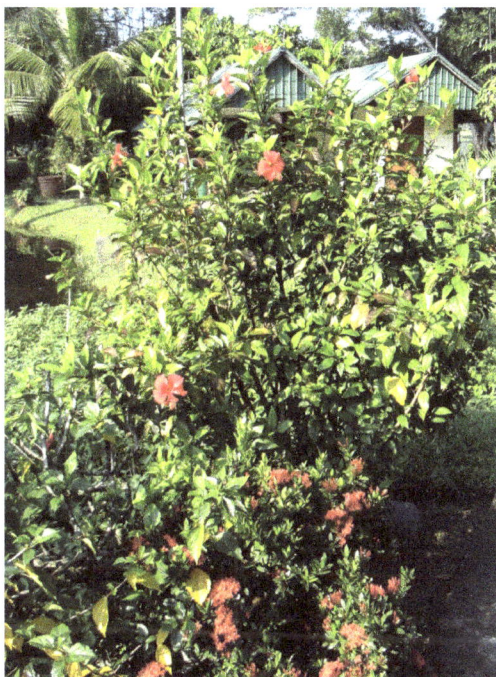

Hibiscus rosa-sinensis shrub

Description: *Hibiscus rosa-sinensis* L. is a small evergreen perennial tree that can grow up to 3.6 m tall. It produces flower all year round. The large glossy leaves are ovate, alternate and vary in colour from pale green to dark green, with serrated edges. The flowers have five petals, and their colour varies but are mostly red.[1,2]

Origin: Native to South Eastern Asia and cultivated throughout the world as decorative plants.[3]

Phytoconstituents: Gossypetin, anthocyanin, myristic acid, palmitic acid, ambrettolide, campesterol, methyl sterculate, malvalate and others.[4–6]

Traditional Medicinal Uses: In Suquang, Papua New Guinea, the whole plant juice is applied directly to sores, to treat headaches and irregular periods.[7] In Central Province, the plant is used to treat eye sores whereas in Northern Province and North Solomons Province, it is used to induce labour. In Indonesia, it is used as a purgative, an abortifacient and to regulate menstruation.[7] The juice also provides a soothing effect on mucous membrane that line the respiratory and digestive tracts.[5] The flowers and leaves are used to treat skin diseases, mumps, to relieve fever,[2,4] as well as to be used as emollient, anodyne and laxative.[8] The flower is also used as an astringent,[5] for excessive menstruation, fever and skin diseases. Its roots are used to treat gonorrhoea.[8] In Finschhafen, the roots and leaves are crushed and the juice is drunk to treat diarrhoea.[7]

Pharmacological Activities: Antianxiety,[9] Anticancer,[10] Anticonvulsant,[9] Antifertility,[11–14] Antioxidant,[15] Hepatoprotective,[16] Hypoglycaemic,[17–19] Hypolipidaemic[18,20] and Wound healing.[21]

Dosage: No information as yet.

Adverse Reactions: No information as yet.

Toxicity: No information as yet.

Contraindications: Not to be taken by small children, and during pregnancy and lactation.[22]

Drug-Herb Interactions: No information as yet.

34. *Hibiscus tiliaceus* L. (Malvaceae)

Linden Hibiscus, Sea Hibiscus, Mahoe

Leaves of *Hibiscus tiliaceus*

Hibiscus tiliaceus flowers

Description: *Hibiscus tiliaceus* L. is an evergreen coastal tree that can grow up to 10 m tall. Leaves are heart-shaped, 10–15 cm long and wide, leathery and green. The large showy flowers are yellow with maroon centre. Fruit is capsular, subglobose to ovoid, about 2–3 cm long and wide, surrounded by the calyx. Seeds are reniform, smooth and glabrous.[1–4]

Origin: Native to the Pantropic.[4]

Phytoconstituents: Hibiscones A–D, hibiscusin, hibiscusamide, friedelin, epifriedelanol, pachysandiol A and others.[2,5–7]

Traditional Medicinal Uses: In Malaysia and Indonesia, the leaves are considered cooling and are used to treat fever. They are also used as a soothing agent and to remove phlegm from the respiratory passages.[2] The flowers are widely used as a means of birth control in traditional medicine in Asian and African countries.[8] In the Philippines, the mucilaginous water obtained by soaking the fresh bark in water is prescribed for dysentery[2] while the root is an ingredient of embrocations and is used as a febrifuge.[8]

Pharmacological Activities: Tyrosinase inhibitory,[9] Antioxidant[10,11] and Cytotoxic.[5]

Dosage: No information as yet.

Adverse Reactions: No information as yet.

Toxicity: No information as yet.

Contraindications: No information as yet.

Drug-Herb Interactions: No information as yet.

35. *Impatiens balsamina* L. (Balsaminaceae)

Balsam Plant

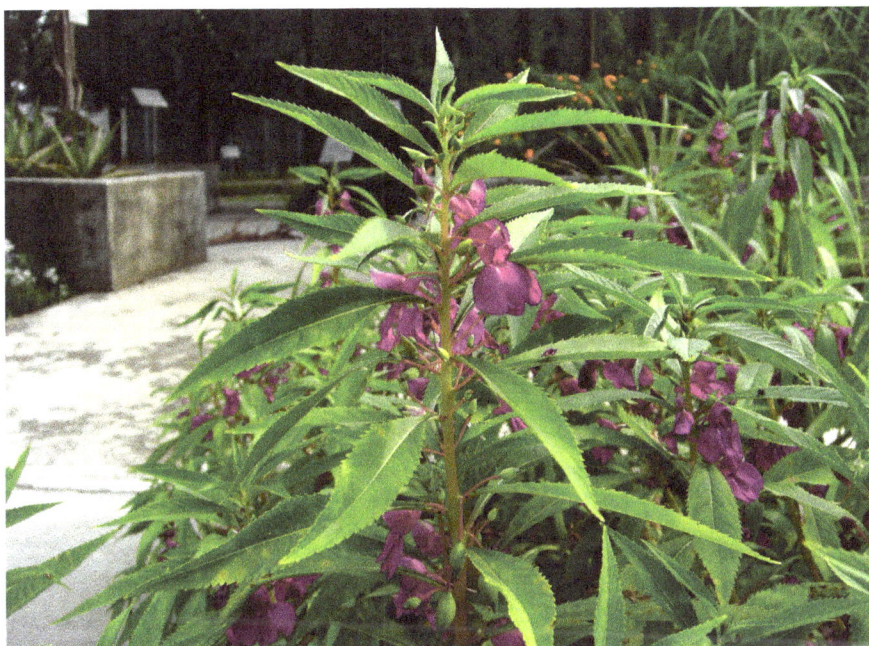

Impatiens balsamina plants

Description: *Impatiens balsamina* L. is an annual, erect, succulent plant, which is about 30–90 cm tall. Leaves are simple, alternate, 15 cm long, serrated and arranged spirally. Flowers are white, pink, purple or variegated and usually in leaf axils. Fruits are capsular, tomentose and contain several seeds.[1–5]

Origin: Native to India.[5]

Phytoconstituents: Impatienol, pelargonidine, delphinidine, cyanidine, balsaminones A and B, hosenkosides F–O and others.[4,6–10]

Traditional Medicinal Uses: A lotion of fresh leaves is used to treat eczema, itches and insect bites.[4] In Vietnam, decoctions of leaves are used to stimulate growth and to wash hair.[4] The juice is also used for warts, cancer treatment and expectorant.[11] A decoction of flowers is taken for infections, vomiting, urine retention and as a tonic. In India, flowers are regarded as cooling, tonic and useful when applied to burns and scalds.[4] The flowers are also used for lumbago and intercostal neuralgia, snakebite, improves circulation and relieves stasis.[11] In Japan, juice squeezed from the white flower petals are applied on the skin to alleviate dermatitis.[6] In China, the seeds are prescribed for difficult labour, puerperal pains, difficult menstruation, cough, hiccups and poisonings. The seeds are mixed with arsenious acid for removing teeth.[4] In Malaysia, the seeds are taken for gastrointestinal tract cancer, and to dislodge fish or chicken bones in throat.[4]

Pharmacological Activities: Antibacterial,[12–15] Antihypotensive,[16,17] Antifungal,[18] Anti-inflammatory[4,8] and Antipruritic.[6,19]

Dosage: No information as yet.

Adverse Reactions: No information as yet.

Toxicity: No information as yet.

Contraindications: No information as yet.

Drug-Herb Interactions: No information as yet.

36. *Imperata cylindrica* (L.) P. Beauv. (Gramineae)

Lalang, Alang-alang, Speargrass

Imperata cylindrica

Infructescence of *Imperata cylindrica*

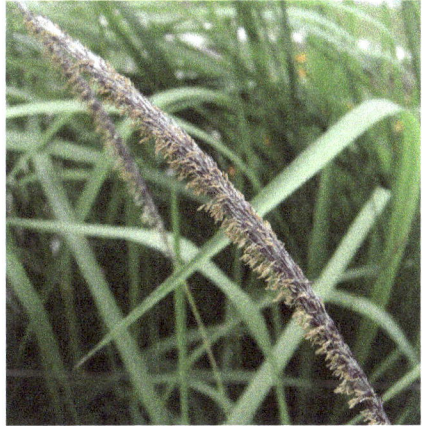

Inflorescence of *Imperata cylindrica*

Description: *Imperata cylindrica* (L.) P. Beauv. is a perennial erect grass about 0.6–1.5 m tall. Leaf blades are linear to lanceolate, long, 0.5–1 cm by 15–30 cm, narrow, with sharp margins and prominent nerves. Inflorescences are white and fluffy. Rhizome is hard, coriaceous and deep within the soil.[1]

Origin: Native to China and Africa but this widespread weed can be found throughout the world.[2]

Phytoconstituents: Arborinione, arundoin, anemonin, isoorientin, imperanene, cylindol A and B, graminones A and B, cylindrene, isoarborinol, impecyloside and others.[3–10]

Traditional Medicinal Uses: Rotted grass from thatch boiled with wine is used to treat bug bite, haemoptysis, severe constipation and vaginismus. It is also used for drug withdrawal symptoms.[3] Besides being used to quench thirst, its flowers and rhizome are also used to treat blood in the sputum, nose bleeds, lung and kidney diseases, jaundice,[11] haemorrhage, wounds, haemoptysis, epistaxis, haematemesis, haematuria, nephritic oedema, high fever, and urinary tract infections.[3] The roots are used to treat fever, cough with phlegm,[11] asthma, cancer, dropsy, epistaxis, haematuria, jaundice, nephritis, diarrhoea, gonorrhoea[3] and dysuria.[12]

Pharmacological Activities: Antidiuretic,[12] Anti-inflammatory,[6] Neuroprotective[9] and Antibacterial.[13]

Dosage: Rhizome is used for the treatment of urodynia, pollakiuria, haematuria and fever. The recommended usual daily dose is 10 to 40 g in the form of a decoction, to be administered orally.[1]

Adverse Reactions: Allergic contact dermatitis[14] and sensitivity to grass pollen (Type I allergy).[15]

Toxicity: No information as yet.

Contraindications: No information as yet.

Drug-Herb Interactions: No information as yet.

37. *Ipomoea pes-caprae* (L.) Sweet (Convolvulaceae)

Beach Morning Glory, Goat's Foot Creeper, Bayhops

Ipomoea pes-caprae plants

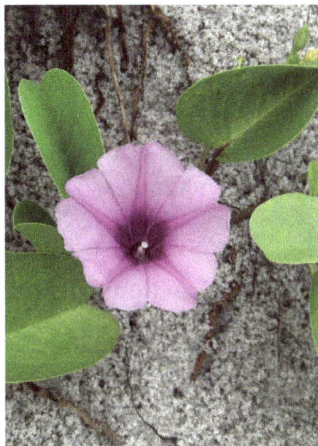

Ipomoea pes-caprae flower

Description: *Ipomoea pes-caprae* (L.) Sweet is a seaside trailing herb. The stems are long and the roots are found at the nodes. Leaves are simple, ovate, quadrangular or rounded, 2.5 cm by 10 cm, with slender petioles. Cymes are 1–3 flowered; the 5 sepals are ovate and blunt. The flower funnel is about 3–5 cm long. The plant bears dehiscent capsules (1.2 cm) containing hairy seeds that break up easily.[1–4]

Origin: Native to the Pantropic.[5]

Phytoconstituents: Pescaproside A & B, pescapreins I–IX, stoloniferin III, E-phytol, β-damascenone and others.[4,6–12]

Traditional Medicinal Uses: The juice squeezed from the plant is used by the Malays to treat fish stings.[3] In Burma, infusions of the plant with rusted iron are used to cure gynaecologic haemorrhages.[4] In India, the plant is known to be cooling, astringent and laxative. It is used internally as an emollient in Vietnam.[4,13] The plant has also been administered for headache.[14] In Thailand, it is used to treat dermatitis caused by jellyfish (e.g., Portuguese man-of-war).[15] The leaves are used in Indonesia to hasten the bursting of boils and used externally for injured feet. Sap from the young leaves are boiled in coconut oil and used to treat sores and ulcers. In the Philippines, the boiled leaves are used to treat rheumatism and fungal infection.[3,4,13] In Papua New Guinea, decoctions of the leaves are applied to sores. On the Solomon Islands, preparations of the leaves are ingested for digestive troubles.[4] The seeds are chewed with areca nut to sooth abdominal pains and cramps or consumed alone as a purgative.[3,13] Decoctions of the tubers are also ingested to treat bladder problems.[4]

Pharmacological Activities: Analgesic,[16] Antidiabetic,[17] Anti-inflammatory,[1,16,18] Antiplatelet aggregation,[14] Antiviral,[4] Antivenom,[4,19] Collagenase inhibitory activity,[12] Muscle relaxant[4,15] and Prostaglandin synthesis inhibition.[9]

Dosage: No information as yet.

Adverse Reactions: No information as yet.

Toxicity: No information as yet.

Contraindications: No information as yet.

Drug-Herb Interactions: No information as yet.

38. *Ixora chinensis* Lam. (Rubiaceae)

Chinese Ixora

Ixora chinensis flowers

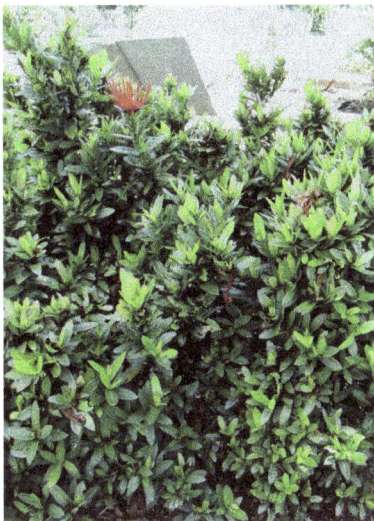

Ixora chinensis shrub

Description: *Ixora chinensis* Lam. is a small, dense shrub that can grow up to 2 m tall. Leaves are short-stalked, obovate-oblong, waxy, 6–10 cm long. Flowers are densely arranged, with 4 petals in bright red.[1]

Origin: This common garden flower is native to China and Thailand.[1]

Phytoconstituents: Ixoric acid, ixoroside, ixoside, geniposidic acid and others.[2–4]

Traditional Medicinal Uses: It is used for hypertension.[5] The whole plant is used to treat rheumatism, abscesses, bruises, and wounds. It is also used as an anodyne, and resolvent. It is beneficial to bone marrow and pregnant uterus as well.[6,7]

Pharmacological Activities: No information as yet.

Dosage: No information as yet.

Adverse Reactions: No information as yet.

Toxicity: No information as yet.

Contraindications: No information as yet.

Drug-Herb Interactions: No information as yet.

39. *Jatropha curcas* L. (Euphorbiaceae)

Barbados Nut, Physic-nut

Flowers of *Jatropha curcas*

Jatropha curcas leaf

Jatropha curcas shrub

Description: *Jatropha curcas* L. is a woody shrub, 2–5 m tall. Stems are smooth and leaves are ovate, 5–15 cm long, green and glabrous. The blade is thin and 5-lobed-cordate. The flowers are 6 mm by 7 mm, hermaphrodite, and with fragrance. Fruit is 3-lobed fleshy capsule of about 2.5 cm in length, containing black seeds.[1,2]

Origin: Native to Mexico,[2] widely cultivated and naturalised in New and Old World Tropics.[3]

Phytoconstituents: Curcain, curcasin, curcacycline A, jatropholone A and B, heudelotinone, nobiletin, jatrocurin, curcin, curcusone B and others.[4–14]

Traditional Medicinal Uses: The leaves are used to treat scabies, parasites and as a rubefacient for paralysis and rheumatism.[15] The fruit is used for dropsy and anasarca.[16] The seed oil is emetic, laxative, purgative and for skin ailments.[4] The latex is applied to bee and wasp stings, to dress ulcers, sores and inflamed tongues. It is also used as a haemostatic agent,[15] and to treat whitlow, carbuncle and sores in mouth.[16] The roots are used in the form of a decoction as mouthwash for bleeding gums and toothache.[15]

Pharmacological Activities: Antidiabetic,[17] Anti-inflammatory,[18] Anticancer,[13,14,19] Antiprotozoal,[20] Antiviral,[21,22] Coagulant,[23] Abortifacient,[24] Haemolytic,[25] Lipolytic,[26] Insecticidal,[27–29] Molluscicidal[30] and Wound healing.[31]

Dosage: Two seeds have been used as a purgative.[32]

Adverse Reactions: No information as yet.

Toxicity: Human poisoning may lead to amnesia, convulsions, delirium, diarrhoea, nausea, vertigo, visual disturbances, acute abdominal pain, nausea, depression and collapse may also occur.[32] Intake of 4–5 seeds may cause death. It is toxic to rats, chicks and human.[32–40]

Contraindications: No information as yet.

Drug-Herb Interactions: May have interactions with other drugs that are metabolised by cytochrome P450 enzymes.[20]

[**Authors' Note:** *Jatropha curcas* is also cultivated as a source of biofuel.]

40. *Juniperus chinensis* L. (Cupressaceae)

Chinese Juniper

Leaves of *Juniperus chinensis*

Juniperus chinensis tree

Description: *Juniperus chinensis* L. is conical shaped tree that can grow up to 25 m tall. Leaves are both scale-like and needle-like. Needle-like leaves are present on both young and adult plants, decussate or in whorls of three, loosely arranged, ascending, nearly lanceolate whereby scale-like leaves are present only in the adult. Seed cones are brown when ripe, usually glaucous, subglobose and 4–9 mm in diameter.[1]

Origin: Native to Myanmar and temperate Asia.[2]

Phytoconstituents: Chinensiol, yatein, podophyllotoxin, thujopsenal, widdrol, sandaracopimaric acid, hinokiic acid and others.[3–11]

Traditional Medicinal Uses: The leaves are used as a tonic to treat bleeding resulting from coughs.[12] It is also used for cold and haemorrhage. Liquor brewed from fresh leaves is used as a tonic and to treat hemoptysis. Others include treatment for convulsions, excessive sweating and hepatitis.[3] Its roots are used on burns, scalds and to promote hair growth on scars.[3]

Pharmacological Activities: Antibacterial,[13] Antifungal,[13,14] Antineoplastic,[15] Antioxidant,[6] Hypoglycemic[16] and Hypolipidaemic.[8,16]

Dosage: No information as yet.

Adverse Reactions: No information as yet.

Toxicity: No information as yet.

Contraindications: No information as yet.

Drug-Herb Interactions: No information as yet.

41. *Kaempferia galanga* L. (Zingiberaceae)

Galangal, Sand Ginger, Kencur

Kaempferia galanga herbs

Description: *Kaempferia galanga* L. is a small herb with short underground stems. Leaves are usually in pairs, oval, glabrous, pointed, 6–15 cm long, and spread out above ground with prominent veins. Flowers are in short stalked spikes. The corolla is white or pinkish, with violet spotted lip.[1]

Origin: Native to tropical Asia.[2]

Phytoconstituents: Ethyl cinnamate, 1,8-cineole, δ-3-carene, α-pinene, camphene, borneol, cyene, α-terpineol, α-gurjunene, germacrenes, cadinenes, caryophyllenes and others.[3–8]

Traditional Medicinal Uses: The whole plant is used as a postpartum protective medicine, treatment for stomachache, diarrhoea, dysentery, treatment for rheumatism, swellings, fever, coughs, asthma and as a tonic/lotion.[8] In Malaya, the leaves and rhizomes are chewed to stop cough. In Indonesia, it is used for abdominal pain, for swelling and muscular rheumatism. In the Philippines, the rhizome is used for boils, chills, dyspepsia, headache and malaria. The Indians also use the rhizomes as lotions, poultices for fever, rheumatism, sore eyes, sore throat and swellings.[3] The rhizomes are stimulant, used to treat toothache, chest pains and constipation.[9] They are also used as carminative, prophylactic, stomachic, for dandruff and scabs. A decoction of the rhizome is used for cholera, contusion, dyspepsia, headache, lameness, lumbago, and malaria. It is also roasted and applied to rheumatism and tumours.[3] To facilitate delivery during birth, it is mixed with the juice of *Curcuma montana, C. aromatica* and ginger rhizomes and consumed.[10]

Pharmacological Activities: Antibacterial,[11,12] Antifungal,[13] Antihypertensive,[14] Anti-inflammatory,[8] Antineoplastic,[15–18] Antioxidant,[19] Antiprotozoal,[20] Depressant,[21] Immunomodulatory,[21] Vasorelaxant,[22] Antiallergy,[23] Insect repellent,[24,25] Insecticidal[25–28] and Wound healing.[29]

Dosage: Oral doses range from 3 to 6 g of the rhizome per day, administered in the form of decoction, powder or pill for the treatment of pectoral and abdominal pains, headache, toothache and cold.[30]

Adverse Reactions: No information as yet.

Toxicity: The ethanolic extract injected intraperitoneally in increasing doses of 25, 100, 250 and 800 mg/kg body weight resulted in a decrease in motor activity, respiratory rate, loss of screen grip and analgesia in rats. A dose of 2000 mg/kg body weight was observed to be lethal. Acute and subacute oral toxicity test of *Kaempferia galanga* produced neither mortality nor significant differences in the body and organ weights between controls and treated rats.[21]

Contraindications: No information as yet.

Drug-Herb Interactions: No information as yet.

42. *Lantana camara* L. (Verbenaceae)

Common Lantana, Bunga Tahi Ayam, Wild Sage

Lantana camara flowers

Description: *Lantana camara* L. is a small shrub with long, weak branches. Leaves are opposite, ovate, 5–12 cm long with a pungent scent. Flowers are in dense spikes, with salver-shaped corolla, 1–1.2 cm long, orange, red, pink or variegated. Fruit is bluish, globose and 4 mm in diameter.[1]

Origin: Native to Mexico and Southern America.[2]

Phytoconstituents: Lantadene A and B, lantanic acid, lantanilic acid, icterogenin, lantanose A & B, lamiridoside, geniposide, δ-guaiene, camarinic acid and others.[3–14]

Traditional Medicinal Uses: The whole plant is used as a bath for scabies and leprosy.[3,15] The entire plant is also a carminative, diaphoretic, vulnerary, used for fistulae, pustules, tumours, treating malaria, rheumatism, tetanus and spasms.[3] In Budiope county, Uganda, the leaves are used for treating malaria.[16] The root and leaves are decocted to treat fever, mumps, neurodermatitis and traumatic injuries.[3,16]

Pharmacological Activities: Anthelmintic,[7,17] Antimutagenic,[13] Antibacterial,[13,18–21] Anticancer/Antineoplastic,[22,23] Antifertility,[24] Anti-inflammatory,[25] Antimicrobial,[26–28] Antimotility,[29] Antioxidant,[30] Antithrombotic,[31] Antiviral,[32] Antifilarial,[33] Insect repellent,[34,35] Insecticidal,[36–39] Larvicidal,[40] Molluscicidal[10,41] and Spermatotoxic.[42]

Dosage: No information as yet.

Adverse Reactions: No information as yet.

Toxicity: Ingestion of green berries had been reported to cause human fatalities.[43] *L. camara* is harmful to animals[44–58] and causes teratogenicity in rats.[59]

Contraindications: No information as yet.

Drug-Herb Interactions: No information as yet.

43. *Lonicera japonica* Thunb. (Caprifoliaceae)

Japanese Honeysuckle, Jin Yin Hua

Lonicera japonica shrub

Lonicera japonica leaves

Description: *Lonicera japonica* Thunb. is a climbing shrub having tomentose young leaves and stems. Leaves are simple, opposite and exstipulate. Blade is elliptic, 3–8 cm by 2–3 cm, truncate at base, obtuse and chartaceous. Flowers are axillary, white, and turns yellow upon maturity. Fruits are globose and black.[1]

Origin: A native of East Asia, widely cultivated and naturalised throughout the world.[2]

Phytoconstituents: Linalool, luteolin, geraniol, aromadendrene, eugenol, loniceroside A, B, C, L-phenylalaninosecologanin, (Z)-aldosecologanin, (E)-aldosecologanin and others.[3–8]

Traditional Medicinal Uses: In China, the flowers are used for influenza, boils and carbuncle.[3] In Malaysia, decoctions of dried flowers are used for cooling, flu, fever, headache, and boils. Distilled flowers are used to produce a medicine for treating postprandial stomachaches.[3] Flower tea is prescribed to treat fever, sore throat, mouth sores, headache, conjunctivitis, keratitis, corneal ulcers, breast infections, muscle and joint pain, stomach problems, diarrhoea, and painful urination.[9] They are used in the treatment of arthritis and inflammation.[10] Flower buds are used in infusions for cutaneous infections, scabies, as diuretic and treat bacterial infection. Decoction is used for bacterial dysentery, cold, enteritis, infected boils, laryngitis, lymphadenitis, rheumatism and sores. The flowers and stems are regarded as cooling and are used to treat aching bones and boils. Other uses include intestinal inflammation, stomach ulcers, painful haemorrhoids, sore throats and intoxication.[3]

Pharmacological Activities: Antibacterial,[11,12] Anticancer/Antineoplastic,[12–15] Antifungal,[16] Antihypertensive,[17] Anti-inflammatory,[9,18–25] Antioxidant,[7] Antiplatelet,[26] Antiviral,[27] Hepatoprotective[28] and Antiatherogenic.[29]

Dosage: In China, 10–60 g of dried floral buds are used for decoction. A combination of 10 g of honeysuckle, 10 g of forsythia, a little mint and bamboo leaf is a prescription for a bad cold from a drugstore in China.[30] About 9–15 g dried flowers has been used in decoction, pills, powder or poultice of the powder.[31] The recommended daily dose is 4 to 8 g of flowers or 10 to 20 g of stems and leaves in the form of a decoction, infusion, extract or alcoholic maceration for the treatment of boils, impetigo, urticaria, allergic rhinitis, fever, malaria, erythema, measles, diarrhoea, dysentery, syphilis, rheumatism and lichen tropicus.[32] For carbuncles, boils, erysipelas, acute dysentery, pharyngitis, upper respiratory infection and epidemic febrile diseases, 6–15 g of dried flowers are used.[33]

Adverse Reactions: Linear, itchy, raised blisters on the skin may occur on contact.[34]

Toxicity: No sign of acute and subacute toxicity was observed.[35]

Contraindications: No information as yet.

Drug-Herb Interactions: No information as yet.

44. *Mangifera indica* L. (Anacardiaceae)

Mango, Mangga

Mangifera indica tree

Fruits of *Mangifera indica*

Description: *Mangifera indica* L. is a fruit tree which grows up to 8 m high. Bark is grey and fissured. Leaves are simple, 12–30 cm by 4–9 cm, narrowly elliptic, pointed, and slightly leathery with wavy edges. Flowers are very small, greenish-yellow or white, fragrant and arranged in panicles. The fruit is kidney-shaped and the yellow to orange flesh is edible. Seed is elongated, fibrous and flattened.[1]

Origin: Native to India and Indochina.[2]

Phytoconstituents: Mangiferin, ambolic acid, ambonic acid, arabinan, mangiferonic acid, quercitin, violaxanthin and others.[3–7]

Traditional Medicinal Uses: The leaves are used in the form of ashes for burns and scalds; chewed to strengthen the gums, while the burning leaf smoke is inhaled for hiccups and other throat ailments. They are also used for skin ailments, asthma and cough.[4] Paste of leaves is applied to warts and used as styptic ointments.[3] The leaves are also used in the treatment of malaria in Budiope county, Uganda.[8] The fruits are used to treat pain in abdomen, diarrhoea, and to quench thirst (with Aegle and salt).[9] Pulp of the fruit is used in China to promote blood circulation, the fruit rind as tonic in Burma and dried slices of young fruits are used for septicaemia in Palau.[3] Its seeds are used for stubborn colds, coughs, diarrhoea, vermifuge and menorrhagia.[4] They are also used for asthma.[9] In the Philippines, raw seeds are used to expel worms and the roasted ones are for diarrhoea.[3] The bark is used for the treatment of fever or sunstroke, cholera, rheumatism, sty in eye, ulcerated tongue (with roots of *Ichnocarpus* and bark of *Zizyphus rugosa*), haemiplegia, diarrhoea (with bark of *Streblus asper*, roots of *Oroxylum indicum* and *Helianthus annuus*), dysentery (with bark of *Streblus* and *Spondias pinnata*), poisoning, uterine haemorrhage and jaundice.[9]

Pharmacological Activities: Analgesic,[10] Anthelmintic,[11] Antibacterial,[12–14] Anticonvulsant,[15] Antidiarrhoeal,[13] Antifungal,[16] Anti-inflammatory,[10,11,17–19] Anticancer,[20-23] Antioxidant,[24–38] Antiviral,[39,40] Gastroprotective,[41,42] Hepatoprotective,[43,44] Antiprotozoal,[15] Hypoglycaemic,[45–47] Hypolipidaemic,[46,48] Immunomodulatory,[49] Larvicidal[50] and Radioprotective.[51,52]

Dosage: As an anthelmintic, it is recommended to take 20–30 g of powdered seed.[53]

Adverse Reactions: Anaphylactic reaction following ingestion of fruit.[54]

Toxicity: Intraperitoneal LD_{50} (50% ethanolic extract) is > 1000 mg/kg in mice.[55]

Contraindications: No information as yet.

Drug-Herb Interactions: Modulation of P450 isozymes.[56,57]

45. *Manihot esculenta* Crantz (Euphorbiaceae)

Tapioca, Cassava

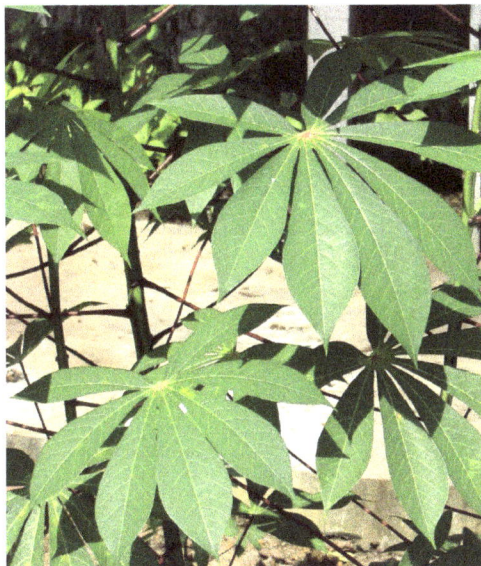

Leaves of *Manihot esculenta*   *Manihot esculenta* shrub

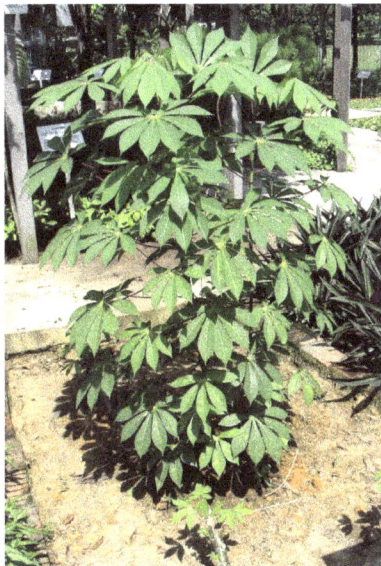

Description: *Manihot esculenta* Crantz is an erect woody shrub that can grow up to 5 m tall. Leaf blades are palmately 3–9-lobed, 5–20 cm. Lobes are oblanceolate to narrowly elliptic, entire, 8–18 cm by 1.5–4 cm. Root tubers beneath the ground yield tapioca, which is fleshy, starchy and edible.[1,2]

Origin: Native to Brazil, cultivated throughout the tropics.[2]

Phytoconstituents: Linamarin, esculentoic acid A and B, esculentin, esculin, scopoletin, scopolin, oxalic acid, saponins and others.[3–8]

Traditional Medicinal Uses: Rhizome is made into a poultice and applied to sores.[5]

Pharmacological Activities: Antifungal,[6] Antineoplastic,[7] Hepatoprotective,[9] Hypercholesterolaemic,[8] Antithyroidal,[10] Neurotoxic[11] and Superoxide dismutase inhibition.[12]

Dosage: No information as yet.

Adverse Reactions: Several minor acute intoxications were seen, with complaints of nausea, vomiting, abdominal pain and headache following a meal of cassava.[12]

Toxicity: Hydrocyanic acid — oral, human LD_{50} 570 μg/kg; Oxalic acid — oral, human LD_{50} 700 mg/kg; Saponin — oral, mouse LD_{50} 3000 mg/kg; Tryptophane — oral, rat TD_{50} 1100 mg/kg. Bitter cassava juice can cause death due to cyanide poisoning.[3] Haemorrhage, necrosis, fibrosis and atrophy of the acinar tissue and fibrosis of the islets of Langerhans of the pancreas occurred in dogs fed on cassava.[13] Epidemic spastic paraparesis (konzo) is known to be due to long-term intake of cassava (*M. esculenta*).[10] Tropical ataxic neuropathy, a polyneuropathy with sensorineural hearing loss and optic atrophy can also result from intake of cassava in humans.[14]

Contraindications: No information as yet.

Drug-Herb Interactions: Co-administration of cassava rich diet and alcohol is found to reduce the alcohol induced toxicity in rats.[7]

46. *Melaleuca cajuputi* Roxb. (Myrtaceae)

Gelam, Paper-bark Tree, Kayu Puteh

Leaves of *Melaleuca cajuputi*

Melaleuca cajuputi flowers

Melaleuca cajuputi tree

Description: *Melaleuca cajuputi* Roxb. is a medium sized tree with an often twisted trunk when the tree is very old. Bark is light brown, flaky and peeling. Leaves are simple, slightly curved, with 5–7 longitudinal veins. Flowers are white and borne in dense spikes. Seeds are borne in a capsule.[1–3]

Origin: Native to tropical Asia and Australasia.[4]

Phytoconstituents: Cajeputol, cineole, β-pinene, eugenol, phellandrene, α-terpineol, eugenetin, isoeugenetin and others.[2,5,6]

Traditional Medicinal Uses: In Malaysia, it is used for the treatment of colic and cholera. It is also used externally for thrush, vaginal infection, acne, athlete's foot, verruca, warts, insect bites, cold sore and nits.[5] Cajuput oil is distilled from the leaves and used by the Burmese to treat gout. The Indochinese uses cajeput oil for rheumatism and pain in the joints and as an analgesic.[2] The oil is used externally in Indonesia for burns, colic, cramps, earache, headache, skin diseases, toothache and wounds. When administered internally, it can induce sweating and act as a stimulant and antispasmodic. In the Philippines, the leaves are used to treat asthma.[2,6]

Pharmacological Activities: Antibacterial.[6]

Dosage: For treatment of coryza, influenza, cough, asthma, dyspepsia, earache, toothache, rheumatism, osteodynia, neuralgia, wounds, burns, post partum haematometra. The reported dose for the above ailments is 20 to 40 g of fresh leaves or 5 to 10 g of dried leaves in the form of a decoction or infusion.[7]

Adverse Reactions: Contact dermatitis may occur. Glottal spasms or bronchial spasms or asthma-like attacks may occur if the oil is applied to the facial areas of infants and small children.[8]

Toxicity: Due to the high cineole content, life-threatening poisonings can occur with overdoses of cajuput oil (more than 10 g). Symptoms of poisoning are reduction in blood pressure, circulatory disorders, collapse and respiratory failure.[8]

Contraindications: Do not consume orally in severe liver diseases and presence of inflammatory condition of gastrointestinal tract or of the biliary ducts. Cajuput oil preparations are not to be applied to the faces of infants or small children.[8]

Drug-Herb Interactions: No information as yet.

47. *Melastoma malabathricum* L. (Melastomaceae)

Sendudok, Singapore Rhododendron

Melastoma malabathricum flower

Fruits of *Melastoma malabathricum*

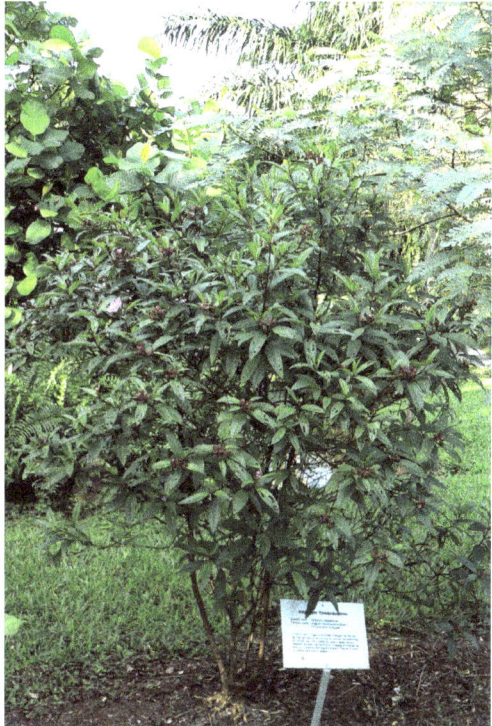

Melastoma malabathricum shrub

Description: *Melastoma malabathricum* L. is a small shrub that can grow up to 1 m tall, densely clothed with tiny scales. Leaves are simple, ovate, elliptic, 4–14 cm by 1.7–3.5 cm, and stiffly papery. Flowers are purple with five petals and fruits are urceolate-globular, 6–15 mm by 6–12 mm and succulent.[1,2]

Origin: Native to tropical and temperate Asia and the Pacific Islands.[2]

Phytoconstituents: Malabathrins A–F, nobotanins B, G, H and J, kaempferol and others.[3–9]

Traditional Medicinal Uses: In Indonesia and Malaysia, the leaves and roots are used as a remedy for diarrhoea and dysentery. The leaves are also used to wash ulcers, to treat piles and for small pox pustules to prevent development of scars.[10] Its bark is used to treat skin diseases.[11]

Pharmacological Activities: Analgesic,[12] Anti-inflammatory,[13] Antineoplastic,[14] Antinociceptive,[13] Antiviral,[14,15] Antioxidant[15] and Antiplatelet.[16]

Dosage: No information as yet.

Adverse Reactions: No information as yet.

Toxicity: *M. malabathricum* was reported to be an aluminium (Al) accumulator (with up to 10 g Al/kg of dry weight in old leaves and up to 7 g Al/kg dry weight in new leaves) and could lead to accumulation of aluminium in the bone and brain, leading to neurotoxicity, when consumed in large amount.[17]

Contraindications: No information as yet.

Drug-Herb Interactions: No information as yet.

48. *Mimosa pudica* L. (Leguminosae)

Touch-me-not, Sensitive Plant, Rumput Simalu

Mimosa pudica leaves and flowers

Description: *Mimosa pudica* L. is a prickly, herbaceous weed. Leaves bipinnate, very sensitive, fold together when touched, in rain or at night. Leaflets are 15–20 pairs, small oblong, nearly sessile. Flowers are pink and fruits are flat pods covered with bristles.[1–3]

Origin: Native to Brazil.[3]

Phytoconstituents: Mimosine, 2-Hydroxymethyl-chroman-4-one and others.[4–6]

Traditional Medicinal Uses: The plant is used on cuts and wounds.[4,7] It is also used for childbirth and infertility in Trinidad and Tobago.[8] Bath with plant decoction relieves insomnia.[4] The leaves are used for hydrocele, dressing for sinus, sores, piles and swelling of feet. In Mexico, aqueous extracts from dried leaves are employed to alleviate depression.[9] In the Philippines, the leaves soaked in coconut oil is used for ulcers.[4]

Pharmacological Activities: Anthelmintic.[10] Antibacterial,[11] Anticonvulsant,[12] Antidepressant,[9] Antifertility,[13] Antifungal,[5] Hyperglycaemic,[14] Antioestrogenic[15] and Antivenom.[16,17]

Dosage: No information as yet.

Adverse Reactions: No information as yet.

Toxicity: No information as yet.

Contraindications: No information as yet.

Drug-Herb Interactions: No information as yet.

49. *Mirabilis jalapa* L. (Nyctaginaceae)

Four O'Clock Flower, Bunga Pukul Empat

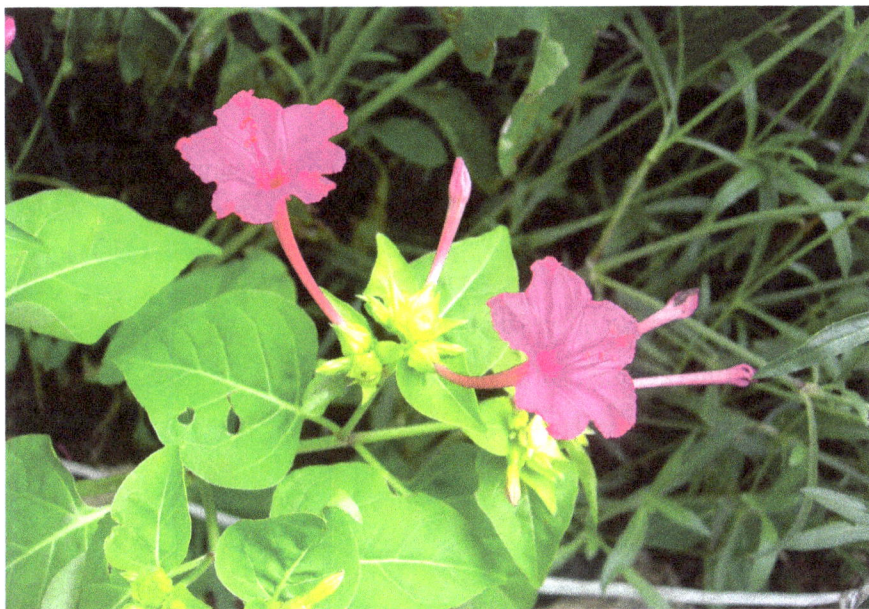

Flowers of *Mirabilis jalapa*

Description: *Mirabilis jalapa* L. is an erect herb that can grow up to 1 m tall. Leaves are simple, heart-shaped, 3–12 cm long, opposite, tapering to a pointed end. Flowers are bisexual, red, pink, yellow or white, with perianth distinctly constricted above, and they bloom late in the afternoon. Fruits are black and globose, 5–8 mm in diameter.[1,2]

Origin: Native to tropical America, introduced in China and in many tropical areas.[2]

Phytoconstituents: Trigonellin, 2'-*O*-methylabronisoflavone, 6-methoxy-boeravinone C and betaxanthins.[3,4]

Traditional Medicinal Uses: Its leaves are used as a decoction for abscesses, juice for wounds and cooked with pork as tonic.[5] The leaves are also placed on boils, blisters, and to relieve urticaria.[6] In Indochina, the seeds are used as a purgative.[3] The flowers release a strong odour at night which will stupefy or drive away mosquitoes.[7] The roots are used as a laxative in a decoction, with or without pork, for colds, inflammation and leucorrhoea.[5] In Malagassy, Madagascar, they are used to treat intestinal pains. In South Africa, the roots are used as purgative agents.[7]

Pharmacological Activities: Antibacterial,[8,9] Antifungal,[4,10] Antineoplastic and Abortifacient.[11]

Dosage: Approximately 8–10 g of roots are taken as a purgative.[12]

Adverse Reactions: No information as yet.

Toxicity: No information as yet.

Contraindications: No information as yet.

Drug-Herb Interactions: No information as yet.

50. *Morinda citrifolia* L. (Rubiaceae)

Mengkudu, Indian Mulberry, Noni

Flowers and fruits of *Morinda citrifolia*

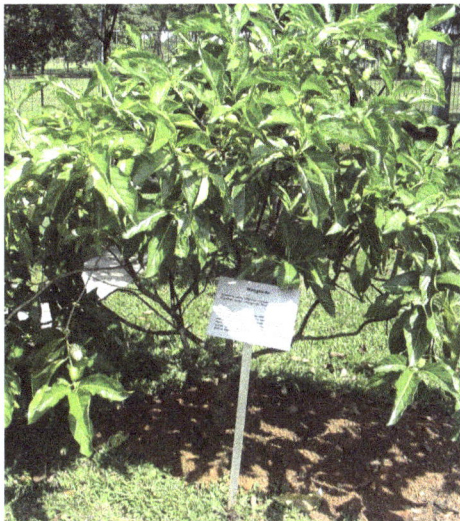

Morinda citrifolia tree

Description: *Morinda citrifolia* L. is a tree that grows up to 9 m. Leaves are simple, large, 10–15 cm by 20–30 cm, decussate and stipulate. Blade is broadly elliptic to obovate, glossy, soft and succulent. Flowers are small, sessile, white, 1.5–2 cm across, terminal and axillary. The fruit is succulent, oval, 5–7 cm across, light greyish green and turns yellow upon maturity.[1–5]

Origin: Native to tropical and temperate Asia and Australasia.[6]

Phytoconstituents: Morintrifolins A and B, morindin, morindone, rubiadin, morindadiol, morindicone, morinthone, morindicinone, morindicininone, noniosides E–H, morinaphthalenone and others.[1,4,7–28]

Traditional Medicinal Uses: The whole plant is used to treat aching bones and arthritis. In Malaysia, the heated leaves are applied to the chest and abdomen to treat coughs, nausea, colic and enlarged spleen. In Japanese and Chinese medicine, *M. citrifolia* is used to treat fever and as a tonic whereas in Indochina, the fruit is prescribed for lumbago, asthma and dysentery.[1] A decoction of the leaves taken orally is effective for the treatment of fever, dysentery and diarrhoea. Poultice of fresh leaves can be used to cure furunculosis. The fruit when consumed together with a little salt is stomachic, aperient, and active on dysentery, uterine haemorrhage, cough, coryza, oedema and neuralgia. The root bark has beneficial effects in hypertension, osteodynia and lumbago.[7] The fruit is also used for throat and gum complaints, dysentery and leucorrhoea while the root is used as a cathartic and febrifuge.[29] The processed fruit juice is also in great demand for various ailments such as diabetes, high blood pressure, headaches, heart disease, AIDS, cancers, gastric ulcers, mental depression, senility, poor digestion, atherosclerosis and drug addiction.[30]

Pharmacological Activities: Analgesic,[30] Anthelmintic,[30,31] Antiangiogenic,[32] Antibacterial,[11,30,33] Anticancer/Antineoplastic,[34-37] Antitumour,[30] Antihypertensive,[30] Antiviral,[30] Anti-inflammatory,[24,25,38] Antioxidant,[14,16,39-44] Antiprotozoal,[45] Antidiabetic,[26] Sedative/Anxiolytic,[25] Chemopreventive,[24] Insecticidal,[9] Wound healing[46] and Hepatoprotective.[47]

Dosage: For treatment of hypertension, osteodynia and lumbago, 10 to 20 g of rootbark is prescribed daily as a decoction or alcoholic maceration of torrefied material.[7]

Adverse Reactions: Sedation, nausea, vomiting, anorexia, hypersensitivity, hyperkalemia may occur.[48, 49]

Toxicity: The lyophilised aqueous extract of *M. citrifolia* roots did not exhibit any toxic effects but did show a significant, dose-related, central analgesic activity in the writhing and hotplate tests.[50] Chemical analysis and genotoxicity tests revealed that noni juice does not have a genotoxic potential and that genotoxic anthraquinones do not exist in noni juice.[51] However, according to some, anthraquinones are the most likely hepatotoxic components found in *M. citrifolia*.[52] Hepatitis induced by Noni juice had been reported.[53,54]

Contraindications: Should not be used during pregnancy and lactation. Contraindicated in people with hyperkalemia or with hypersensitivity to *M. citrifolia*.[48]

Drug-Herb Interactions: No information as yet.

51. *Nelumbo nucifera* Gaertn. (Nymphaeaceae)

Sacred Lotus, East Indian Lotus, Oriental Lotus

Nelumbo nucifera plants with flowers

Description: *Nelumbo nucifera* Gaertn. is an aquatic plant that grows in shallow waters. Leaves are green, round, 30–60 cm across and with long petiole. Flowers are pink, white or red, 10–30 cm and solitary. Fruits are non-edible and non-fleshy.[1]

Origin: Native to tropical and temperate Asia, Australia and Eastern Europe.[2]

Phytoconstituents: Nuciferin, nornuciferin, nelumboroside A & B, nelumstemine, dotriacontane, ricinoleic, roemerin, liensinine, neferine, lotusine, liriodenine, asimilobin, pronuciferine and others.[3–8]

Traditional Medicinal Uses: The leaves are used to treat sunstroke, diarrhoea, dysentery, fever, dizziness and vomiting of blood.[9] The plant is used as an antidote for mushroom poisoning[9] and for smallpox.[10] In Ayurveda,

105

the plant is used to treat cholera, diarrhoea, worm infestation, vomiting, exhaustion and intermittent fever.[3] The fruits are used in decoction for agitation, fever, heart and haematemesis while the stamens are used to "purify the heart, permeate the kidneys, strengthen virility, to blacken the hair, for haemoptysis and spermatorrhoea".[10] They are also used to treat premature ejaculation,[9] as astringent for bleeding,[3] excessive bleeding from the uterus,[9] abdominal cramps, bloody discharges, metrorrhagia, non-expulsion of the amniotic sac,[10] and as cooling agent during cholera.[11] The seeds are believed to promote virility, for leucorrhoea and gonorrhoea.[10] Powdered beans are used in treating digestive disorders, particularly diarrhoea.[3] They are also used as a tonic,[9] for enteritis, insomnia, metrorrhagia, neurasthenia, nightmare, spermatorrhoea, splenitis and seminal emissions.[10] The roots are for the treatment of diarrhoea, dysentery, dyspepsia, ringworm and other skin ailments and as a tonic as well.[10,11]

Pharmacological Activities: Antianxiety,[12] Antiarrhythmic,[13] Antibacterial,[4] Anticonvulsant,[14] Antidiarrhoeal,[15] Anti-inflammatory,[16] Hepatoprotective,[18] Antioxidant,[8,17–20] Antiplatelet,[21] Antiproliferative,[22] Antipyretic,[23,24] Antiviral,[25,26] Hypoglycaemic,[27,28] Hypolipidaemic,[29] Immunomodulatory[30] and Insecticidal.[4]

Dosage: A daily dose of 10 to 30 g of the ripe seeds as a decoction or powder is used for the treatment of neurasthenia, spermatorrhoea and metrorrhoea. The pericarps should be removed before using the seeds. Decoctions of 15 to 20 g and 2 to 4 g of dried leaves and seed cores respectively have been used for treating insomnia, haemorrhage and haematemesis. 6 to 12 g of plumules, 5 to 10 g of filaments, or 15 to 30 g of the receptacles in the form of a decoction, are used in the treatment of bloody stools, haematuria, uterine haemorrhage and haematemesis.[31] A daily dose of 5 to 8 g of rhizome or seeds taken by mouth is recommended.[32]

Adverse Reactions: Stamen, receptacle, rhizome node, leaf, and embryo can be safely consumed with proper usage.[33] No known side effects with appropriate therapeutic dosages.[3]

Toxicity: No information as yet.

Contraindications: Seed is contraindicated in constipation and stomach distention.[33]

Drug-Herb Interactions: No information as yet.

52. *Nephelium lappaceum* L. (Sapindaceae)

Rambutan, Hairy Lychee

Nephelium lappaceum fruits

Nephelium lappaceum tree

Description: *Nephelium lappaceum* L. is an evergreen tree that can grow up to 10 m tall. Leaves are pinnate, 25–40 cm long, compound with 2–4 pairs of leaflets. Blades are elliptic or obovate, 6–18 cm by 4–7.5 cm, thinly leathery, glabrous, with 7–9 pairs of lateral veins. Flowers are greenish white, fragrant and with no petals. The fruit is hairy, turning from green to red when mature and contains one hard seed.[1–3]

Origin: Native to Indonesia, Malaysia, Philippines and Thailand.[3]

Phytoconstituents: Type II and III cyanolipids, β-caryophyllene, monoterpene lactones 1 & 2, paullinic acid and others.[4–7]

Traditional Medicinal Uses: The fruits are used to treat diarrhoea and fever,[2] and also for the treatment of dysentery and dyspepsia.[8] Its seeds are narcotic[8] and the bark is used to treat disease of the tongue while the roots are used for fever.[2]

Pharmacological Activities: Antiviral,[9] Antifungal,[6] Antibacterial,[10] Antioxidant[10–13] and Cytotoxic.[12,13]

Dosage: No information as yet.

Adverse Reactions: No information as yet.

Toxicity: No information as yet.

Contraindications: No information as yet.

Drug-Herb Interactions: No information as yet.

53. *Nerium oleander* L. (Apocynaceae)

Oleander

Flowers of *Nerium oleander*

Nerium oleander shrub

Description: *Nerium oleander* L. is a small shrub up to 2 m high. Leaves are very narrowly elliptic, 5–21 by 1–3.5 cm, dark green, without stipules, leathery and arranged in whorls of three. Flowers are showy and fragrant. Sepals are narrowly triangular to narrowly ovate, 3–10 mm. Corolla is purplish red, pink, white, salmon, or yellow. Fruits consist of cylindrical follicles, 12–23 cm. Seeds are oblong, coma, about 0.9–1.2 cm.[1,2]

Origin: Native to southern Europe,[2] and widely cultivated and naturalised in Asia, Europe and North America.[1]

Phytoconstituents: Oleandroside, kaneroside, neriaside, nerigoside, neriumoside, neridiginoside, nerizoside, neritaloside, proceragenin, neridienone A, cardenolides N-1 to N-4 and others.[3–11]

Traditional Medicinal Uses: The plant is used in Ayurveda to treat scabies, eye disease and haemorrhoids.[3] It is used to treat parasitic infection in Calabria (Southern Italy).[12] Leaf decoction is used to treat diabetes in southeastern Morocco.[13] Bark, leaf, flower are used medicinally as a cardio-tonic and diuretic.[14]

Pharmacological Activities: Analgesic,[15] Anti-inflammatory,[8,9,15] Antibacterial,[16] Anticancer/Antineoplastic,[8,9,17–22] Antifungal,[23] Depressant,[24,25] Antimitotic,[26] Insecticidal,[27] Larvicidal,[28] Muscle stimulatory[29–30] and inhibits Nuclear factor-kappa B (NF-κB) activation.[31]

Dosage: No information as yet.

Adverse Reactions: Depression, dizziness, stupor, headache, nausea, vomiting, anorexia, abdominal cramps, spontaneous abortion, hypersensitivity, contact dermatitis, hyperkalemia and tachypnea.[32]

Toxicity: Toxic to humans[33–39] and animals.[40–48] The plant contains numerous toxic compounds, many of which can be deadly to people.[33–39] Ingestion can cause both gastrointestinal and cardiac effects and also affect the central nervous system.[33–39]

Contraindications: Should not be used during pregnancy and lactation, in children and in persons with hypersensitivity to oleander.[32] Should not be taken internally.[49]

Drug-Herb Interactions: Fatal digitalis toxicity can occur with concurrent usage of cardiac glycosides such as digoxin and digitoxin.[32] Concurrent use of quinidine, calcium salts, saluretics, laxatives or glucocorticids increases efficacy as well as side effects.[3]

54. *Ophiopogon japonicus* Ker-Gawl. (Liliaceae)

Dwarf Lilyturf, Mondo Grass, Mai Men Dong

Ophiopogon japonicus herb

Description: *Ophiopogon japonicus* Ker-Gawl. is an evergreen, stemless, rhizomatous herb. Leaves are sessile, long, 10–50 cm by 2–4 mm, linear, grass-like, 3–7 veined, and have pointed tips. Flowers are either solitary or paired.[1,2]

Origin: Native to Japan and Korea.[2]

Phytoconstituents: Ophiopogonin D and E, ophiopogonin C′ and D′, bornanol, ophiopogonanone A, C, E and F, ophiopojaponin D and others.[3–9]

Traditional Medicinal Uses: The roots are used to cool the body system, as a tonic, purgative, thirst quencher, treatment for sore throat, cough and fever. [10] In China, the roots are also used to treat bronchitis, cold, dysuria, haemoptysis, laryngitis, restlessness, thirst, tuberculosis, stress, as an aphrodisiac, promoting fertility and memory, also as a sialogogue, to treat cancer and frequently included in polyherbal prescriptions for diabetes mellitus. The Indochinese uses the rhizomes to treat fever and inflammation, as a febrifuge, galactagogue, and also for intestinal, kidney and liver ailments.[3]

Pharmacological Activities: Antiarrhythmic,[11] Anti-inflammatory,[12,13] Antithrombotic,[14,15] Immunomodulatory,[16] Immunostimulatory,[17,18] Cardioprotective[19,20] and Chemoprotective.[21]

Dosage: The daily dose of tuberous roots of 6 to 20 g in the form of a decoction, pills or syrup consumed as an expectorant or antitussive.[22]

Adverse Reactions: No information as yet.

Toxicity: LD_{50} value of *O. japonicus* in mice was more than 2 g/kg intraperitoneally.[12]

Contraindications: No information as yet.

Drug-Herb Interactions: No information as yet.

55. *Peltophorum pterocarpum* Backer ex K. Heyne (Leguminosae)

Jemerlang Laut, Yellow Flame, Yellow Flamboyant

Peltophorum pterocarpum tree

Peltophorum pterocarpum flowers

Fruits of *Peltophorum pterocarpum*

Description: *Peltophorum pterocarpum* Backer ex K. Heyne is a large tree with dome-shaped crown that can grow up to 24 m tall. The main rachis are 15–30 cm long and the pinnae are 6–20 paired with each pinna having 20–30 pairs of oblong leaflets. Flowers are yellow and in large bunches. The tree bears pods which are oblong, 5–10 cm long, flat and thin.[1–3]

Origin: Native to Malaysia, Ceylon, the Andamans and North Australia.[3]

Phytoconstituents: Rhamnetin, hirusitidin, bergenin and others.[4]

Traditional Medicinal Uses: The bark extract is used internally to cure dysentery and externally as a lotion to treat sprains, muscular aches, ulcers, and as an eye lotion, gargle and tooth powder.[2]

Pharmacological Activities: Antibacterial[5–7] and Antifungal.[7]

Dosage: No information as yet.

Adverse Reactions: No information as yet.

Toxicity: No information as yet.

Contraindications: No information as yet.

Drug-Herb Interactions: No information as yet.

56. *Persicaria hydropiper* L. (Polygonaceae)

Water Pepper, Laksa Plant

Persicaria hydropiper herbs

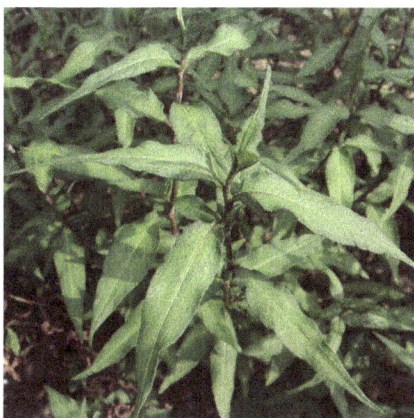

Persicaria hydropiper top view

Description: *Persicaria hydropiper* L. is an erect herb that can grow up to 30–50 cm tall. Leaves are oblong-lanceolate, short petioled, tapering to a pointed end, ciliate on the under surface. Flowers are white, in sparse, thin and with hanging false ears. Fruit is black, nut-like and has a flat and domed side.[1]

Origin: Native to Northern Africa, tropical and temperate Asia, Australia and Europe.[2]

Phytoconstituents: Rhamnazin, hydropiperoside, polygoidal, warburganal, isopolygodial, isodrimeninol, drimenol, confertifolin and others.[1,3–5]

Traditional Medicinal Uses: The liquid extract of the plant is used as a contraceptive and a haemostatic.[6] The plant is also used alone or with other herbs decocted for diarrhoea, dyspepsia, dysentery, enteritis, diuretic, expelling worms, heat stroke, itching skin, haemorrhage, jaundice and cancer as well.[7] In folk medicine, it is used internally for uterine bleeding, menstrual bleeding, bleeding of haemorrhoids, gastrointestinal bleeding, rheumatic pain, as a diuretic, for bladder and kidney disease, and gout. It is used externally for poorly healing wounds, sprains and contusions.[1] The leaves are pounded and applied to skin diseases, for uterine disorders,[7] while its seeds are used as carminative, diuretic and stimulant.[7]

Pharmacological Activities: Analgesic,[8] Anthelmintic,[9] Antifertility,[10,11] Antifungal,[12–14] Antineoplastic,[15] Antioxidant,[12] Antimutagenic[16] and Insect repellent.[17,18]

Dosage: A tea prepared by pouring 0.25 L of hot water over 1 heaped teaspoon of the dried plant extract and strained after 10 min is to be drunk 3 times a day.[1] Homeopathically, it is used to treat varicose veins, 5 drops, 1 tablet or 10 globules are to be taken every 30 to 60 min in acute cases, or 1 to 3 times daily for chronical cases.[1]

Adverse Reactions: No known side effects with therapeutic dosages. Larger amounts can cause gastroenteritis. Skin irritation may occur if applied externally.[1]

Toxicity: Toxic to animals[19–21] and the LD_{50} for its chloroform leaf extract was 758.58 mg/kg in male albino mice.[22]

Contraindications: No information as yet.

Drug-Herb Interactions: No information as yet.

57. *Phyllanthus amarus* Schum. & Thonn.
(Euphorbiaceae)

Pick-a-back, Carry Me Seed, Ye Xia Zhu

Phyllanthus amarus herbs

Description: *Phyllanthus amarus* Schum. & Thonn. is a small herbaceous annual plant that can grow up to 60 cm tall. Leaves are simple, alternate, green and stipulate. The blade is 3–8 mm by 2–4.5 mm and oblong-elliptic. The fruits are green, depressed globose in shape, 3-lobed and smooth. Both the flowers and fruits are borne under the branches.[1,2]

Origin: Native to Mexico and South America.[3]

Phytoconstituents: Phyllanthusin D, geraniin, corilagin, elaeocarpusin, amariin, amariinic acid, amarosterol-A and B, phyllantin, hypophyllantin and others.[4–11]

Traditional Medicinal Uses: The aerial part of the plant is used for various conditions. In Chinese medicine, the plant is made into a tea to cure kidney problems, venereal diseases, stones in the kidneys and bladder. The Malays use it to increase menstrual flow, reduce fever and cure colic. It is used by the Indians as a fish poison.[5] Indians also use the plant as liver tonic to treat liver ailments, ascites, jaundice, diarrhoea, dysentery, intermittent fever, conditions of the urogenital tract, eye disease, scabies, ulcers and wounds.[12] In Vietnam, it is used to induce sweating, and increase menstrual flow. It is also prescribed for toothache, muscle spasms and gonorrhoea. It is considered a diuretic, colic remedy and abortifacient in Southeast Asia.[5] It is also commonly used in Benin, Africa, as folk medicine against malaria.[13]

Pharmacological   Activities: Analgesic,[14]   Antibacterial,[15,16] Antidiarrhoeal,[17] Antifertility,[18] Antifungal,[19] Anti-inflammatory,[20–24] Antineoplastic,[25–29] Antioxidant,[30–32] Antiplasmodial,[13,33] Antiviral,[34–39] Diuretics,[40] Hepatoprotective,[41–43] Hypoglycaemic,[11,30,44] Inhibition of gastric lesion,[22] Antimutagenic,[27,45] Insecticidal[46] and Radioprotective.[47]

Dosage: A decoction may be prepared with 10 plants in 1 L of water.[12]

Adverse Reactions: No known side effects with therapeutic dosages.[12]

Toxicity: Non-toxic to mice at a dose of 100 mg/kg body weight[18] but at doses of 400 mg/kg, 800 mg/kg and 1000 mg/kg body weight in rats.[48]

Contraindications: No information as yet.

Drug-Herb Interactions: An alcoholic extract of *P. amarus* was found to inhibit cytochrome P450 enzymes both *in vivo* as well as *in vitro*.[48]

58. *Piper nigrum* L. (Piperaceae)

Pepper, Lada

Leaves and fruits of *Piper nigrum*   *Piper nigrum* herbs

Description: *Piper nigrum* L. is a perennial woody vine with aerial roots at the stem nodes. Leaves are shiny, broad, alternate and heart-shaped. Flowers are small and white. The fruits are globular berries, about 5 mm in diameter, green when unripe but turn bright red upon maturity and each bears a single seed.[1]

Origin: Native to Southeast Asia.[2]

Phytoconstituents: Piperine, sabinene, nigramides A–S, pipertipine, pipercitine; pellitorine, guineensine, piperettine, pipericine, dipiperamides A–C, pipnoohine, pipyahyine and others.[3–10]

Traditional Medicinal Uses: The plant is used in many Asian countries as a stimulant, for the treatment of colic, rheumatism, headache, diarrhoea, dysentery, cholera, menstrual pains, removing excessive gas and increasing the flow of urine.[11] It is also used in folk medicine for stomach disorders and digestive problems, neuralgia and scabies. In Ayurveda, it is used for arthritis, asthma, fever, cough, catarrh, dysentery, dyspepsia, flatulence, haemorrhoids, urethral discharge and skin damage. In Chinese medicine, it is used for vomiting, diarrhoea and gastric symptoms. Homeopathically, it is used for irritation of mucous membrane and galactorrhoea.[3] A heavy dose of pepper with wild bamboo shoots is said to cause abortion.[4] In Assam, a method of birth control include *Cissampelos pareira* in combination with *Piper nigrum*, root of *Mimosa pudica* and *Hibiscus rosa-sinensis*.[12] The fruits are used to remove excessive gas in system, increase flow of urine, treat colic, rheumatism, headache, diarrhoea, dysentery, cholera, and menstrual pains.[13] White pepper is used for cholera, malaria, stomachache while black pepper is used for abdominal fullness, adenitis, cancer, cholera, cold, colic, diarrhoea, dysentery, dysmenorrhoea, dysuria, furuncles, headache, gravel, nausea, poisoning due to fish, mushrooms or shellfish.[4]

Pharmacological Activities: Antibacterial,[14–16] Antidiabetic,[17] Antifungal,[18] Anti-inflammatory,[19] Antineoplastic,[20] Antioxidant,[21–26] Gastroprotective,[27] Hepatoprotective,[28,29] Hypolipidaemic,[30,31] Antimutagenic,[32] Antithyroidal[33] and Insecticidal.[5,6,9,34–38]

Dosage: A single dose ranges from 300–600 mg of the berries and is taken internally for stomach disorder.[3] Homeopathic recommendations for irritation of mucous membranes and galactorrhoea are 5–10 drops, 1 tablet or 5–10 globules 1–3 times daily.[3] For the treatment of haemorrhoids, 5–15 whole peppercorns are recommended to be taken.[39] For congestion, cold, head cold, chicken soup with black pepper can be taken.[40] The average daily dose of the berries is stated to be 1 to 3 g as a decoction, powder or pills, for the treatment of dyspepsia, vomiting, diarrhoea and colic resulting from cold.[41]

Adverse Reactions: No known side effects with appropriate therapeutic dosages.[3] Hypersensitivity may occur in certain people. The fruit can be safely consumed with proper usage.[42] However, when taken in large amounts in children, apnoea can occur.[43]

Toxicity: *P. nigrum* did not cause any adverse effects when tested with rats at doses 5 to 20 times normal human intake.[44]

Contraindications: Not to be taken during pregnancy and lactation or by children and people who are allergic to it.[43]

Drug-Herb Interactions: *P. nigrum* extracts inhibits the activity of cytochrome P450 enzymes.[8,45–49] Piperine enhanced the bioavailability of beta lactam antibiotics, amoxycillin trihydrate and cefotaxime sodium significantly in rats.[50] Piperine potentiated pentobarbitone sleeping time in a dose dependant manner, with peak effect at 30 min.[19] Piperine enhanced the bioavailability of phenytoin significantly, possibly by increasing the absorption.[51] Piperine from black peppers increased the AUC of phenytoin, propranolol and theophylline in healthy volunteers and plasma concentrations of rifampicin in patients with pulmonary tuberculosis.[52] Avoid concurrent use with drugs metabolised by cytochrome P450 enzymes.[43] Avoid concurrent use with anticoagulant agents as well.[53]

59. *Piper sarmentosum* Roxb. (Piperaceae)

Wild Pepper, Kadok, Sirih Tanah

Piper sarmentosum herb

Close up of *Piper sarmentosum* flowers and fruits

Description: *Piper sarmentosum* Roxb. is a climbing herb that can grow up to 10 m long with long runners that can develop into plantlets. Leaves are alternate, simple, 7–14 cm by 6–13 cm, heart-shaped and young leaves have a waxy surface. Flowers are bisexual or unisexual, in terminal or leaf opposite spikes. Fruit is small, dry, with several rounded bulges. Plant has a characteristic pungent odour.[1–5]

Origin: Native to Cambodia, India, Indonesia, Laos, Malaysia, Philippines and Vietnam.[5]

Phytoconstituents: Sarmentosine, sarmentine, (+)-sesamin, horsfieldin, brachystamide B, sarmentamide A, B, and C, (+)-asarinin, methyl piperate and others.[6–11]

Traditional Medicinal Uses: The whole plant can be used to treat fever and aids digestion.[3] The fruit is used as an expectorant[6] while the roots are used to treat toothache, fungal dermatitis on the feet, coughing, asthma and pleurisy.[6,12]

Pharmacological Activities: Antibacterial,[8,10,13] Antineoplastic,[6] Antiprotozoal,[10,14,15] Antipsychotic,[16] Hypoglycaemic,[17] Antifungal,[10] Insecticidal,[18,19] Antituberculosis,[11] Antimalarial[14] and Antioxidant.[20]

Dosage: No information as yet.

Adverse Reactions: No information as yet.

Toxicity: No information as yet.

Contraindications: No information as yet.

Drug-Herb Interactions: No information as yet.

60.   *Plantago major* L. (Plantaginaceae)

Common Plantain, Whiteman's Foot, Daun Sejumbok

Plantago major herbs

Description: *Plantago major* L. is a small perennial herb. Leaves are nearly all basal, exstipulate, lanceolate to ovate, 5–20 cm long and rosette. Flowers are small, white, in dense spike-like inflorescence. Sepals are broadly elliptic, oblong to rounded obtuse or subacute and corolla are greenish or yellowish, with four lobed and imbricate. Seeds are dull black and endospermous.[1,2]

Origin: It is found in Europe, Northern and Central Asia, and introduced all over the world.[2]

Phytoconstituents: Aucubin, catalpol, scutellarein, nepetin, chlorogenic acid, neochlorogenic acid, hispidulin, homoplantaginin, nepitrin, ursolic acid and others.[3,4]

Traditional Medicinal Uses: The Greeks and Romans used it as an astringent, to heal wounds, asthma, fever and eye disorders.[5] In Brazil, it has been used to treat skin ulceration (cutaneous leishmaniasis) caused by *Leishmania braziliensis*.[6] *P. major* has been used in Turkey in the treatment of ulcers by taking the powdered dried leaves together with honey daily before breakfast.[7] Infusion of the leaf has been taken for diarrhoea, ulcers, bloody urine, digestive disorders, and excess mucous discharge. The American Indian groups make use of a poultice of the leaves for pain, swelling, cuts, wounds, sores, infections, blisters, insect bites, snakebites and haemorrhoids.[5] Its seeds are used to induce sweating, increase flow of urine, treat diarrhoea, dysentery, rheumatism, malaria, asthma, kidney problems, bladder diseases, gonorrhoea and piles.[3] Its roots are used to treat fever, respiratory infections and constipation.[5] The Commision E approved the internal use of plantain for catarrhs of the respiratory tract and inflammatory alterations of the oral and pharyngeal mucosa while its external application is approved for inflammatory reactions of the skin.[8]

Pharmacological Activities: Analgesic,[9] Antibacterial,[10,11] Antidiarrhoeal,[12] Anti-inflammatory,[4,13–15] Anticancer,[10,16–18] Antioxidant,[19] Antiprotozoal,[20,21] Antiviral,[22] Immunomodulatory,[23,24] Immunostimulatory,[25,26] Proliferative,[10] Antiulcerogenic,[6] Antimutagenic,[27,28] Uterotonic[29] and Wound healing.[30]

Dosage: A daily dose of 8 to 16 g of the whole plant or seeds in the form of a decoction or extract is used to treat oedema, dysuria, haematuria, persistent cough, bronchitis and ophthalmia.[31] Approximately 2 to 4 ml of the fluid extract taken orally three times daily serves for general well being.[32] As a rinse or gargle, 1.4 g of cut herb is immersed in 150 ml of cold water for 1 to 2 hours.[8] For internal use, 1.4 g of herb is immersed in 150 ml of boiled water for 10–15 min, drunk as infusion, for 3 to 4 times daily.[8]

Adverse Reactions: Nausea, vomiting, diarrhoea, anorexia, bloating, hypersensitivity and dermatitis may arise. Life threatening anaphylaxis may occur in more serious cases.[32]

Toxicity: The 70% ethanol extract was found to be toxic to shrimps[33] but *P. major* possesses a low toxicity in rats after oral and intraperitoneal administration.[34]

Contraindications: Should not be used during pregnancy and lactation. Should not be used in persons with intestinal obstruction or those who developed hypersensitivity to plantain.[32]

Drug-Herb Interactions: Decreases the effects of carbamazepine and enhances the effects of cardiac glycosides, β-blockers and calcium channel blockers.[32]

61. *Punica granatum* L. (Punicaceae)

Pomegranate, Delima

Punica granatum fruit

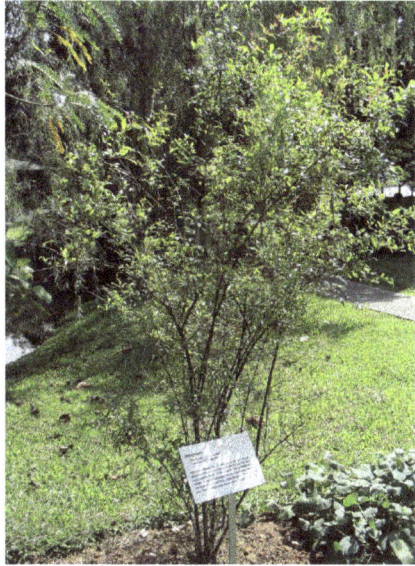

Punica granatum tree

Description: *Punica granatum* L. is a small woody shrub that can grow up to 3–4.5 m tall. Leaves are opposite, oblong-lanceolate, 3–8 cm and branches are spiny. Flowers are large, trumpet-shaped, and bright orange-red in colour. Fruit is globose, with waxy surface, tough leathery skin, turns deep pink or red upon maturity, and contains numerous seeds with fleshy covering.[1]

Origin: Native to temperate and tropical Asia.[2]

Phytoconstituents: Punicafolin, punicalagin, friedelin, betulic acids, estrone, estradiol, piperidine, pomegranatate, pseudopelletierine and others.[3–7]

Traditional Medicinal Uses: The plant is used in folk medicine to expel worms, for diarrhoea, dysentery, as an abortifacient, astringent, for haemorrhoids (when used externally) and as a gargle for sore throat. In Ayurveda it is used for diarrhoea, dysentery, vomiting, and eye pain. Used in Chinese medicine for chronic diarrhoea and dysentery, it is also used to treat blood in stools, worm infestation, and anal prolapse. Homeopathically, it is used for gastrointestinal disturbances.[5] The leaves are used to relieve itch. The pericarp is used to treat diarrhoea, rectocele and to expel pinworms.[8] The dried pericarp is decocted with other herbs for colic, colitis, diarrhoea, dysentery, leucorrhoea, menorrhagia, oxyuriasis, paralysis and rectocele,[3] to treat piles, yellowish discharge from the vagina, sore throat, bad breath and nose bleed.[8] The stem bark is used as an emmenagogue and for dysentery.[3]

Pharmacological Activities: Anthelmintic,[3,9,10] Antibacterial,[11–21] Anticancer/ Antineoplastic,[22–36] Antidiabetic/Hypoglycaemic,[37–41] Antidiarrhoeal,[42] Antifertility,[43] Antifungal,[44,45] Anti-inflammatory,[46,47] Antimalarial,[48] Antioxidant,[34,48–62] Antiviral,[63–65] Gastroprotective,[66,67] Hepatoprotective,[68] Hypolipidaemic,[41,69] Immunomodulatory,[70,71] Neuroprotective,[72–75] Antiatherogenic,[76–78] Wound healing,[79,80] Larvicidal[81] and Molluscicidal.[82,83]

Dosage: Daily dose for the treatment of tapeworm is one part pericarp, root, or stem bark boiled with 5 parts of water. Bark juice extract is recommended for tapeworm at a single dose of 20 g. Otherwise, 250 parts powdered bark are boiled in 1500 parts water for 30 minutes.[5] Doses of 4–5 g of powdered dried flower are used in haematuria, haemorrhoids, dysentery, chronic diarrhoea, and bronchitis. Either 1.5–3 g of root and bark powder, 100–200 ml of bark decoction (for children, 28–56 ml) is used for anthelmintic purposes.[84] 7 g flower in 300 ml water is prescribed for inflamed mouth and throat; 5–12 g root or stem bark boiled in 240 cm^3 water until 1/3 has evaporated and taken on a 3 hourly basis on empty stomach 2 hours after taking 40 ml castor oil.[85] A daily dose of 20 to 50 g of dried root bark or stem bark as a decoction is used in the treatment of taeniasis (tapeworm infection). Daily dose of 15 to 20 g fruit rind in the form of a decoction is taken to treat dysentery and diarrhoea.[86]

Adverse Reactions: No known side effects with appropriate therapeutic dosages. In some cases, gastrointestinal disturbances may occur due to high tannin content.[5] Pomegranate fruits have rarely been reported to cause immediate hypersensitivity.[87]

Toxicity: The bark[5,9,88,89] and peel[90,91] are toxic.

Contraindications: Contraindicated in the elderly, children and pregnant women.[92] Pericarp is contraindicated in diarrhoea. Should not be taken with fats or oils when it is used for killing parasites.[88] Should not be used by patients with hepatic diseases or asthma, and who are hypersensitive to pomegranate.[91]

Drug-Herb Interactions: Inhibition of cytochrome P450 enzymes,[93,94] interaction with carbamazepine[95] and interaction with tolbutamide.[96]

62. *Rhodomyrtus tomentosa* (Ait.) Hassk
(Myrtaceae)

Rose Myrtle

Rhodomyrtus tomentosa plants

Flowers of *Rhodomyrtus tomentosa*

Description: *Rhodomyrtus tomentosa* (Ait.) Hassk is a bushy shrub with woolly young parts. Leaves are simple, oblong, 4–10 cm, opposites with three main longitudinal veins. Flowers are large, showy and pink, 3–4 cm wide, solitary and auxillary. Berries are oblong, 1–1.25 cm long, dark purple, fleshy and aromatic.[1–3]

Origin: Native to tropical and temperate Asia.[4]

Phytoconstituents: Rhodomyrtone, casuariin, castalagin, friedelin and others.[5–11]

Traditional Medicinal Uses: The leaves are used by the Chinese as a pain-killer. They are also used in Indonesia to heal wounds.[1] In Malaysia, the plant, including the roots, are decocted to treat diarrhoea and heartburn.[1,5] The buds and young leaves are used for treatment of colic, diarrhoea, dysentery, abscesses, furunculosis and haemorrhage. Concentrated decoction of leaves is used as a disinfectant for wounds, impetigo and abscesses.[12] The seeds are used as a digestive tonic and to treat snake bites.[1]

Pharmacological Activities: Antibacterial.[6]

Dosage: Daily dose of 10 to 30 g of fresh buds or young leaves in the form of an extracted juice or dried for use as a powder or in a decoction is used for the treatment of colic, diarrhoea, dysentery, abscesses, furunculosis and haemorrhage.[12]

Adverse Reactions: No information as yet.

Toxicity: No information as yet.

Contraindications: No information as yet.

Drug-Herb Interactions: No information as yet.

63. *Rhoeo spathacea* (Sw.) Stearn (Commelinaceae)

Purple-leaved Spider Wort, Moses-in-the-Cradle, Oyster Plant

Rhoeo spathacea plants

Flowers of *Rhoeo spathacea*

Description: *Rhoeo spathacea* (Sw.) Stearn is a short-stemmed herb with underground rhizome that can grow up to 90 cm tall. Leaves pointed upwards, fleshy, hard, long, narrow, 15–25 cm by 3–5 cm, pointed, dark green on the adaxial surface and purple on abaxial surface. Flowers are small, white, and concealed within a boat-shaped envelope of two bracts.[1,2]

Origin: Native to Central America and West Indies, widely cultivated as ornamental plants in most tropical countries.[2]

Phytoconstituents: No information as yet.

Traditional Medicinal Uses: They are used to treat intestinal bleeding, blood in the sputum and dysentery.[3]

Pharmacological Activities: Anti-inflammatory,[4] Antiadrenergic[5] and Uterostimulatory.[6]

Dosage: No information as yet.

Adverse Reactions: No information as yet.

Toxicity: No information as yet.

Contraindications: No information as yet.

Drug-Herb Interactions: No information as yet.

64. *Ricinus communis* L. (Euphorbiaceae)

Castor Oil Plant, Castor Bean

Ricinus communis leaf

Fruits of *Ricinus communis*

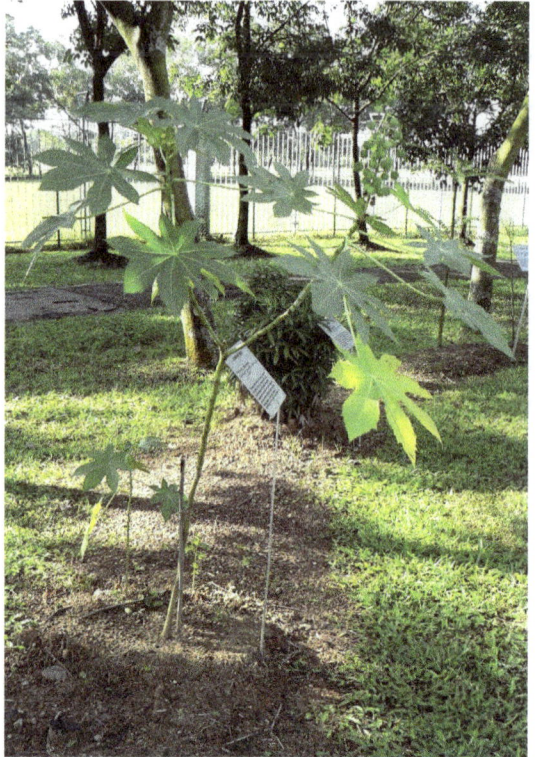
Ricinus communis plant

Description: *Ricinus communis* L. is an erect herb, growing up to 3.6 m high, having pinkish succulent stem and large alternate palmate leaves that are green or reddish brown. Leaves are lobed, consisting of 6–8 radiating leaflets with serrated edges and prominent central veins. Flowers are green, pink or red and inconspicuous, with no petals. The fruits are capsular, with three lobes, prickly and green, containing three seeds.[1]

Origin: Native to Africa, naturalised throughout tropics and subtropics.[2]

Phytoconstituents: Ricin, ricinoleic acid, ricinine, p-coumaric acid, ferulic acid, o-coumaric acids, syringic acid, cinnamic acids, stigmasterol, fucosterol and others.[3–8]

Traditional Medicinal Uses: Its leaf poultice is applied to boils and sores in India; to treat headaches and fever in Hawaii.[9] The leaves and roots are used in a decoction for anal prolapse, arthritis, constipation, facial palsy, lymphadenopathy, strabismus, uteral prolapse, cough, and also as a discutient and expectorant. The heated leaves are applied to gout and swellings as well.[5] The leaves and oil are used for dermatological purposes in Nigeria.[10] Its seeds are used to treat abscesses and skin eruptions, deafness, headache, skin problems, bleeding, constipation, boils, piles and to promote labour.[4] They are rubbed on the temple for headache, powdered for abscesses, boils, and carbuncles. The plant is also used for dogbite, scrofula and several skin infections. The Chinese rub the oil on the body for skin ailments. The seeds are crushed and made into a pulp and rubbed into the palms for palsy, introduced into the urethra in stricture and rubbed on the soles of feet of parturient woman to hasten birth or expulsion of the placenta. The seeds are also used to treat colic, diarrhoea, dysentery, enteritis, acute constipation, for itching, ringworms, warts, dandruff, hair loss and haemorrhoids. It is also used as a laxative before X-ray examination of bowels.[3,5] Midwives sometimes use castor oil to induce labour.[11]

Pharmacological Activities: Antifertility,[12–18] Antioxidant,[19] Antipsychotic,[20] Antiviral,[21] Anti-inflammatory,[19] Convulsant,[20] Hepatoprotective,[22] Filaricidal,[23] Haemaglutination[6] and Insecticidal.[24]

Dosage: For internal use in the treatment of constipation or to expel worms, at least five 2 g capsules or ten 1 g capsules are recommended.[3] Externally, a paste made of ground seeds may be applied to affected skin areas 2 times daily, up to 15 days.[3] Homeopathically, 5 drops, 1 tablet or 10 globules are recommended every 30–60 min for acute treatment or 1–3 times daily for chronic cases.[3] Standard adult dose is reported to be 5.0–20.0 ml/dose, not exceeding 60.0 ml per day. Standard child dose is reported to be 4.0–12.0 ml/dose.[11] The seed oil is recommended for use as a laxative in a dose of 2 to 5 ml and as a purgative in a dose of 20 to 30 ml. 15 seeds may be crushed and applied to the plantar surface of the feet for the induction of labour at term and placenta delivery in cases of placenta retention.[25] The woman's feet are advised to be thoroughly washed after delivery of foetus and placenta.[25] For constipation, it is reported that 15 to 60 ml of castor oil may be taken orally a day.[26]

Adverse Reactions: Gastric irritation, nausea, vomiting, colic, and severe diarrhoea may occur. Long term use may result in loss of electrolytes, especially potassium, and causes intestinal motility inhibition.[3]

Toxicity: Seeds are highly poisonous due to the toxic lectins, principally the albumin and ricin.[27] The seeds are toxic to both animals and humans.[3,12,26-35]

Contraindications: Not to be taken by children under 12 years of age, patients with acute inflammatory intestinal diseases, inflammatory bowel disease and appendicitis.[3] contraindicated in pregnancy, intestinal obstruction and abdominal pain.[3,11] Not to be given with potentially toxic oil-soluble anthelmintics.[36]

Drug-Herb Interactions: Effect of cardioactive steroids may be increased.[3] To prevent decreased absorption of castor, it is advised not to take within one hour of antacids, other drugs and milk.[37]

65. *Ruta graveolens* L. (Rutaceae)

Herb of Grace, Common Rue

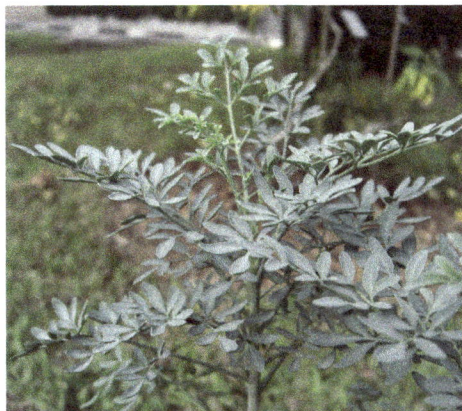

Ruta graveolens herb   *Ruta graveolens* leaves

Description: *Ruta graveolens* L. is a glabrous herb with stem that can grow up to 14–45 cm. Lower leaves are more or less long-petiolate with ultimate segments 2–9 mm wide, lanceolate to narrowly oblong. Inflorescence is rather lax; pedicels are as long as or longer than the capsule; bracts are lanceolate, leaf-like. Sepals are lanceolate and acute. Petals are oblong-ovate, denticulate and undulate. Capsule is glabrous; segments somewhat narrowed above to an obtuse apex.[1]

Origin: Native to Europe.[2]

Phytoconstituents: Rutoside, rutaverine, arborinine, rutin, elemol, pregeijerene, geijerene, furocoumarins, bergapten, xanthotoxin, fagarine, graveolinine and others.[3–5]

Traditional Medicinal Uses: It is frequently used to treat worm and parasitic infection.[6] It has been commonly used for the treatment of psoriasis and vitiligo due to the psoralens and methoxypsoralens present.[7] It is also used to relieve muscle spasms, as carminative, emmenagogue, haemostat, uteronic, vermifuge, to treat hepatitis, dyspepsia, diarrhoea, bug bite, cancer, cold, fever, snakebite, earache, toothache and as an antidote especially in malarial poisoning.[3,4,8] It is also used as an abortifacient to terminate pregnancy.[9] The plant has been used for pain relief in Mexico.[10]

Pharmacological Activities: Analgesic,[11] Antibacterial,[12] Anticancer,[13–17] Antifertility,[18–20] Anti-inflammatory,[5,21] Antimicrobial,[22–24] Antiprotozoal[25] and Cytotoxic.[24]

Dosage: For the treatment of earache, the oil is poured on cotton and inserted into the affected ear. For oral administration, 1 capsule or 0.5 to 1 teaspoon of extract is taken three times daily with meals. The topical cream may be applied to the affected area when necessary.[26] For delayed menstruation, 2 cups of infusion per day is used.[8]

Adverse Reactions: Hypotension, hypersensitivity, rash, erythema and blisters may occur when applied topically.[26] Therapeutic dosages could bring about melancholic moods, sleep disorders, tiredness, dizziness and spasms.[27] Misuse as an abortive during pregnancy can lead to vomiting, epigastric pain, kidney damage, depression, sleep disorders, feelings of vertigo, delirium, fainting, tremor, spasm and sometimes may end up with fatal outcome.[8]

Toxicity: Toxic to humans[27–33] and animals.[34]

Contraindications: Should not be used during pregnancy.[9] Should not be used during lactation, in children and in person with hypersensitivity to rue. Patients with heart diseases should use it with caution.[26] Contraindicated in patients with poor renal function.[9]

Drug-Herb Interactions: Concurrent use of cardiac glycosides (e.g. digoxin, digitoxin) with rue may cause increased inotropic effects. Concurrent use of antihypertensives with rue may cause increased vasodilation.[26]

66. *Saccharum officinarum* L. (Gramineae)

Sugarcane, Tebu

Saccharum officinarum plants

Saccharum officinarum stems

Description: *Saccharum officinarum* L. is a perennial plant that forms tall clumps that can grow up to 6 m tall. Stems are greenish, yellowish or dark purplish and juicy. Leaf blades are broadly linear, glabrous and 80–150 cm by 4–6 cm. Inflorescence is a large silky panicle.[1,2]

Origin: Native to Southeast Asia and Pacific Islands, widely cultivated elsewhere.[2]

Phytoconstituents: Octacosanol, policosanol, orientin, tricin-7-*O*-glycoside, palmitic acid, oleic acid and linolenic acid and others.[3–9]

Traditional Medicinal Uses: Consumption of beets (*Beta vulgaris*) combined with molasses from *S. officinarum*, is used by Dominican healers to shrink fibroids or to "strengthen and fortify the uterus after the fibroid had been drained from the body".[10] The cane juice promotes expulsion of phlegm from the respiratory passages, stimulates gastric activities, treats wounds, ulcers and boils.[11]

137

Pharmacological Activities: Analgesic,[3] Anticancer,[6] Anti-inflammatory,[3,9,12] Antiosteoporotic,[5] Antioxidant,[6,13,14] Antiplatelet,[15–17] Antithrombotic,[18] Hypocholesterolaemic,[19–24] Immunomodulatory,[25,26] Immunostimulatory,[27] Antiatherogenic[28] and Myocardial protective.[4,17]

Dosage: Short-term studies have shown the efficacy and tolerability of policosanol at 10 mg/day on hypercholesterolaemia in obese patients with Type 2 diabetes.[19,20]

Adverse Reactions: Pollen extract of *S. officinarum* showed strong sensitising potential which induces allergy.[29]

Toxicity: Studies on the toxicity of higher aliphatic primary acids (D003) and higher aliphatic primary alcohols (policosanol) isolated from *S. officinarum* did not show any toxic effects.[30–33]

Contraindications: No information as yet.

Drug-Herb Interactions: Pretreatment with high doses of policosanol significantly increased propranolol-induced hypotensive effects, while the effects of nifedipine remained unchanged. Policosanol does not antagonise the hypotensive effect of β-blockers but it can increase the hypotensive effect of beta-blockers without modifying cardiac frequency.[34] Pretreatment with single doses of policosanol significantly increased the nitroprusside-induced hypotensive effect.[35] Warfarin alone and the combination of policosanol and warfarin induced a moderate, but significant prolongation of the bleeding time.[36]

[**Authors' Note:** Freshly squeezed sugarcane juice is commonly consumed and sugar is commercially obtained from sugarcane.]

67. *Sauropus androgynus* (L.) Merr.
(Euphorbiaceae)

Sweet Leaf Bush, Cekup Manis, Daun Katuk

Leaves of *Sauropus androgynus*

Sauropus androgynus plant

Description: *Sauropus androgynus* (L.) Merr. is a small shrub that can grow up to 3 m tall. Leaf blade is ovate-lanceolate, oblong-lanceolate, or lanceolate, 3–10 cm by 1.5–3.5 cm and thinly papery. Flowers are small and borne in axillary clusters. Mature fruit is about 5 cm wide. Young branches and leaves are used as vegetable.[1,2]

Origin: Native to China and tropical Asia.[3]

Phytoconstituents: Sauroposide, (−)-isolariciresinol, corchoionoside C and others.[4–10]

Traditional Medicinal Uses: The fresh leaves or roots possess uterotonic activity and are used for the treatment of retained placenta. A mouth wash made from the juice of the fresh leaves and honey, and applied to the tongue and gums cures thrush of the tongue in infants. The leaves are used for erythema, measles and dysuria. Its roots serve as a diuretic and relieve congestion.[4] The root decoction is used for fever and urinary bladder complaints.[11]

Pharmacological Activities: Anthelmintic.[12]

Dosage: For the treatment of retained placenta, a dose of 40 g, in the form of an extracted juice is administered in 2 subdoses at 10 minutes interval.[4]

Adverse Reactions: Excessive consumption of the leaf reportedly caused dizziness, drowsiness and constipation.[13]

Toxicity: Reported to cause bronchiolitis obliterans,[14–16] ischaemic bronchi necrosis,[17] irreversible obstructive ventilatory defect,[18,19] breathing difficulties[20,21] and is cytotoxic.[22]

Contraindications: No information as yet.

Drug-Herb Interactions: No information as yet.

68. *Sesbania grandiflora* Pers. (Leguminosae)

Scarlet Wisteria Tree, Red Wisteria, Daun Turi

Sesbania grandiflora leaves

Fruit of *Sesbania grandiflora*

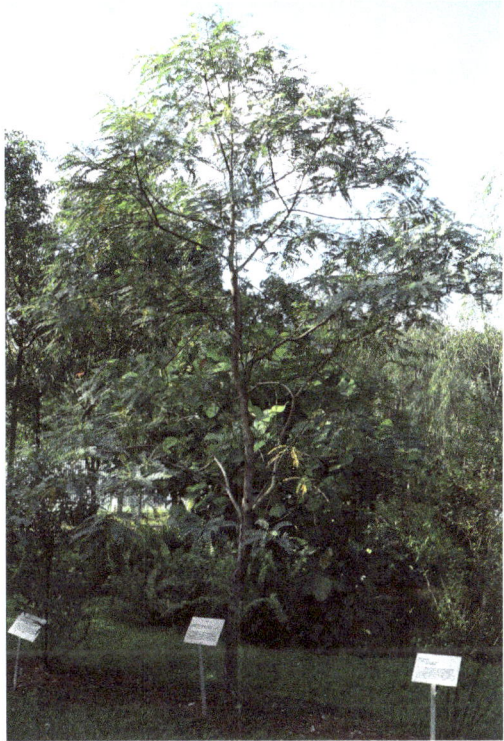
Sesbania grandiflora tree

Description: *Sesbania grandiflora* Pers. is a tree that can grow to 8–10 m in height. The compound leaves are about 30 cm long with 12 to 20 pairs of rounded, narrow, oblong leaflets, 3–4 cm by 1 cm. Flowers are 5–10 cm by 3 cm, in pale pink, red, purple or white. The pods are 25–50 cm, slender, and cylindrical with many light brown to red brown seeds.[1]

Origin: Native to Malesia and cultivated in the tropics.[2]

Phytoconstituents: Grandiflorol, (+)-leucocyanidin, oleanolic acid, lutein, beta-carotene, violaxanthin, neoxanthin, zeaxanthin and others.[3–9]

Traditional Medicinal Uses: In the Philippines, the plant is used for its hypotensive properties.[10] It is used in Indian folk medicine for the treatment of liver disorders.[11] The juice of the leaves and flowers are popularly used for nasal catarrh and headache when taken as snuff. Various leaf preparations are used to treat epileptic fits. Applied externally for treatment of leprous eruptions. A poultice of the leaves is used for bruises. The leaf juice is mixed with honey for congenital bronchitis or cold in babies.[12]

Pharmacological Activities: Antibacterial,[13] Anticonvulsant,[14] Anti-inflammatory,[15] Anxiolytic,[14] Depressant,[15] Diuretic,[15] Hepatoprotective,[11] Hypoglycaemic,[16] Hypotensive[15] and Haemolytic.[17,18]

Dosage: No information as yet.

Adverse Reactions: No information as yet.

Toxicity: No information as yet.

Contraindications: No information as yet.

Drug-Herb Interactions: No information as yet.

69. *Solanum nigrum* L. (Solanaceae)

Black Nightshade, Terong Meranti, Poison Berry

Fruits of *Solanum nigrum*

Solanum nigrum flower

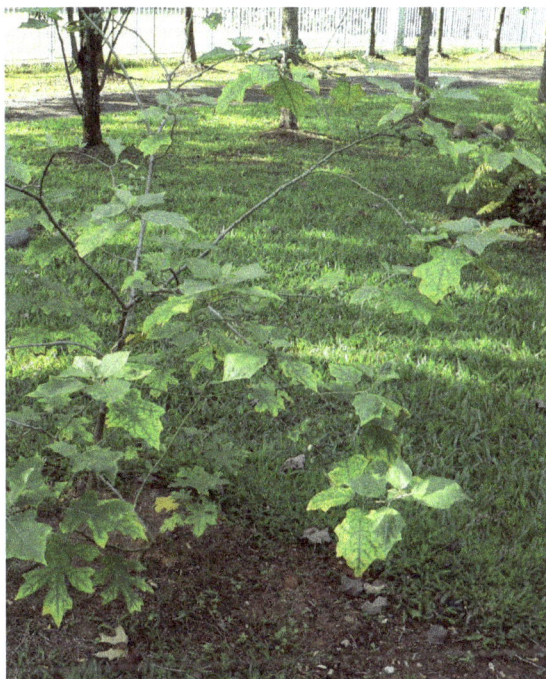
Solanum nigrum shrub

Description: *Solanum nigrum* L. is a small herb, up to 1.5 m tall. Leaves are ovate, ovate-oblong, glabrous, hairy, 1–16 cm by 0.25–12 cm. Inflorescence of 2–10 in an extra-axillary cluster, with white or purple corolla and yellow central protrusion. Fruit is globose, black in colour but is green when immature, 0.5 cm in diameter, with many seeds.[1]

Origin: Native to Southwest Asia, Europe, India and Japan.[2]

Phytoconstituents: Solanidine, α-,β-,γ-chaconine, desgalactotigonin, α-,β-solamargine, diosgenin, solanadiol, α-,β-,γ-solanines, soladulcidine, solanocapsine, α-,β-solansodamine, solasodine, α-solasonine, tigogenin, tomatidenol, uttronins A and B, uttrosides A and B, solanigroside A–H and others.[3–9]

Traditional Medicinal Uses: The stem, leaves and roots are used as a decoction for wounds, tumours and cancerous growths, sores and as an astringent.[3,5] They are also used as a condiment, stimulant, tonic, for treatment of piles, dysentery, abdominal pain, inflammation of bladder, relief of asthma, bronchitis, coughs, eye ailments, itch, psoriasis, skin diseases, eczema, ulcer, relief of cramps, rheumatism, neuralgia and expulsion of excess fluids. The roots are used as an expectorant.[5] The plant has yielded medicines for sore throats, coughs and digestive problems. It has also been used as an agricultural insecticide.[10] Europeans in Africa used the plant to treat convulsions. It is used by the Africans for treating headache, ulcers and as a sedative. The whole plant is used for the treatment of dermatitis, inflammation, heavy female discharge, diarrhoea and dysentery.[11] It is also used as a diuretic and febrifuge. Whole plant is decocted for abscesses, cancer of the cervix, inflammation, leucorrhoea and open sores. Young shoots are consumed as virility tonic for men and to treat dysmenorrhoea in females.[3] In Indochina, the leaves are used as purgative and high blood pressure lowering agents while the fruits are used as laxatives.[12]

Pharmacological Activities: Antibacterial,[13] Anticancer/antineoplastic,[8,14–33] Antiulcerogenic,[36,37] Antinociceptive,[15] Anti-inflammatory,[15] Antioxidant,[34,35] Antiviral,[38] Depressant,[39] Hepatoprotective,[40–43] Hypolipidaemic,[44] Antimutagenic,[45] Enzyme modulation,[46] Larvicidal,[47] Molluscicidal[47–49] and Parasiticidal.[50]

Dosage: 10 drops of extract is taken internally 2 to 3 times a day or 5 to 10 g of tincture may be taken daily for gastric irritation, cramps and whooping cough. For external use as a rinse of moist compress, it is boiled in 1 L water for 10 minutes before usage for psoriasis, haemorrhoids, abscesses, eczema and bruising.[4]

Adverse Reactions: No known side effects with appropriate therapeutic dosage.[4]

Toxicity: Harmful to rats.[51] It is toxic to cattle as it can cause acute nitrate toxicity which leads to death in cattle. In chronic cases, decreased milk yield, abortion, impaired vitamin A and iodine nutrition can occur. The proposed LD_{50} for nitrate toxicity is 160–224 mg/NO_3/kg for cattle.[52] Overdoses can lead to headache, queasiness and vomiting, due to high alkaloid content. Mydriasis may also occur, although rare.[4,52] Solanine, in doses of 200–400 mg, may cause gastroenterosis, tachycardia, dyspnea, vertigo, sleepiness, lethargy, twitching of the extremities and cramps. It is also teratogenic.[52]

Contraindications: No information as yet.

Drug-herb Interactions: No information as yet.

70. *Swietenia macrophylla* King (Meliaceae)

Honduras Mahogany, Broad-leaved Mahogany

A fruiting *Swietenia macrophylla* tree

Seeds and fruits of *Swietenia macrophylla*

Description: *Swietenia macrophylla* King is an evergreen tree, up to 30–35 m tall. Bark is grey and smooth when young, turning dark brown, ridged and flaky when old. Leaves are up to 35–50 cm long, alternate, glabrous with 4–6 pairs of leaflets. Each leaflet is 9–18 cm long. Flowers are small and white; and the fruit is dehiscent, usually 5-lobed capsule, erect, 12–15 cm long, grayish brown, smooth or minutely verrucose. The seed is woody, glossy and possesses wing-like structure at the base that aids its dispersion by wind.[1]

Origin: Native to South America, cultivated in the Asia-Pacific and the Pacific for its quality wood.[1]

Phytoconstituents: Swietenine, swietenolide, andirobin, khayasin T, swietemahonins E–G, swietenine acetate, swietenolide tiglate and others.[2–7]

Traditional Medicinal Uses: The seeds of *Swietenia macrophylla* are traditionally used in several indigenous systems of medicine for the treatment of various ailments such as hypertension, diabetes and malaria.[9] The local folks of Malaysia believe that the seeds are capable of "curing" hypertension and diabetes. The seeds are usually consumed raw by chewing.[8] A decoction of seeds of *Swietenia macrophylla* is reported to treat malaria in Indonesia.[9] Among the Amazonian Bolivian ethnic groups, the seeds are traditionally used to induce abortion by drinking a decoction of the seeds and to heal wounds and various ailments of the skin via external application of the mashed seeds.[10]

Pharmacological Activities: Antimalarial,[11] Antihypertensive[8] and Antidiarrhoeal.[12]

Dosage: No information as yet.

Adverse Reactions: No information as yet.

Toxicity: Uterine haemorrhage.[13]

Contraindications: No information as yet.

Drug-Herb Interactions: No information as yet.

71. *Terminalia catappa* L. (Combretaceae)

Indian Almond, Katapang

Fruits of *Terminalia catappa*

Terminalia catappa tree

Description: *Terminalia catappa* L. is a tall tree, up to 25 m tall. Branches are horizontally whorled, giving it a pagoda shape. Leaves are shiny, obovate, 10–25 cm long, tapering to a short thick petiole. Leaves are yellow that turn red before shedding. Flowers are small and white. Fruits have smooth outer coat, 3–6 cm long, flattened edges, with a pointed end. Pericarp is fibrous and fleshy.[1–3]

Origin: Native to tropical and temperate Asia, Australasia, the Pacific and Madagascar.[4]

Phytoconstituents: Catappanin A, chebulagic acid, 1-desgalloylleugeniin, geraniin, granatin B, punicalagin, punicalin, tercatain, terflavins A & B, tergallagin, euginic acid and others.[2,5–13]

Traditional Medicinal Uses: *Terminalia catappa* has been used to treat dysentery in a number of Southeast Asian countries. In Indonesia, the leaves are used as a dressing for swollen rheumatic joints while in the Philippines, they are used to expel worms.[2] In Karkar Island, New Guinea, juice from the squeezed leaves is applied to sores and the sap from the white stem pith is squeezed and drunk to relieve cough. In Nasingalatu, Papua New Guinea, the flower is crushed, mixed with water and drunk to induce sterility. In New Britain, the old yellow leaves are crushed in water and drunk to sooth sore throat. In Bougainville, the leaves are heated and placed on pimples and the bark is applied to sores. In Tonga, the juice from pounded leaves and bark is applied to mouth sore. In Irian Jaya, the leaves are applied to wounds and burns while in Somoa, it is used to cure cough and sore throat.[14] The fruits are used after childbirth to strengthen the back. An enema made from the crushed fruit mixed with *Trigonella foenum-graecum*, animal fat and warm water is administered to the new mother after childbirth.[15] The leaves are used for the treatment of scabies and skin diseases while the juice is used to treat headache and colic.[13,14] The bark is used as a diuretic, cardiotonic and for dysentery.[16] The leaves of this plant have also been used as a folk medicine for treating hepatitis in India and Philippines.[5]

Pharmacological   Activities: Antimicrobial,[17–19]   Anticancer/ Antineoplastic,[20,21]   Anti-inflammatory,[5,22]   Hypoglycaemic,[8]   Anti- oxidant,[6,10,11,23,24] Hepatoprotective,[7,11,25–29] Antiviral,[30] Chemopreventive,[31] Aphrodisiac[32] and Antimutagenic.[33]

Dosage: 2 tablespoons of a decoction (few leaves in 200 ml water) is given every 2 hours to stop diarrhoea.[34]

Adverse Reactions: No information as yet.

Toxicity: Preliminary oral LD_{50} doses of petroleum ether, methanol and aqueous extracts of *T. catappa* in mice were found to be 343, 195, and 210 mg/kg respectively.[8] Rats fed on *T. catappa* diet maintained their body weight but suffered from stomach, small intestine and pancreas hypertrophies as well as spleen atrophy.[35] Larger doses enhanced liver damage.[36]

Contraindications: No information as yet.

Drug-Herb Interactions: No information as yet.

72. *Thevetia peruviana* (Pers.) K. Schum.
(Apocynaceae)

Yellow Oleander, Trumpet Flower

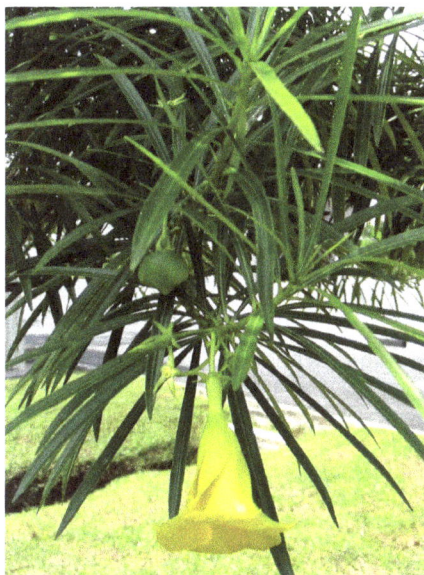

Thevetia peruviana flower and fruit

Thevetia peruviana shrub

Description: *Thevetia peruviana* (Pers.) K. Schum. is a shrub, up to 6 m tall. All parts contain highly poisonous milky latex. Leaves are simple, few, exstipulate and spirally arranged. Blade is linear, 7–13 cm by 0.5–1 cm and glossy. Flowers are large, yellow, 5 cm across, gathered in few flowered terminal cymes. Fruits are green, shiny, globose, 4–5.5 cm across with 4 or less poisonous seeds.[1]

Origin: Native to Central and South America.[2]

Phytoconstituents: Thevetins A and B, thevetosides, acetylperuvoside, epiperuviol, perusitin, theveneriin, thevebioside, thevefolin, pervianoside I–III and others.[3–6]

Traditional Medicinal Uses: Used as an abortifacient, to treat congestive heart failure, malaria, leprosy, indigestion, ringworm, venereal disease and even as a suicide instrument.[7] Used in India as an astringent to the bowel, useful in urethral discharge, worms, skin diseases, wounds, piles, eye problems and itch. Used in continental Europe and is considered particularly useful in mild myocardial insufficiency and digitalis intolerance.[5] Its bark is used as an emetic, febrifuge, insecticidal, poison and for reviving patients with heart failure.[4]

Pharmacological Activities: Antiarrhythmic,[5] Antifungal,[8] Hepatotoxicity and Nephrotoxicity,[5] Larvicidal, [9] Molluscicidal,[10] and Cardiotonic.[5]

Dosage: No information as yet.

Adverse Reactions: Vomiting, dizziness, and cardiac dysrhythmias such as conduction block affecting the sinus and AV nodes may occur.[11]

Toxicity: Toxic to humans,[3,11–16] fishes[17–19] and animals.[5,20] Ingestion of half to fifteen seeds may cause a varying degree of vomiting, palpitation, arrhythmia including sinus bradycardia and could also lead to death.[21] The kernels of about 10 fruits may be fatal to adults while the kernel of one fruit may be fatal to children.[22]

Contraindications: No information as yet.

Drug-Herb Interactions: No information as yet.

[**Authors' Note:** The purified glucoside thevetin, extracted from the seeds, is prescribed as a cardiotonic drug in a 0.1% solution orally, in a dose of 1 to 2 ml daily, or in 1 mg/2 ml ampoules parenterally. 1 to 2 ampoules have been given daily for its cardiotonic property.[23] A dose of 2.4 mg followed by average maintenance dose of 600 μg has been shown to be effective in 22 patients with congestive heart failure.[5] The herb has been administered as a thevetin tablet of 0.25 mg/tablet. A maintenance dose consists of 0.25 mg/day. For injection administration, an injection ampoule is prepared from 0.25 mg/ml diluted in a 5% glucose solution.[24] However, note that thevetin is not conventionally used in clinical practice and its dosage is not found in pharmacopoeial monographs.]

73. *Tinospora crispa* (L.) Diels (Menispermaceae)

Akar Putarwali, Batang Wali

Tinospora crispa shrub

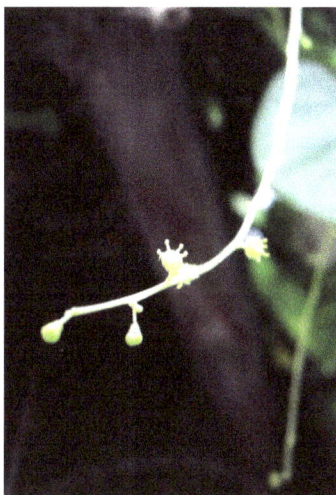

Flowers of *Tinospora crispa*

Description: *Tinospora crispa* (L.) Diels is a woody climber with numerous protrusions on the stem. Leaves are oblong-ovate, cordate, 8–9 cm by 7–8 cm and tapering to a pointed end. Flowers are small, with 6 petals, 2 mm in length and 8–27 cm racemes. Male flowers have yellow sepals whereas female flowers have green sepals. Drupelets are red, juicy and 7–8 mm long.[1,2]

Origin: Native to Malesia, Indochina, Indian subcontinent and China.[3]

Phytoconstituents: Boropetol B, borapetoside B, C & F, jatrorhizine, magnoflorine, palmatine, protoberberine, tembolarine, diosmetin, cycloeucalenol, cycloeucalenone and others.[4–8]

Traditional Medicinal Uses: It is used for hypertension, diabetes mellitus,[9] to treat malaria, remedy for diarrhoea and as vermifuge.[6] In Malaysia, *T. crispa* extract is taken orally by Type 2 (non-insulin-dependent) diabetic patients to treat hyperglycaemia.[10]

Pharmacological Activities: Anti-inflammatory,[11] Antioxidant,[7] Antimalarial,[12,13] Antiprotozoal[14] and Hypoglycaemic.[10,15,16]

Dosage: No information as yet.

Adverse Reactions: The plant may result in an increased risk of hepatic dysfunction due to marked elevation of liver enzymes but is reversible upon discontinuation of *T. crispa*.[17]

Toxicity: No information as yet.

Contraindications: No information as yet.

Drug-Herb Interactions: *T. crispa* extract showed 61.3% increase of cytochrome P450 3A4 (CYP3A4) enzyme inhibition after 20 min of preincubation with human liver microsomes.[18]

[**Authors' Note:** In a randomised double blind placebo controlled trial[17] to determine the efficacy of *T. crispa* as an additional treatment in patients with Type 2 diabetes mellitus who did not respond to oral hypoglycaemic drugs and refused insulin injection, 20 patients received *T. crispa* powder in capsule form at a dose of 1 gram three times daily for six months. Twenty patients received a placebo. The results showed no evidence to support the use of *T. crispa* for additional therapy in such patients. Two patients had elevated levels of liver enzymes, which was reversible on discontinuing *T. crispa*. Patients receiving *T. crispa* had significant weight reduction and cholesterol elevation. Note that the study only included patients that did not respond to oral hypoglycaemic therapy.]

74.   *Vitex rotundifolia* L. f. (Verbenaceae)

Round Leaf Chastetree, Beach Vitex

Vitex rotundifolia leaves

Vitex rotundifolia tree

Description: *Vitex rotundifolia* L. f. is an evergreen woody tree, densely covered with short hairs. Leaves are opposite, simple, ovate, broadly oblong-elliptic, 2–5 cm long by 1.5–3 cm wide, rounded or abruptly acute at the base. Inflorescence panicles are at the terminal, densely flowered, 4–7 cm long with purple corolla. Fruits are globose, 5–7 mm.[1]

Origin: Native to Temperate and Tropical Asia, Australasia and Pacific.[2]

Phytoconstituents: Rotundifuran, prerotundifuran, vitexilactone, previtexilactone, vitexicarpin, vitricine, vitetrifolins D–G, vitexifolins A–E, isoambreinolide and others.[1,3–9]

Traditional Medicinal Uses: In Malaysia, various parts of the plants are considered panacea for illnesses ranging from headache to tuberculosis.[10] In China, the plant has been used for the treatment of cancer.[11] A poultice of the leaves is used to treat rheumatism, contusions, swollen testicles and as a discutient in sprains. In Indonesia, leaves have been used in medicinal baths, as a tincture or for intestinal complaints.[10] In Papua New Guinea, sap from crushed heated leaves is diluted with water and drunk to relieve headaches. The fruits are used to expel worms and in Vietnam, a decoction of dried fruits has been used to treat colds, headache, watery eyes and mastitis.[10] In Thailand, fruits have been used for asthmatic cough and haemorrhoids.[10] Infusion of the boiled roots is regarded as diaphoretic and diuretic, drunk for fever, after child-birth and for liver diseases.[10]

Pharmacological Activities: Analgesic,[12] Antibacterial,[13,14] Antifungal,[15] Anti-inflammatory,[16] Antineoplastic,[11,15,17–20] Antinociceptive,[8] Antioxidant,[6] Antiprotozoal,[21] Hypotensive,[22] Immunomodulatory,[17] Antimutagenic[23] and Insect repellant.[15]

Dosage: No information as yet.

Adverse Reactions: *V. rotundifolia* may trigger various allergic reactions such as sneezing, respiratory problems, dizziness, headache and nausea.[10]

Toxicity: No information as yet.

Contraindications: No information as yet.

Drug-Herb Interactions: No information as yet.

75. *Zingiber officinale* Roscoe (Zingiberaceae)

Common Ginger

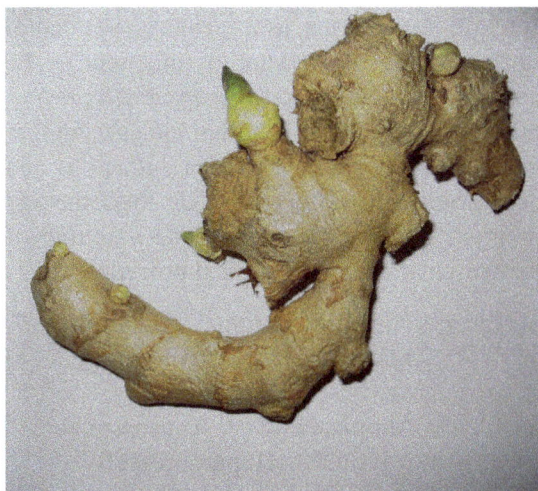

Zingiber officinale rhizome

Zingiber officinale plant

Description: *Zingiber officinale* Roscoe is a herbaceous plant that grows up to 1.2 m high and with an underground rhizome. The stem grows above ground and leaves are narrow, long, lanceolate, with distinct venation pattern and pointed apex. Flowers are white or yellowish-green, streaked with purple and fragrant.[1]

Origin: Originate from tropical Asia, widely cultivated in the tropics.[2]

Phytoconstituents: Gingerol, zingiberene, farnesene, camphene, neral, nerol, 1,8-cineole, geranial, geraniol, geranyl acetate and others.[3–7]

Traditional Medicinal Uses: Ginger is the folk remedy for anaemia, nephritis, tuberculosis, and antidote to *Arisaema* and *Pinellia*.[8] Sialogogue when chewed, causes sneezing when inhaled and rubefacient when applied externally. Antidotal to mushroom poisoning, ginger peel is used for opacity of the cornea. The juice is used as a digestive stimulant and local application in ecchymoses.[8] Underground stem is used to treat stomach upset, nausea, vomiting, nose bleeds, rheumatism, coughs, blood in stools, to improve digestion, expel intestinal gas, and stimulate appetite.[9] The rhizomes are used to treat bleeding, chest congestion, cholera, cold, diarrhoea, dropsy, dysmenorrhoea, nausea, stomachache, and also for baldness, cancer, rheumatism, snakebite and toothache.[8] It is also used as postpartum protective medicine, treatment for dysentery, treatment for congestion of the liver, complaints with the urino-genital system/female reproduction system and sinus.[10] Besides that, it is used to alleviate nausea, as a carminative, circulatory stimulant and to treat inflammation and bacterial infection.[3] The Commision E approved the internal use of ginger for dyspepsia and prevention of motion sickness.[11] The British Herbal Compendium indicates ginger for atonic dyspepsia, colic, vomiting of pregnancy, anorexia, bronchitis and rheumatic complaints.[12] European Scientific Cooperative on Phytotherapy (ESCOP) indicates its use for prophylaxis of the nausea and vomiting of motion sickness and to alleviate nausea after minor surgical procedures.[13]

Pharmacological Activities: Analgesic,[14–16] Anthelmintic,[17] Antiarthritic,[18] Anticancer,[19–33] Antidiabetic,[34,35] Antidiarrhoeal,[36] Antiemetic,[37–53] Antihyperlipidaemic,[54] Antihypertensive,[55] Anti-inflammatory,[14,15,56–66] Antimicrobial,[67–78] Antioxidant,[6,7,79–82] Antiplatelet,[83–87] Antispasmodic,[88] Antiulcer,[89,90] Antiviral,[91–93] Anxiolytic,[94] Hepatoprotective,[93–97] Hypocholesterolaemic,[57,98,99] Hypoglycaemic,[14,100,101] Hypolipidaemic,[102] Hypotensive,[103] Immunomodulatory,[104–106] Neuroprotective,[107] Insect repellent[69,108] and Radioprotective.[109–111]

Dosage: A tea is prepared by pouring boiling water over 0.5 to 1 g of the coarsely powdered ginger for 5 min and passing through a tea strainer, taken to prevent vomitting.[112]

Adverse Reactions: Fresh rhizome can be safely consumed with proper usage.[113] Contact dermatitis of the fingertips has been reported in sensitive patients.[114]

Toxicity: It is nontoxic when tested in rats[115] but overdose may cause cardiac arrhythmia and CNS depression.[116]

Contraindications: Consult physician before using ginger preparations in patients with blood coagulation disorders, taking anticoagulant drugs or with gallstones. Avoid dried rhizomes during pregnancy.[113]

Drug-Herb Interactions: Interacts with anticoagulants such as heparin, warfarin, drugs used in chemotherapy and ticlopidine. Ginger taken prior to 8-MOP (treatment for patients undergoing photopheresis) may substantially reduce nausea caused by 8-MOP.[117] Ginger appears to increase the risk of bleeding in patients taking warfarin.[118] However, ginger at recommended doses does not significantly affect clotting status, the pharmacokinetics or pharmacodynamics of warfarin in healthy subjects.[119] Ginger also significantly decreased the oral bioavailability of cyclosporine.[120]

[**Authors' Note:** Ginger is widely eaten as a food ingredient and used in many different cultures as traditional medicine. Ongoing scientific research has shown diverse pharmacological activities.]

Appendix

List of Native and Non-native Plants

Native	Non-native	
Ardisia elliptica	Abrus precatorius	Juniperus chinensis
Asplenium nidus	Adiantum capillus-veneris	Kaempferia galanga
Barringtonia asiatica	Allamanda cathartica	Lantana camara
Barringtonia racemosa	Aloe vera	Lonicera japonica
Calophyllum inophyllum	Andrographis paniculata	Mangifera indica
Cerbera odollam	Areca catechu	Manihot esculenta
Crinum asiaticum	Aster tataricus	Mimosa pudica
Eurycoma longifolia	Azadirachta indica	Mirabilis jalapa
Hibiscus tiliaceus	Bauhinia purpurea	Morinda citrifolia
Ipomoea pes-caprae	Bixa orellana	Nelumbo nucifera
Melaleuca cajuputi	Cananga odorata	Nerium oleander
Melastoma malabathricum	Capsicum annuum	Ophiopogon japonicus
Nephelium lappaceum	Cassia fistula	Persicaria hydropiper
Peltophorum pterocarpum	Catharanthus roseus	Phyllanthus amarus
Piper sarmentosum	Celosia argentea	Piper nigrum
Rhodomyrtus tomentosa	Centella asiatica	Plantago major
Terminalia catappa	Cissus quadrangularis	Punica granatum
	Cocos nucifera	Rhoeo spathacea
	Coix lacryma-jobi	Ricinus communis
	Cymbopogon citratus	Ruta graveolens
	Dolichos lablab	Saccharum officinarum
	Elephantopus scaber	Sauropus androgynus
	Euphorbia hirta	Sesbania grandiflora
	Hibiscus mutabilis	Solanum nigrum
	Hibiscus rosa-sinensis	Swietenia macrophylla
	Impatiens balsamina	Thevetia peruviana
	Imperata cylindrica	Tinospora crispa
	Ixora chinensis	Vitex rotundifolia
	Jatropha curcas	Zingiber officinale

Note: Native plants listed above refer to those that are indigenous to Singapore while the non-native plants are those that originate elsewhere and have been subsequently introduced to Singapore. Many of the non-native plants are now growing well in Singapore.

List of Plants by Scientific Name

Scientific Name	Common Name
Abrus precatorius	Rosary Pea, Indian Licorice, Precatory Bean
Adiantum capillus-veneris	Black Maidenhair Fern, Southern Maidenhair Fern, Venus Hair Fern
Allamanda cathartica	Allamanda, Common Allamanda, Golden Trumpet
Aloe vera	Aloe, Lidah Buaya
Andrographis paniculata	Hempedu Bumi, Sambiloto, Chuan Xin Lian
Ardisia elliptica	Mata Pelanduk/Ayam, Sea-Shore Ardisia, Shoebutton Ardisia
Areca catechu	Betel Nut Palm, Areca Nut, Pinang
Asplenium nidus	Bird's Nest Fern
Aster tataricus	Tatarian Aster, Tatarian Daisy
Azadirachta indica	Neem
Barringtonia asiatica	Beach Barringtonia, Fish-Killer Tree, Putat Laut
Barringtonia racemosa	Putat Kampong, Samundrapandu
Bauhinia purpurea	Butterfly Tree
Bixa orellana	Annatto, Lipstick Tree
Calophyllum inophyllum	Indian Laurel, Penaga Laut, Borneo Mahogany
Cananga odorata	Kenanga, Ylang-Ylang
Capsicum annuum	Chilli, Red Pepper
Cassia fistula	Golden Shower Tree, Indian Laburnum, Purging Cassia
Catharanthus roseus	Madagascar Periwinkle, Rose Periwinkle
Celosia argentea	Feather Cockscomb, Red Spinach
Centella asiatica	Indian Pennywort, Asiatic Pennywort
Cerbera odollam	Pong Pong Tree, Indian Suicide Tree, Sea Mango
Cissus quadrangularis	Grape Leaf, Veld Grape
Cocos nucifera	Coconut Palm, Kelapa
Coix lacryma-jobi	Job's Tears, Adlay
Crinum asiaticum	Crinum Lily, Spider Lily, Bawang Tanah
Cymbopogon citratus	Lemon Grass
Dolichos lablab	Lablab Bean, Hyacinth Bean
Elephantopus scaber	Elephant's Foot, Tutup Bumi, Tapak Sulaiman
Euphorbia hirta	Asthma Weed
Eurycoma longifolia	Tongkat Ali, Ali's Umbrella, Pasak Bumi

Scientific Name	Common Name
Hibiscus mutabilis	Cotton Rose, Chinese Rose
Hibiscus rosa-sinensis	Hawaiian Hibiscus, China Rose, Bunga Raya
Hibiscus tiliaceus	Linden Hibiscus, Sea Hibiscus, Mahoe
Impatiens balsamina	Balsam Plant
Imperata cylindrica	Lalang, Alang-Alang, Speargrass
Ipomoea pes-caprae	Beach Morning Glory, Goat's Foot Creeper, Bayhops
Ixora chinensis	Chinese Ixora
Jatropha curcas	Barbados Nut, Physic-Nut
Juniperus chinensis	Chinese Juniper
Kaempferia galanga	Galangal, Sand Ginger, Kencur
Lantana camara	Common Lantana, Bunga Tahi Ayam, Wild Sage
Lonicera japonica	Japanese Honeysuckle, Jin Yin Hua
Mangifera indica	Mango, Mangga
Manihot esculenta	Tapioca, Cassava
Melaleuca cajuputi	Gelam, Paper-Bark Tree, Kayu Puteh
Melastoma malabathricum	Sendudok, Singapore Rhododendron
Mimosa pudica	Touch-Me-Not, Sensitive Plant, Rumput Simalu
Mirabilis jalapa	Four O'clock Flower, Bunga Pukul Empat
Morinda citrifolia	Mengkudu, Indian Mulberry, Noni
Nelumbo nucifera	Sacred Lotus, East Indian Lotus, Oriental Lotus
Nephelium lappaceum	Rambutan, Hairy Lychee
Nerium oleander	Oleander
Ophiopogon japonicus	Dwarf Lilyturf, Mondo Grass, Mai Men Dong
Peltophorum pterocarpum	Jemerlang Laut, Yellow Flame, Yellow Flamboyant
Persicaria hydropiper	Water Pepper, Laksa Plant
Phyllanthus amarus	Pick-A-Back, Carry Me Seed, Ye Xia Zhu
Piper nigrum	Pepper, Lada
Piper sarmentosum	Wild Pepper, Kadok, Sirih Tanah
Plantago major	Common Plantain, Whiteman's Foot, Daun Sejumbok
Punica granatum	Pomegranate, Delima
Rhodomyrtus tomentosa	Rose Myrtle
Rhoeo spathacea	Purple-Leaved Spider Wort, Moses-In-The-Cradle, Oyster Plant
Ricinus communis	Castor Oil Plant, Castor Bean
Ruta graveolens	Herb of Grace, Common Rue
Saccharum officinarum	Sugarcane, Tebu

Scientific Name	Common Name
Sauropus androgynus	Sweet Leaf Bush, Cekup Manis, Daun Katuk
Sesbania grandiflora	Scarlet Wisteria Tree, Red Wisteria, Daun Turi
Solanum nigrum	Black Nightshade, Terong Meranti, Poison Berry
Swietenia macrophylla	Honduras Mahogany, Broad-Leaved Mahogany
Terminalia catappa	Indian Almond, Katapang
Thevetia peruviana	Yellow Oleander, Trumpet Flower
Tinospora crispa	Akar Putarwali, Batang Wali
Vitex rotundifolia	Round Leaf Chastetree, Beach Vitex
Zingiber officinale	Common Ginger

List of Plants by Common Name

Common Name	Scientific Name
Akar Putarwali	*Tinospora crispa*
Alang-Alang	*Imperata cylindrica*
Ali's Umbrella	*Eurycoma longifolia*
Allamanda	*Allamanda cathartica*
Aloe	*Aloe vera*
Annatto	*Bixa orellana*
Areca Nut	*Areca catechu*
Asiatic Pennywort	*Centella asiatica*
Asthma Weed	*Euphorbia hirta*
Balsam Plant	*Impatiens balsamina*
Barbados Nut	*Jatropha curcas*
Batang Wali	*Tinospora crispa*
Bawang Tanah	*Crinum asiaticum*
Bayhops	*Ipomoea pes-caprae*
Beach Barringtonia	*Barringtonia asiatica*
Beach Morning Glory	*Ipomoea pes-caprae*
Beach Vitex	*Vitex rotundifolia*
Betel Nut Palm	*Areca catechu*
Bird's Nest Fern	*Asplenium nidus*
Black Maidenhair Fern	*Adiantum capillus-veneris*
Black Nightshade	*Solanum nigrum*
Borneo Mahogany	*Calophyllum inophyllum*
Broad-Leaved Mahogany	*Swietenia macrophylla*
Bunga Pukul Empat	*Mirabilis jalapa*
Bunga Raya	*Hibiscus rosa-sinensis*
Bunga Tahi Ayam	*Lantana camara*
Butterfly Tree	*Bauhinia purpurea*
Carry Me Seed	*Phyllanthus amarus*
Cassava	*Manihot esculenta*
Castor Bean	*Ricinus communis*
Castor Oil Plant	*Ricinus communis*
Cekup Manis	*Sauropus androgynus*
Chilli	*Capsicum annuum*
China Rose	*Hibiscus rosa-sinensis*

Common Name	Scientific Name
Chinese Ixora	*Ixora chinensis*
Chinese Juniper	*Juniperus chinensis*
Chinese Rose	*Hibiscus mutabilis*
Chuan Xin Lian	*Andrographis paniculata*
Coconut Palm	*Cocos nucifera*
Common Allamanda	*Allamanda cathartica*
Common Ginger	*Zingiber officinale*
Common Lantana	*Lantana camara*
Common Plantain	*Plantago major*
Common Rue	*Ruta graveolens*
Cotton Rose	*Hibiscus mutabilis*
Crinum Lily	*Crinum asiaticum*
Daun Katuk	*Sauropus androgynus*
Daun Sejumbok	*Plantago major*
Daun Turi	*Sesbania grandiflora*
Delima	*Punica granatum*
Dwarf Lilyturf	*Ophiopogon japonicus*
East Indian Lotus	*Nelumbo nucifera*
Elephant's Foot	*Elephantopus scaber*
Feather Cockscomb	*Celosia argentea*
Fish-Killer Tree	*Barringtonia asiatica*
Four O'Clock Flower	*Mirabilis jalapa*
Galangal	*Kaempferia galanga*
Gelam	*Melaleuca cajuputi*
Goat's Foot Creeper	*Ipomoea pes-caprae*
Golden Shower Tree	*Cassia fistula*
Golden Trumpet	*Allamanda cathartica*
Grape Leaf	*Cissus quadrangularis*
Hairy Lychee	*Nephelium lappaceum*
Hawaiian Hibiscus	*Hibiscus rosa-sinensis*
Hempedu Bumi	*Andrographis paniculata*
Herb of Grace	*Ruta graveolens*
Honduras Mahogany	*Swietenia macrophylla*
Hyacinth Bean	*Dolichos lablab*
Indian Almond	*Terminalia catappa*
Indian Laburnum	*Cassia fistula*
Indian Laurel	*Calophyllum inophyllum*

Common Name	Scientific Name
Indian Licorice	*Abrus precatorius*
Indian Mulberry	*Morinda citrifolia*
Indian Pennywort	*Centella asiatica*
Indian Suicide Tree	*Cerbera odollam*
Japanese Honeysuckle	*Lonicera japonica*
Jemerlang Laut	*Peltophorum pterocarpum*
Jin Yin Hua	*Lonicera japonica*
Job's Tears	*Coix lacryma-jobi*
Kadok	*Piper sarmentosum*
Katapang	*Terminalia catappa*
Kayu Puteh	*Melaleuca cajuputi*
Kelapa	*Cocos nucifera*
Kenanga	*Cananga odorata*
Kencur	*Kaempferia galanga*
Lablab Bean	*Dolichos lablab*
Lada	*Piper nigrum*
Laksa Plant	*Persicaria hydropiper*
Lalang	*Imperata cylindrica*
Lemon Grass	*Cymbopogon citratus*
Lidah Buaya	*Aloe vera*
Linden Hibiscus	*Hibiscus tiliaceus*
Lipstick Tree	*Bixa orellana*
Madagascar Periwinkle	*Catharanthus roseus*
Mahoe	*Hibiscus tiliaceus*
Mangga	*Mangifera indica*
Mango	*Mangifera indica*
Mata Pelanduk/Ayam	*Ardisia elliptica*
Mengkudu	*Morinda citrifolia*
Mondo Grass	*Ophiopogon japonicus*
Moses-In-The-Cradle	*Rhoeo spathacea*
Neem	*Azadirachta indica*
Noni	*Morinda citrifolia*
Oleander	*Nerium oleander*
Oriental Lotus	*Nelumbo nucifera*
Oyster Plant	*Rhoeo spathacea*
Paper-Bark Tree	*Melaleuca cajuputi*
Pasak Bumi	*Eurycoma longifolia*

Common Name	Scientific Name
Penaga Laut	*Calophyllum inophyllum*
Pepper	*Piper nigrum*
Physic-Nut	*Jatropha curcas*
Pick-A-Back	*Phyllanthus amarus*
Pinang	*Areca catechu*
Poison Berry	*Solanum nigrum*
Pomegranate	*Punica granatum*
Pong Pong Tree	*Cerbera odollam*
Precatory Bean	*Abrus precatorius*
Purging Cassia	*Cassia fistula*
Purple-Leaved Spider Wort	*Rhoeo spathacea*
Putat Kampong	*Barringtonia racemosa*
Putat Laut	*Barringtonia asiatica*
Rambutan	*Nephelium lappaceum*
Red Pepper	*Capsicum annuum*
Red Spinach	*Celosia argentea*
Red Wisteria	*Sesbania grandiflora*
Rosary Pea	*Abrus precatorius*
Rose Myrtle	*Rhodomyrtus tomentosa*
Rose Periwinkle	*Catharanthus roseus*
Round Leaf Chastetree	*Vitex rotundifolia*
Rumput Simalu	*Mimosa pudica*
Sacred Lotus	*Nelumbo nucifera*
Sambiloto	*Andrographis paniculata*
Samundrapandu	*Barringtonia racemosa*
Sand Ginger	*Kaempferia galanga*
Scarlet Wisteria Tree	*Sesbania grandiflora*
Sea Hibiscus	*Hibiscus tiliaceus*
Sea Mango	*Cerbera odollam*
Sea-Shore Ardisia	*Ardisia elliptica*
Sendudok	*Melastoma malabathricum*
Sensitive Plant	*Mimosa pudica*
Shoebutton Ardisia	*Ardisia elliptica*
Singapore Rhododendron	*Melastoma malabathricum*
Sirih Tanah	*Piper sarmentosum*
Southern Maidenhair Fern	*Adiantum capillus-veneris*
Speargrass	*Imperata cylindrica*

Common Name	Scientific Name
Spider Lily	*Crinum asiaticum*
Sugarcane	*Saccharum officinarum*
Sweet Leaf Bush	*Sauropus androgynus*
Tapak Sulaiman	*Elephantopus scaber*
Tapioca	*Manihot esculenta*
Tatarian Aster	*Aster tataricus*
Tatarian Daisy	*Aster tataricus*
Tebu	*Saccharum officinarum*
Terong Meranti	*Solanum nigrum*
Tongkat Ali	*Eurycoma longifolia*
Touch-Me-Not	*Mimosa pudica*
Trumpet Flower	*Thevetia peruviana*
Tutup Bumi	*Elephantopus scaber*
Veld Grape	*Cissus quadrangularis*
Venus Hair Fern	*Adiantum capillus-veneris*
Water Pepper	*Persicaria hydropiper*
Whiteman's Foot	*Plantago major*
Wild Pepper	*Piper sarmentosum*
Wild Sage	*Lantana camara*
Ye Xia Zhu	*Phyllanthus amarus*
Yellow Flamboyant	*Peltophorum pterocarpum*
Yellow Flame	*Peltophorum pterocarpum*
Yellow Oleander	*Thevetia peruviana*
Ylang-Ylang	*Cananga odorata*

List of Abbreviations and Symbols

α	alpha
Al	aluminium
AUC	area under the curve
β	beta
cm^3	cubic centimetre
δ	delta
EC_{50}	effective concentration, i.e. minimum concentration required for 50% of the test subjects to exhibit the desired response
γ	gamma
g	gramme
IgE	immunoglobulin E
kg	kilogramme
L	litre
LD_{50}	lethal dose, i.e. minimum concentration required to kill 50% of the test subjects
mg	milligramme
ml	millimetre
mg/kg	milligrammes per kilogram of body weight
mg/ml	milligrammes per millimetre of fluid
mg/m^2	milligrammes per square metre of body surface area
μg	microgramme
NO_3	nitrate
ppm	parts per million
%	percent
TD_{50}	toxic dose, i.e. minimum concentration required to induce toxicity in 50% of the test subjects

References

Abrus precatorius L.

1. Hsuan, K. 1990. The Concise Flora of Singapore, Gymnosperms and Dicotyledons. Singapore University Press: National University of Singapore.
2. Nguyen V. D. 1993. Medicinal Plants of Vietnam, Cambodia and Laos. Nguyen Van Duong: Vietnam.
3. Wee, Y. C. and Hsuan, K. 1990. An illustrated Dictionary of Chinese Medicinal Herbs. Times Edition and Eu Yan Seng Holdings Ltd: Singapore.
4. eFloras.org 2008 http://www.efloras.org/florataxon.aspx?flora_id=5&taxon_id=200011844 Date Accessed: 30th March 2008
5. Duke, J. A. and Ayensu, E. S. 1985. Medicinal Plants of China, volume one and two. Reference Publications, Inc.: United States of America.
6. Nguyen, V. D. and Doan, T. N. 1989. Medicinal Plants in Vietnam. World Health Organisation, Regional Office for the Western Pacific, Manila, Institute of Materia Medica, Hanoi.
7. Saxena, V. K. and Sharma, D. N. 1999. *Fitoterapia* 70: 328–329.
8. Kennelly, E. J., Cai, L. N., Kim, N. C. and Kinghorn, A. D. 1996. *Phytochemistry* 41: 1381–1383.
9. Tahirov, T. H., Lu, T. H., Liaw, Y. C., Chen, Y. L. and Lin, J. Y. 1995. *Journal of Molecular Biology* 250: 354–367.
10. Wang, J. P., Hsu, M. F., Chang, L. C., Kuo, J. S. and Kuo, S. C. 1995. *European Journal of Pharmacology* 273: 73–81.
11. Lin, J. Y., Lee, T. C. and Tung, T. C. 1978. *International Journal of Peptide and Protein Research* 12: 311–317.
12. Lin, J. Y., Lee, T. C., Hu, S. T. and Tung, T. C. 1981. *Toxicon* 19: 41–51.
13. Bhardwaj, D. K., Bisht, M. S. and Mehta, C. K. 1980. *Phytochemistry* 19: 2040–2041.
14. Ghosal, S. and Dutta, S. K. 1971. *Alkaloids of Abrus Precatorius. Phytochemistry* 10: 195–198.
15. Kim, N. C., Kim, D. S. and Kinghorn, A. D. 2002. *Natural Product Letters* 16: 261–266.
16. Yadava, R. N. and Reddy, V. M. 2002. *Journal of Asian Natural Product Research* 4: 103–107.
17. Limmatvapirat, C., Sirisopanaporn, S. and Kittakoop, P. 2004. *Planta Medica* 70: 276–278.
18. Singh, R. B. and Shelley. 2007. *Journal of Environmental Biology* 28: 461–464.
19. Jain, S. K. and DeFilipps, R. A. 1991. Medicinal Plants of India, volume one and two. Reference Publications Inc.: United States of America.
20. Zore, G. B., Awad, V., Thakre, A. D., Halde, U. K., Meshram, N. S., Surwase, B. S. and Karuppayil, S. M. 2007. *Natural Product Research* 21: 933–940.
21. Mølgaard, P., Nielsen, S. B., Rasmussen, D. E., Drummond, R. B., Makaza, N. and Andreassen, J. 2001. *Journal of Ethnopharmacology* 74: 257–264.
22. Anam, E. M. 2001. *Phytomedicine* 8: 24–27.
23. Kuo, S. C., Chen, S. C., Chen, L. H., Wu, J. B., Wang, J. P. and Teng, C. M. 1995. *Planta Medica* 61: 307–312.
24. Moriwaki, S. O. H., Bakalova, R., Yasuda, S. and Yamasaki, N. 2004. *Toxicology and Applied Pharmacology* 195: 182–193.
25. Ramnath, V., Kuttan, G. and Kuttan, R. 2002. *Indian Journal of Physiology and Pharmacology* 46: 69–77.
26. Ohba, H., Moriwaki, S., Bakalova, R., Yasuda, S. and Yamasaki, N. 2004. *Toxicology and Applied Pharmacology* 195: 182–193.
27. Ghosh, D. and Maiti, T. K. 2007. *Immunobiology* 212: 589–599.
28. Ghosh, D. and Maiti, T. K. 2007. *Immunobiology* 212: 667–673.
29. Bhutia, S. K., Mallick, S. K., Stevens, S. M., Prokai, L., Vishwanatha, J. K. and Maiti, T. K. 2008. *Toxicology in Vitro* 22: 344–351.
30. Bhutia, S. K., Mallick, S. K., Maiti, S. and Maiti, T. K. 2008. *Chemico-Biological Interactions*, 174: 11–18.

31. Schottelius, J., Marinkelle, C. J. and Gomez-Leiva, M. A. 1986. *Tropical Medicine and Parasitology* 37: 54–58.
32. Ramnath, V., Kuttan, G. and Kuttan, R. 2002. *Indian Journal of Experimental Biology* 40: 910–913.
33. Tripathi, S. and Maiti, T. K. 2003. *International Immunopharmacology* 3: 375–381.
34. Ramnath, V., Kuttan, G. and Kuttan, R. 2006. *Immunopharmacology and Immunotoxicology* 28: 259–268.
35. Tripathi, S. and Maiti, T. K. 2005. *International Journal of Biochemistry and Cell Biology* 37: 451–462.
36. Singh, S. and Singh, D. K. 1999. *Phytotherapy Research* 13: 210–213.
37. Duke, J. A., Bogenschutz-Godwin, M. J., duCellier, J. and Duke, P. A. K. 2002. Handbook of medicinal herbs. CRC Press: Boca Raton, FL.
38. Fernando, C. 2001. *Anaesthesia.* 56: 1178–1180.
39. Bradberry, S. 2007. *Medicine* 35: 576–577.
40. Sinha, R. 1990. *Journal Ethnopharmacology* 128: 173–181.
41. Sinha, S. and Mathur, R. S. 1990. *Indian Journal of Experimental Biology* 28: 752–756.
42. Ratnasooriya, W. D., Amarasekera, A. S., Perera, N. S. D. and Premakumara, G. A. S. 1991. *Journal of Ethnopharmacology* 33: 85–90.

Adiantum capillus-veneris L.

1. Duke, J. A. and Ayensu, E. S. 1985. Medicinal Plants of China, volume one and two. Reference Publications, Inc.: United States of America.
2. Jain, S. K. and DeFilipps, R. A. 1991. Medicinal Plants of India, volume one and two. Reference Publications Inc.: United States of America.
3. Foster, S. and Duke, J. A. 1990. A Field Guide to Medicinal Plants, Eastern and Central North America. Houghton Mifflin Company: New York.
4. Schauenberg, P. and Paris, F. 1990. Guide to Medicinal Plants (translated from French by Maurice Pugh Jones). Keats Publishing Inc.: New Canaan, Connecticut.

5. Bianchini, F. and Corbetta, F. 1977. Health Plants of the World, Atlas of Medicinal Plants. Newsweek Books: New York.
6. eFloras.org 2008 http://www.efloras.org/florataxon.aspx?flora_id=1&taxon_id=200003518 Date Accessed: 30th March 2008
7. Imperato, F. 1982. *Phytochemistry* 21: 2158–2159.
8. Imperato, F. 1982. *Phytochemistry* 21: 2717–2718.
9. Berti, G., Bottari, F. and Marsili, A. 1969. *Tetrahedron* 25: 2939–2947.
10. Nakane, T., Maeda, Y., Ebihara, H., Arai, Y., Masuda, K., Takano, A., Ageta, H., Shiojima, K., Cai, S. Q. and Abdel-Halim, O. B. 2002. *Chemical and Pharmaceutical Bulletin (Tokyo)* 50: 1273–1275.
11. Marino, A., Elberti, M. G. and Cataldo, A. 1989. *Bollettino della Società italiana di biologia sperimentale* 65: 461–463.
12. Singh, M., Singh, N., Khare, P. B. and Rawat, A. K. 2008. *Journal of Ethnopharmacology* 115: 327–329.
13. Fleming, T. (Ed.). 2000. PDR for Herbal Medicines, 2nd edition. Medical Economics Company: New Jersey.

Allamanda cathartica L.

1. Hsuan, K. 1990. The Concise Flora of Singapore, Gymnosperms and Dicotyledons. Singapore University Press: National University of Singapore.
2. Wiart, C. 2006. Medicinal Plants of the Asia-Pacific: Drugs for the future? World Scientific Publishing Co. Pte. Ltd.: Singapore.
3. Wee, Y. C. 1992. A Guide to Medicinal Plants. Singapore Science Centre Publication: Singapore.
4. eFloras.org 2008 http://www.efloras.org/florataxon.aspx?flora_id=2&taxon_id=210000031 Date Accessed: 30th March 2008
5. Carballeira, N. M. and Cruz, C. 1998. *Phytochemistry* 49: 1253–1256.
6. Coppen, J. J. W. 1983. *Phytochemistry* 22: 179–182.

7. Tiwari, T. N., Pandey, V. B. and Dubey, N. K. 2002. *Phytotherapy Research* 16: 393–394.
8. Agrawal, P. K. and Sharma, M. P. 1983. *Indian Journal of Pharmaceutical Sciences* 45: 246–248.
9. De Melo, S. J. and De Mello, J. F. 1997. *Fitoterapia* 68: 478.
10. Abdel-Kader, M. S., Wisse, J., Evans, R., van der Werff, H. and Kingston, D. G. 1997. *Journal of Natural Products* 60: 1294–1297.
11. Jain, S. K. and DeFilipps, R. A. 1991. Medicinal Plants of India, volume one and two. Reference Publications Inc.: United States of America.
12. Alen, Y., Nakajima, S., Nitoda, T., Baba, N., Kanzaki, H. and Kawazu, K. 2000. *Zeitschrift für Naturforschung [C]* 55: 295–299.
13. Dixit, S. N., Tripathi, S. C. and Ojaha, T. N. 1982. *Bokin Bobai* 10: 197–199.
14. Connér, D. E. and Beuchat, L. R. 1984. *Journal of Food Science* 49: 429–434.
15. Tripathi, S. C. and Dixit, S. N. 1977. *Experientia* 33: 207–208.
16. Tiwari, T. N., Dubey, N. K. and Pandey, V. B. 1998. *Indian Drugs* 35: 415–416.
17. Otero, R., Núñez, V., Barona, J., Fonnegra, R., Jiménez, S. L., Osorio, R. G., Saldarriaga, M. and Díaz, A. 2000. *Journal of Ethnopharmacology* 73: 233–241.
18. Nayak, S., Nalabothu, P., Sandiford, S., Bhogadi, V. and Adogwa, A. 2006. *BMC Complementary and Alternative Medicine* 6: 12.
19. Apollo, M., Dash, S. K. and Padhy, S. 2006. *Journal of Human Ecology* 19: 239–248.

Aloe vera Mill.

1. Hsuan, K. 1992. Orders and Families of Malayan Seed Plants. Singapore University Press: Singapore.
2. Wee, Y. C. 1992. A Guide to Medicinal Plants. Singapore Science Centre Publication: Singapore.
3. eFloras.org 2008 http://www.efloras.org/florataxon.aspx?flora_id=1&taxon_id=200027555 Date Accessed: 30th March 2008
4. van Wyk, B. E. and Wink, M. 2004. Medicinal plants of the world: an illustrated scientific guide to important medicinal plants and their uses. Timber Press: Portland.
5. Goh, S. H., Chuah, C. H., Mok, J. S. L. and Soepadmo, E. 1995. Malaysian Medicinal Plants for the Treatment of Cardiovascular Diseases. Pelanduk Publications (M) Sdn Bhd: Malaysia.
6. Li, T. S. C. 2000. Medicinal Plants. Culture, Utilisation and Phytopharmacology. Technomic Publishing Co., Inc.: Lancaster.
7. Djeraba, A. and Quere, P. 2000. *International Journal of Immunopharmacology* 22: 365–372.
8. Okamura, N., Hine, N., Tateyama, Y., Nakazawa, M., Fujioka, T., Mihashi, K. and Yagi, A. 1998. *Phytochemistry* 49: 219–223.
9. Pugh, N., Ross, S. A., ElSohly, M. A. and Pasco, D. S. 2001. *Journal of Agriculture and Food Chemistry* 49: 1030–1034.
10. Wang, H. M., Shi, W., Xu, Y. K., Liu, Y., Lu, M. J. and Pan, J. Q. 2003. Spectral study of a new dihydroisocoumarin. *Magnetic Resonance in Chemistry* 41: 718–720.
11. Zhang, X. F., Wang, H. M., Song, Y. L., Nie, L. H., Wang, L. F., Liu, B., Shen, P. P. and Liu, Y. 2006. *Bioorganic and Medicinal Chemistry Letters* 16: 949–953.
12. Esua, M. F. and Rauwald, J. W. 2006. *Carbohydrate Research* 341: 355–364.
13. Wee, Y. C. and Hsuan, K. 1990. An illustrated Dictionary of Chinese Medicinal Herbs. Times Edition and Eu Yan Seng Holdings Ltd: Singapore.
14. Goh, S. H. 1988. Proceedings: Malaysian Traditional Medicine, Kuala Lumpur, pp. 7–26. Institute of Advanced Studies, University of Malaya, Kuala Lumpur.
15. Jaganath, I. B. and Ng, L. T. 2000. Herbs: The green pharmacy of Malaysia. Vinpress Sdn. Bhd: Kuala Lumpur, in collaboration with the Malaysian Agricultural Research and Development (MARDI).
16. Jain, S. K. and DeFilipps, R. A. 1991. Medicinal Plants of India, volume one and two. Reference Publications Inc.: United States of America.
17. Duke, J. A. and Ayensu, E. S. 1985. Medicinal Plants of China, volume one and two. Reference Publications, Inc.: United States of America.
18. Keys, J. D. 1976. Chinese herbs: their botany, chemistry, and pharmacodynamics.

Charles E. Tuttle Co. of Rutland: Vermont & Tokyo, Japan.

19. Grover, J. K., Yadav, S. and Vats, V. 2002. *Journal of Ethnopharmacology* 81: 81–100.

20. Moon, E. J., Lee, Y. M., Lee, O. H., Lee, M. J., Lee, S. K., Chung, M. H., Park, Y. I., Sung, C. K., Choi, J. S. and Kim, K. W. 2000. *Angiogenesis* 3: 117–123.

21. Jasso de Rodríguez, D., Hernández-Castillo, D., Rodríguez-García, R. and Angulo-Sánchez, J. L. 2005. *Industrial Crops and Product* 21: 81–87.

22. Stuart, R. W., Lefkowitz, D. L., Lincoln, J. A., Howard, K., Gelderman, M. P. and Lefkowitz, S. S. 1997. *International Journal of Immunopharmacology* 19: 75–82.

23. Rosca-Casian, O., Parvu, M., Vlase, L. and Tamas, M. 2007. *Fitoterapia* 78: 219–222.

24. Rajasekaran, S., Ravi, K., Sivagnanam, K. and Subramanian, S. 2006. *Clinical and Experimental Pharmacology and Physiology* 33: 232–237.

25. Tanaka, M., Misawa, E., Ito, Y., Habara, N., Nomaguchi, K., Yamada, M., Toida, T., Hayasawa, H., Takase, M., Inagaki, M. and Higuchi, R. 2006. *Biological and Pharmaceutical Bulletin* 29: 1418–1422.

26. Duansak, D., Somboonwong, J. and Patumraj, S. 2003. *Clinical Hemorheology and Microcirculation* 29: 239–246.

27. Langmead, L., Makins, R. J. and Rampton, D. S. 2004. *Alimentary Pharmacology and Therapeutics* 19: 521–527.

28. Lee, C. K., Han, S. S., Shin, Y. K., Chung, M. H., Park, Y. I., Lee, S. K. and Kim, Y. S. 1999. *International Journal of Immunopharmacology* 21: 303–310.

29. Prabjone, R., Thong-Ngam, D., Wisedopas, N., Chatsuwan, T. and Patumraj, S. 2006. *Clinical Hemorheology and Microcirculation* 35: 359–366.

30. Habeeb, F., Stables, G., Bradbury, F., Nong, S., Cameron, P., Plevin, R. and Ferro, V. A. 2007. *Methods* 42: 388–393.

31. Lee, K. H., Kim, J. H., Lim, D. S. and Kim, C. H. 2000. *Journal of Pharmacy and Pharmacology* 52: 593–598.

32. Pecere, T., Gazzola, M. V., Mucignat, C., Parolin, C., Dalla, V. F., Cavaggioni, A., Basso, G., Diaspro, A., Salvato, B., Carli, M.

and Palu, G. 2000. *Cancer Research* 60: 2800–2804.

33. Acevedo-Duncan, M., Russell, C., Patel, S. and Patel, R. 2004. *International Immunopharmacology* 4: 1775–1784.

34. Akev, N., Turkay, G., Can, A., Gurel, A., Yildiz, F., Yardibi, H., Ekiz, E. E. and Uzun, H. 2007. *European Journal of Cancer Prevention* 16: 151–157.

35. Akev, N., Turkay, G., Can, A., Gurel, A., Yildiz, F., Yardibi, H., Ekiz, E. E. and Uzun, H. 2007. *Phytotherapy Research* 21: 1070–1075.

36. Chen, S. H., Lin, K. Y., Chang, C. C., Fang, C. L. and Lin, C. P. 2007. *Food and Chemical Toxicology* 45: 2296–2303.

37. Esmat, A. Y., El-Gerzawy, S. M. and Rafaat, A. 2005. *Cancer Biology and Therapy* 4: 108–112.

38. Nićiforović, A., Adzić, M., Spasić, S. D. and Radojcić, M. B. 2007. *Cancer Biology and Therapy* 6: 1200–1205.

39. Habeeb, F., Shakir, E., Bradbury, F., Cameron, P., Taravati, M. R., Drummond, A. J., Gray, A. I. and Ferro, V. A. 2007. *Methods* 42: 315–320.

40. Yagi, A., Kabash, A., Mizuno, K., Moustafa, S. M., Khalifa, T. I. and Tsuji, H. 2003. *Planta Medica* 69: 269–271.

41. Saada, H. N., Ussama, Z. S. and Mahdy, A. M. 2003. *Pharmazie* 58: 929–931.

42. Rajasekaran, S., Sivagnanam, K. and Subramanian, S. 2005. *Pharmacological Reports* 57: 90–96.

43. Gupta, R. and Flora, S. J. 2005. *Phytotherapy Research* 19: 23–28.

44. Wu, J. H., Xu, C., Shan, C. Y. and Tan, R. X. 2006. *Life Sciences* 78: 622–630.

45. Kardosová, A. and Machová, E. 2006. *Fitoterapia* 77: 367–373.

46. Yoo, E. J. and Lee, B. M. 2005. *Journal of Toxicology and Environmental Health, Part A* 68: 1841–1860.

47. Maze, G., Terpolilli, R. N. and Lee, M. K. 1997. *Medical Science Research* 25: 765–766.

48. Eamlamnam, K., Patumraj, S., Visedopas, N. and Thong-Ngam, D. 2006. *World Journal of Gastroenterology* 12: 2034–2039.

49. Etim, O. E., Farombi, E. O., Usoh, I. F. and Akpan, E. J. 2006. *Pakistan Journal of Pharmaceutical Sciences* 19: 337–340.

50. Chandan, B. K., Saxena, A. K., Shukla, S., Sharma, N., Gupta, D. K., Suri, K. A., Suri, J., Bhadauria, M. and Singh, B. 2007. *Journal of Ethnopharmacology* 111: 560–566.
51. Lin, H. J., Lai, C. C., Lee Chao, P. D., Fan, S. S., Tsai, Y., Huang, S. Y., Wan, L. and Tsai, F. J. 2007. *Pharmacology and Therapeutics* 23: 152–171.
52. Geremias, R., Pedrosa, R. C., Locatelli, C., de Fávere, V. T., Coury-Pedrosa, R. and Laranjeira, M. C. 2006. *Phytotherapy Research* 20: 288–293.
53. Hart, L. A. T., Van Enckevort, P. H., Van Dijk, H., Zaat, R., De Silva, K. T. D. and Labadie, R. P. 1988. *Journal of Ethnopharmacology* 23: 61–71.
54. Hart, L. A. T., Van den Berg, A. J. J., Kuis, L., Van Dijk, H. and Labadie, R. P. 1989. *Planta Medica* 55: 509–512.
55. Chen, X. D., Huang, L. Y., Wu, B. Y., Jiang, Q., Wang, Z. C. and Lin, X. H. 2005. *Zhongguo Wei Zhong Bing Ji Jiu Yi Xue* 17: 296–298.
56. Chen, X. D., Wu, B. Y., Jiang, Q., Huang, L. Y. and Wang, Z. C. 2005. *Zhongguo Zhong Yao Za Zhi* 30: 1944–1946.
57. Liu, C., Leung, M. Y., Koon, J. C., Zhu, L. F., Hui, Y. Z., Yu, B. and Fung, K. P. 2006. *International Immunopharmacology* 6: 1634–1641.
58. Djeraba, A. and Quere, P. 2000. *International Journal of Immunopharmacology* 22: 365–372.
59. Womble, D. and Helderman, J. H. 1988. *International Journal of Immunopharmacology* 10: 967–974.
60. Dutta, A., Mandal, G., Mandal, C. and Chatterjee, M. 2007. *Glycoconjugate Journal* 24: 81–86.
61. Dutta, A., Bandyopadhyay, S., Mandal, C. and Chatterjee, M. 2007. *Journal of Medical Microbiology* 56: 629–636.
62. Kirdpon, S., Kirdpon, W., Airarat, W., Trevanich, A. and Nanakorn, S. 2006. *Journal of the Medical Association of Thailand* 89 (Suppl 2): S9–14.
63. Kirdpon, S., Kirdpon, W., Airarat, W., Thepsuthammarat, K. and Nanakorn, S. 2006. *Journal of the Medical Association of Thailand* 89: 1199–1205.
64. Wang, Z. W., Zhou, J. M., Huang, Z. S., Yang, A. P., Liu, Z. C., Xia, Y. F., Zeng, Y. X. and Zhu, X. F. 2004. *Journal of Radiation Research (Tokyo)* 45: 447–454.
65 Wang, Z. W., Huang, Z, S., Yang, A. P., Li, C. Y., Huang, H., Lin, X., Liu, Z. C. and Zhu, X. F. 2005. *Ai Zheng* 24: 438–442.
66. Abdullah, K. M., Abdullah, A., Johnson Mary, L., Bilski, J. J., Petry, K., Redmer, D. A., Reynolds, L. P. and Grazul-Bilska, A. T. 2003. *Journal of Alternative and Complementary Medicine* 9: 711–718.
67. Chithra, P., Sajithlal, G. B. and Chandrakasan, G. 1998. *Journal of Ethnopharmacology* 59: 179–186.
68. Bezáková, L., Oblozinský, M., Sýkorová, M., Paulíková, I. and Kostálová, D. 2005. *Ceská a Slovenská Farmacie* 54: 43–46.
69. Maenthaisong, R., Chaiyakunapruk, N., Niruntraporn, S. and Kongkaew, C. 2007. *Burns* 33: 713–718.
70. Bradley, P. R. (Ed.). 1992. British herbal compendium. Volume 1. A handbook of scientific information on widely used plant drugs. Companion to Volume 1 of the British Herbal Pharmacopoeia. British Herbal Medicine Association: United Kingdom.
71. Fetrow, C. W. and Avila, J. R. 2000. The complete guide to herbal medicines. Springhouse Corp: Springhouse, PA.
72. Kuhn, M. A. and Winston, D. 2000. Herbal therapy and supplements: a scientific and traditional approach. Lippincott: Philadelphia.
73. Quick access: Professional guide to conditions, herbs & supplements. 2000. Newton, MA: Integrative Medicine Communications.
74. Boudreau, M. D. and Beland, F. A. 2006. *Journal of Environmental Science and Health. Part C, Environmental Carcinogenesis & Ecotoxicology Reviews* 24: 103–154.
75. Bottenberg, M. M., Wall, G. C., Harvey, R. L. and Habib, S. 2007. *The Annals of Pharmacotherapy* 41: 1740–1743.
76. Bisset, N. G. 1994. Herbal drugs and phytopharmaceuticals. CRC Press: Boca Raton, FL.
77. Fleming, T. (Ed.). 2000. PDR for Herbal Medicines, 2nd edition. Medical Economics Company: New Jersey.
78. Vinson, J. A., Al Kharrat, H. and Andreoli, L. 2005. *Phytomedicine* 12: 760–765.

Andrographis paniculata (Barm.f.) Nees

1. Wiart, C. 2006. Medicinal Plants of the Asia-Pacific: Drugs for the future? World Scientific Publishing Co. Pte. Ltd.: Singapore.
2. USDA, ARS GRIN Taxonomy for Plants 2008 http://www.ars-grin.gov/cgi-bin/npgs/html/taxon.pl?104887 Date Accessed: 30th June 2008
3. Jain, D. C., Gupta, M. M., Saxena, S. and Kumar, S. 2000. *Journal of Pharmaceutical and Biomedical Analysis* 22: 705–709.
4. Cheung, H. Y., Cheung, C. S. and Kong, C. K. 2001. *Journal of Chromatography A* 930: 171–176.
5. Reddy, M. K., Reddy, M. V., Gunasekar, D., Murthy, M. M., Caux, C. and Bodo, B. 2003. *Phytochemistry*, 62: 1271–1275.
6. Koteswara, R. Y., Vimalamma, G., Rao, C. V. and Tzeng, Y. M. 2004. *Phytochemistry* 65: 2317–2321.
7. Chen, L. X., Qu, G. X. and Qiu, F. 2006. *Zhongguo Zhong Yao Za Zhi* 31: 391–395.
8. Chen, L. X., Qu, G. X. and Qiu, F. 2006. *Zhongguo Zhong Yao Za Zhi* 19: 1594–1597.
9. Reddy, V. L., Reddy, S. M., Ravikanth, V., Krishnaiah, P., Goud, T. V., Rao, T. P., Ram, T. S., Gonnade, R. G., Bhadbhade, M. and Venkateswarlu, Y. 2005. *Natural Product Research* 19: 223–230.
10. Pramanick, S., Banerjee, S., Achari, B., Das, B., Sen, A. K. Sr., Mukhopadhyay, S., Neuman, A. and Prangé, T. 2006. *Journal of Natural Products* 69: 403–405.
11. Shen, Y. H., Li, R. T., Xiao, W. L., Xu, G., Lin, Z. W., Zhao, Q. S. and Sun, H. D. 2006. *Journal of Natural Products* 69: 319–322.
12. Li, W., Xu, X., Zhang, H., Ma, C., Fong, H., van Breemen, R. and Fitzloff, J. 2007. *Chemical and Pharmaceutical Bulletin* 55: 455–458.
13. Wu, T. S., Chern, H. J., Damu, A. G., Kuo, P. C., Su, C. R., Lee, E. J. and Teng, C. M. 2008. *Journal of Asian Natural Product Research* 10: 17–24.
14. Chan, W. R., Taylor, D. R., Willis, C. R. and Fehlhaber, H. W. 1968. *Tetrahedron Letters* 9: 4803–4806.
15. Perry, L. M. 1980. Medicinal plants of East and Southeast Asia: Attributed properties and uses. The MIT Press: Cambridge, Massachusetts and London.
16. Natural Medicines Comprehensive Database 2008 http://www.naturaldatabase.com/S(fpd5gq45bydk4145u4wt5dij))/nd/Search.aspx?cs=&s=ND&pt=9&Product=andrographis+paniculata&btnSearch.x=21&btnSearch.y=9 Date Accessed: 23rd June2008
17. Chen, J. H., Hsiao, G., Lee, A. R., Wu, C. C. and Yen, M. H. 2004. *Biochemical Pharmacology* 67: 1337–1345.
18. Singha, P. K., Roy, S. and Dey, S. 2003. *Fitoterapia* 74: 692–694.
19. Limsong, J., Benjavongkulchai, E. and Kuvatanasuchati, J. 2004. *Journal of Ethnopharmacology* 92: 281–289.
20. Zaidan, M. R., Noor Rain, A., Badrul, A. R., Adlin, A., Norazah, A. and Zakiah, I. 2005. *Tropical Biomedicine* 22: 165–170.
21. Kumar, R. A., Sridevi, K., Kumar, N. V., Nanduri, S. and Rajagopal, S. 2004. *Journal of Ethnopharmacology* 92: 291–295.
22. Cheung, H. Y., Cheung, S. H., Li, J., Cheung, C. S., Lai, W. P., Fong, W. F. and Leung, F. M. 2005. *Planta Medica* 71: 1106–1111.
23. Jada, S. R., Hamzah, A. S., Lajis, N. H., Saad, M. S., Stevens, M. F. and Stanslas, J. 2006. *Journal of Enzyme Inhibition and Medicinal Chemistry* 21: 145–155.
24. Zhou, J., Zhang, S., Ong, C. N. and Shen, H. M. 2006. *Biochemical Pharmacology* 72: 132–144.
25. Jada, S. R., Subur, G. S., Matthews, C., Hamzah, A. S., Lajis, N. H., Saad, M. S., Stevens, M. F. G. and Stanslas, J. 2007. *Phytochemistry* 68: 904–912.
26. Sheeja, K. and Kuttan, G. 2007. *Immunopharmacology and Immunotoxicology* 29: 81–93.
27. Sheeja, K. and Kuttan, G. 2007. *Integrative Cancer Therapies* 6: 66–73.
28. Sheeja, K., Guruvayoorappan, C. and Kuttan, G. 2007. *International Immunopharmacology* 7: 211–221.
29. Ji, L., Liu, T., Liu, J., Chen, Y. and Wang, Z. 2007. *Planta Medica* 73: 1397–1401.
30. Jiang, C. G., Li, J. B., Liu, F. R., Wu, T., Yu, M. and Xu, H. M. 2007. *Anticancer Research* 27: 2439–2447.
31. Husen, R., Pihie, A. H. and Nallappan, M. 2004. *Journal of Ethnopharmacology* 95: 205–208.

32. Yu, B. C., Hung, C. R., Chen, W. C. and Cheng, J. T. 2003. *Planta Medica* 69: 1075–1079.

33. Zhang, X. F. and Tan, B. K. 2000. *Clinical and Experimental Pharmacology and Physiology* 27: 358–363.

34. Reyes, B. A., Bautista, N. D., Tanquilut, N. C., Anunciado, R. V., Leung, A. B., Sanchez, G. C., Magtoto, R. L., Castronuevo, P., Tsukamura, H. and Maeda, K. I. 2006. *Journal of Ethnopharmacology* 105: 196–200.

35. Xu, Y., Chen, A., Fry, S., Barrow, R. A., Marshall, R. L. and Mukkur, T. K. 2007. *International Immunopharmacology* 7: 515–523.

36. Umamaheswari, S. and Mainzen Prince, P. S. 2007. *Acta Poloniae Pharmaceutica* 64: 53–61.

37. Akbarsha, M. A. and Murugaian, P. 2000. *Phytotherapy Research* 14: 432–435.

38. Habtemariam, S. 1998. *Phytotherapy Research* 12: 37–40.

39. Hidalgo, M. A., Romero, A., Figueroa, J., Cortes, P., Concha, I. I., Hancke, J. L. and Burgos, R. A. 2005. *British Journal of Pharmacology* 144: 680–686.

40. Shen, Y. C., Chen, C. F. and Chiou, W. F. 2002. *British Journal of Pharmacology* 135: 399–406.

41. Madav, S., Tripathi, H. C., Tandan, S. K., Kumar, D. and Lal, J. 1998. *Indian Journal of Pharmaceutical Sciences* 60: 176–178.

42. Iruretagoyena, M. I., Tobar, J. A., Gonzalez, P. A., Sepulveda, S. E., Figueroa, C. A., Burgos, R. A., Hancke, J. L. and Kalergis, A. M. 2005. *The Journal of Pharmacology and Experimental Therapeutics* 312: 366–372.

43. Ji, L. L., Wang, Z., Dong, F., Zhang, W. B. and Wang, Z. T. 2005. *Journal of Cellular Biochemistry* 95: 970–978.

44. Ko, H. C., Wei, B. L. and Chiou, W. F. 2006. *Journal of Ethnopharmacology* 107: 205–210.

45. Sheeja, K., Shihab, P. K. and Kuttan, G. 2006. *Immunopharmacology and Immunotoxicology* 28: 129–140.

46. Liu, J., Wang, Z. T. and Ji, L. L. 2007. *The American Journal of Chinese Medicine* 35: 317–328.

47. Liu, J., Wang, Z. T., Ji, L. L. and Ge, B. X. 2007. *Molecular and Cellular Biochemistry* 298: 49–57.

48. Liu, J., Wang, Z. T. and Ge, B. X. 2008. *International Immunopharmacology* 8: 951–958.

49. Kamdem, R. E., Sang, S. and Ho, C. T. 2002. *Journal of Agricultural and Food Chemistry* 50: 4662–4665.

50. Tripathi, R. and Kamat, J. P. 2007. *Indian Journal of Experimental Biology* 45: 959–967.

51. Tripathi, R., Mohan, H. and Kamat, J. P. 2007. *Food Chemistry* 100: 81–90.

52. Burgos, R. A., Imilan, M., Sanchez, N. S. and Hancke, J. L. 2000. *Journal of Ethnopharmacology* 71: 115–121.

53. Thisoda, P., Rangkadilok, N., Pholphana, N., Worasuttayangkurn, L., Ruchirawat, S. and Satayavivad, J. 2006. *European Journal of Pharmacology* 553: 39–45.

54. Rahman, N. N. N. A., Furuta, T., Kojima, S., Takane, K. and Mustafa-Ali, M. 1999. *Journal of Ethnopharmacology* 64: 249–254.

55. Zaridah, M. Z., Idid, S. Z., Omar, A. W. and Khozirah, S. 2001. *Journal of Ethnopharmacology* 78: 79–84.

56. Dua, V. K., Ojha, V. P., Roy, R., Joshi, B. C., Valecha, N., Devi, C. U., Bhatnagar, M. C., Sharma, V. P. and Subbarao, S. K. 2004. *Journal of Ethnopharmacology* 95: 247–251.

57. Calabrese, C., Berman, S. H., Babish, J. G., Ma, X., Shinto, L., Dorr, M., Wells, K., Wenner, C. A. and Standish, L. J. 2000. *Phytotherapy Research: PTR* 14: 333–338.

58. Wiart, C., Kumar, K., Yusof, M. Y., Hamimah, H., Fauzi, Z. M. and Sulaiman, M. 2005. *Phytotherapy Research* 19: 1069–1070.

59. Woo, A. Y., Waye, M. M., Tsui, S. K., Yeung, S. T. and Cheng, C. H. 2008. *The Journal of Pharmacology and Experimental Therapeutics* 325: 226–235.

60. Chang, K. T., Lii, C. K., Tsai, C. W., Yang, A. J. and Chen, H. W. 2008. *Food and Chemical Toxicology* 46: 1079–1088.

61. Singh, R. P., Banerjee, S. and Rao, A. R. 2001. *Phytotherapy Research: PTR* 15: 382–390.

62. Visen, P. K., Shukla, B., Patnaik, G. K. and Dhawan, B. N. 1993. *Journal of Ethnopharmacology* 40: 131–136.

63. Trivedi, N. P., Rawal, U. M. and Patel, B. P. 2007. *Integrative Cancer Therapies* 6: 271–280.

64. Zhang, C. Y. and Tan, B. K. 1996. *Clinical and Experimental Pharmacology and Physiology* 23: 675–678.
65. Panossian, A., Davtyan, T., Gukassyan, N., Gukasova, G., Mamikonyan, G., Gabrielian, E. and Wikman, G. 2002. *Phytomedicine* 9: 598–605.
66. Burgos, R. A., Seguel, K., Perez, M., Meneses, A., Ortega, M., Guarda, M. I., Loaiza, A. and Hancke, J. L. 2005. *Planta Medica* 71: 429–434.
67. Mandal, S. C., Dhara, A. K. and Maiti, B. C. 2001. *Phytotherapy Research: PTR* 15: 253–256.
68. Yoopan, N., Thisoda, P., Rangkadilok, N., Sahasitiwat, S., Pholphana, N., Ruchirawat, S. and Satayavivad, J. 2007. *Planta Medica* 73: 503–511.
69. Pharmacopoeia of the People's Republic of China, volume 1. English Edition, 2000. Compiled by the State Pharmacopoeia Commission of People's Republic of China. Chemical Industry Press: Beijing.
70. Mills, S. and Bone, K. 2000. Principles and Practice of Phytotherapy. Churchill Livingstone: New York.
71. Chang, H. and But, P. 1987. Pharmacology and Application of China Materia Medica, volume two. World Scientific Publication: Singapore.
72. Hovhannisyan, A. S., Abrahamyan, H., Gabrielyan, E. S. and Panossian, A. G. 2006. *Phytomedicine* 13: 318–323.

Ardisia elliptica Thunb.

1. Wee, Y. C. 1992. A Guide to Medicinal Plants. Singapore Science Centre Publication: Singapore.
2. Hsuan, K. 1992. Orders and Families of Malayan Seed Plants. Singapore University Press: Singapore.
3. Wee, Y. C. and Corlett, R. 1986. The City and the Forest: Plant Life in Urban Singapore. Singapore University Press: National University of Singapore.
4. USDA, ARS GRIN Taxonomy for Plants 2008 http://www.ars-grin.gov/cgi-bin/npgs/html/taxon.pl?403463 Date Accessed: 30th June 2008

5. Chow, P. W., Sim, K. Y., Lim, P. L. and Chung, V. C. 1991. *Bulletin of the Singapore National Institute of Chemistry* 19: 87–93.
6. Ahmad, S. A., Catalano, S., Marsili, A., Morelli, I. and Scartoni, V. 1977. *Planta Medica* 32: 162–164.
7. Kobayashi, H. and de Mejia, E. 2005. *Journal of Ethnopharmacology* 96: 347–354.
8. Phadungkit, M. and Luanratana, O. 2006. *Natural Product Research* 20: 693–697.
9. Liu, N., Li, Y., Gua, J. X. and Qian, D. G. 1993. *Acta Academiae Medicinae Shanghai* 20: 49–54.
10. Jalil, J., Jantan, I., Shaari, K. and Rafi, I. A. A. 2004. *Pharmaceutical Biology* 42: 457–461.
11. Ibrahim, B. J., Kang, Y. H., Suh, D. Y. and Han, B. H. 1996. *Natural Product Sciences* 2: 86–89.

Areca catechu L.

1. Wee, Y. C. 1992. A Guide to Medicinal Plants. Singapore Science Centre Publication: Singapore.
2. Palaniswamy, U. R. 2003. A guide to Medicinal Plants of Asian origin and culture. CPL Press: United States of America.
3. Hsuan, K., Chin, S. C. and Tan, H. T. W. 1998. The Concise Flora of Singapore volume II: Monocotyledons. Singapore University Press: Singapore.
4. Wagner, W. L., Herbst, D. R. and Sohmer, S. H. 1999. Manual of the flowering plants of Hawaii (Revised edition). University of Hawaii Press: Honolulu.
5. USDA, ARS GRIN Taxonomy for Plants 2008 http://www.ars-grin.gov/cgi-bin/npgs/html/taxon.pl?104887 Date Accessed: 30th June 2008
6. Duke, J. A. and Ayensu, E. S. 1985. Medicinal Plants of China, volume one and two. Reference Publications, Inc.: United States of America.
7. Nguyen, V. D. and Doan, T. N. 1989. Medicinal Plants in Vietnam. World Health Organisation, Regional Office for the Western Pacific, Manila, Institute of Materia Medica, Hanoi.
8. Holdsworth, D. K., Jones, R. A. and Self, R. 1998. *Phytochemistry* 48: 581–582.

9. Woodley, E. (Ed.). 1991. Medicinal plants of Papua New Guinea, part I. Veerlag Josef Margraf Scientific Books: Weikersheim.
10. Keys, J. D. 1976. Chinese herbs: their botany, chemistry, and pharmacodynamics. Charles E. Tuttle Co. of Rutland: Vermont & Tokyo, Japan.
11. Jain, S. K. and DeFilipps, R. A. 1991. Medicinal Plants of India, volume one and two. Reference Publications Inc.: United States of America.
12. Ghelardini, C., Galeotti, N., Lelli, C. and Bartolini, A. 2001. *Il Farmaco* 56: 383–385.
13. Muhammad, M. M., Abdul, M. A., Mohd, A. A., Rozita, M. N., Nashriyah, B. M., Abdul, R. R. and Kazuyoshi, K. 1997. *Pesticide Science* 51: 165–170.
14. Wang, Y. C. and Huang, T. L. 2005. Screening of anti-*FEMS Immunology and Medical Microbiology* 43: 295–300.
15. Chang, Y. C., Tai, K. W., Li, C. K., Chou, L. S. and Chou, M. Y. 1999. *Clinical Oral Investigations* 3: 25–29.
16. Sundqvist, K., Liu, Y., Erhardt, P., Nair, J., Bartsch, H. and Grafstrom, R. C. 1991. *IARC Scientific Publications* 105: 281–285.
17. Lodge, D., Johnston, G. A. R., Curtis, D. R. and Brand, S. J. 1977. *Brain Research* 136: 513–522.
18. Ahsana, D. and Shagufta, K. 1997. *Phytotherapy Research* 11: 174–176.
19. Ahsana, D. and Shagufta, K. 2000. *Pharmacology Biochemistry and Behavior* 65: 1–6.
20. Inokuchi, J., Okabe, H., Yamauchi, T., Nagamatsu, A., Nonaka, G. and Nishioka, I. 1986. *Life Sciences* 38: 1375–1382.
21. Chung, F. M., Shieh, T. Y., Yang, Y. H., Chang, D. M., Shin, S. J., Tsai, J. C., Chen, T. H., Tai, T. Y. and Lee, Y. J. 2007. *Translational Research* 150: 58–65.
22. Chang, M. C., Ho, Y. S., Lee, P. H., Chan, C. P., Lee, J. J., Hahn, L. J., Wang, Y. J. and Jeng, J. H. 2001. *Carcinogenesis* 22: 1527–1535.
23. Owen, P. L., Matainaho, T., Sirois, M. and Johns, T. 2007. *Journal of Biochemical and Molecular Toxicology* 21: 231–242.
24. Jeng, J. H., Ho, Y. S., Chan, C. P., Wang, Y. J., Hahn, L. J., Lei, D., Hsu, C. C. and Chang, M. C. 2000. *Carcinogenesis* 21: 1365–1370.
25. Park, Y. B., Jeon, S. M., Byun, S. J., Kim, H. S. and Choi, M. S. 2002. *Life Sciences* 70: 1849–1859.
26. Lai, Y. L., Lin, J. C., Yang, S. F., Liu, T. Y. and Hung, S. L. 2007. *Journal of Periodontal Research* 42: 69–76.
27. Hung, S. L., Cheng, Y. Y., Peng, J. L., Chang, L. Y., Liu, T. Y. and Chen, Y. T. 2005. *Journal of Periodontal Research* 76: 373–379.
28. Chiang, C. P., Hsieh, R. P., Chen, T. H., Chang, Y. F., Liu, B. Y., Wang, J. T., Sun, A. and Kuo, M. Y. 2002. *Journal of Oral Pathology and Medicine* 31: 402–409.
29. Hung, S. L., Lin, Y. J., Chien, E. J., Liu, W. G., Chang, H. W., Liu, T. Y. and Chen, Y. T. 2007. *Journal of Periodontal Research* 42: 393–401.
30. Chang, L. Y., Wan, H. C., Lai, Y. L., Liu, T. Y. and Hung, S. L. 2006. *Journal of Periodontal Research* 77: 1969–1977.
31. Wang, C. C., Liu, T. Y., Wey, S. P., Wang, F. I. and Jan, T. R. 2007. *Food and Chemical Toxicology* 45: 1410–1418.
32. Lee, K. K., Cho, J. J., Park, E. J. and Choi, J. D. 2001. *International Journal of Cosmetic Science* 23: 341–346.
33. Pithayanukul, P., Ruenraroengsak, P., Bavovada, R., Pakmanee, N., Suttisri, R. and Saen-oon, S. 2005. *Journal of Ethnopharmacology* 97: 527–533.
34. Chang, M. C., Kuo, M. Y., Hahn, L. J., Hsieh, C. C., Lin, S. K. and Jeng, J. H. 1998. *Journal of Periodontal Research* 69: 1092–1097.
35. Jaiswal, P. and Singh, D. K. 2008. *Veterinary Parasitology* 152: 264–270.
36. Chatrchaiwiwatana, S. 2006. *Journal of the Medical Association of Thailand* 89: 1004–1011.
37. Skidmore-Roth, L. 2001. Mosby's handbook of herbs and natural supplements. Mosby, Inc.: St. Louis.
38. Fetrow, C. W. and Avila, J. R. 2000. The complete guide to herbal medicines. Springhouse Corp: Springhouse, PA.
39. Fleming, T. (Ed.). 2000. PDR for Herbal Medicines, 2nd edition. Medical Economics Company: New Jersey.
40. Wu, M. T., Wu, D. C., Hsu, H. K., Kao, E. L. and Lee, J. M. 2003. *British Journal of Cancer* 89: 1202–1204.

41. Kang, I. M., Chou, C. Y., Tseng, Y. H., Huang, C. C., Ho, W. Y., Shih, C. M. and Chen, W. 2007. *Journal of Occupational and Environmental Medicine* 49: 776–769.
42. Lan, T. Y., Chang, W. C., Tsai, Y. J., Chuang, Y. L., Lin, H. S. and Tai, T. Y. 2007. *American Journal of Epidemiology* 165: 677–683.
43. Yang, M. S., Lee, C. H., Chang, S. J., Chung, T. C., Tsai, E. M., Ko, A. M. and Ko, Y. C. 2008. *Drug and Alcohol Dependence* 95: 134–139.
44. López-Vilchez, M. A., Seidel, V., Farré, M., García-Algar, O., Pichini, S. and Mur, A. 2006. *Pediatrics* 117: e129–131.
45. García-Algar, O., Vall, O., Alameda, F., Puig, C., Pellegrini, M., Pacifici, R. and Pichini, S. 2005. *Archives of Diseases in Childhood: Fetal and neonatal edition* 90: F276–277.
46. Chou, W. W., Guh, J. Y., Tsai, J. F., Hwang, C. C., Chen, H. C., Huang, J. S., Yang, Y. L., Hung, W. C. and Chuang, L. Y. 2008. *Toxicology* 243: 1–10.
47. Lai, K. C. and Lee, T. C. 2006. *Mutation Research* 599: 66–75.

Asplenium nidus L.

1. Wee, Y. C. 1992. A Guide to Medicinal Plants. Singapore Science Centre Publication: Singapore.
2. USDA, ARS GRIN Taxonomy for Plants 2008 http://www.ars-grin.gov/cgi-bin/npgs/html/taxon.pl?5590 Date Accessed: 30th June 2008
3. Imperato, F. 1987. *Chemistry & Industry* 14: 487.
4. Perry, L. M. 1980. Medicinal plants of East and Southeast Asia: Attributed properties and uses. The MIT Press: Cambridge, Massachusetts and London.
5. Bourdy, G. and Walter, A. 1992. *Journal of Ethnopharmacology* 37: 179–196.
6. Bourdy, G., Francois, C., Andary, C. and Boucard, M. 1996. *Journal of Ethnopharmacology* 52: 139–143.

Aster tataricus L.f.

1. Hsuan, K. 1992. Orders and Families of Malayan Seed Plants. Singapore University Press: Singapore.

2. Wee, Y. C. and Hsuan, K. 1990. An illustrated Dictionary of Chinese Medicinal Herbs. Times Edition and Eu Yan Seng Holdings Ltd: Singapore.
3. eFloras.org 2008 http://www.efloras.org/florataxon.aspx?flora_id=1&taxon_id=200023501 Date Accessed: 30th March 2008
4. Ng, T. B., Liu, F., Lu, Y. H., Cheng, C. H. K. and Wang, Z. T. 2003. *Comparative Biochemistry and Physiology Part C: Toxicology & Pharmacology* 136: 109–115.
5. Schumacher, K. K., Hauze, D. B., Jiang, J. J., Szewczyk, J., Reddy, R. E., Davis, F. A. and Joullié, M. M. 1999. *Tetrahedron Letters* 40: 455–458.
6. Cheng, D. L. and Shao, Y. 1993. *Phytochemistry* 35: 173–176.
7. Cheng, D. L., Shao, Y., Hartmann, R., Roeder, E. and Zhao, K. 1996. *Phytochemistry* 41: 225–227.
8. Morita, H., Nagashima, S., Shirota, O., Takeya, K. and Itokawa, H. 1993. *Chemistry Letters* 11: 1877–1880.
9. Morita, H., Nagashima, S., Takeya, K., Itokawa, H. and Iitaka, Y. 1995. *Tetrahedron* 51: 1121–1132.
10. Akihisa, T., Kimura, Y., Koike, K., Tai, T., Yasukawa, K., Arai, K., Suzuki, Y. and Nikaido, T. 1998. *Chemical & Pharmaceutical Bulletin* 46: 1824–1826.
11. Akihisa, T., Kimura, Y., Tai, T. and Arai, K. 1999. *Chemical & Pharmaceutical Bulletin* 47: 1161–1163.
12. Perry, L. M. 1980. Medicinal plants of East and Southeast Asia: Attributed properties and uses. The MIT Press: Cambridge, Massachusetts and London.
13. Keys, J. D. 1976. Chinese herbs: their botany, chemistry, and pharmacodynamics. Charles E. Tuttle Co. of Rutland: Vermont & Tokyo, Japan.
14. Rossi, F., Zanotti, G., Saviano, M., Iacovino, R., Palladino, P., Saviano, G., Amodeo, P., Tancredi, T., Laccetti, P., Corbier, C. and Benedetti, E. 2004. *Journal of Peptide Science* 10: 92–102.
15. Cozzolino, R., Palladino, P., Rossi, F., Calì, G., Benedetti, E. and Laccetti, P. 2005. *Carcinogenesis* 26: 733–739.

Azadirachta indica A. Juss.

1. Palaniswamy, U. R. 2003. A guide to Medicinal Plants of Asian origin and culture. CPL Press: United States of America.
2. Hsuan, K. 1990. The Concise Flora of Singapore, Gymnosperms and Dicotyledons. Singapore University Press: National University of Singapore.
3. Sumner J. 2000. The Natural History of Medicinal Plants. Timber Press, Portland.
4. eFloras.org 2008 http://www.efloras.org/florataxon.aspx?flora_id=5&taxon_id=220001427 Date Accessed: 30ᵗʰ March 2008
5. Siddiqui, B. S., Afshan, F. and Faizi, S. 2001. *Tetrahedron* 57: 10281–10286.
6. Siddiqui, B. S., Afshan, F., Ghiasuddin, Faizi, S., Naqvi, S. N. H. and Tariq, R. M. 2000. *Phytochemistry* 53: 371–376.
7. Siddiqui, S., Faizi, S., Mahmood, T. and Siddiqui, B. S. 1986. *Tetrahedron* 42: 4849–4856.
8. Sinniah, D., Goh, S. H. and Baskaran, G. 1988. Proceedings: Malaysian Traditional Medicines, Kuala Lumpur, pp. 65–69. Institute of Advanced Studies, University of Malaya, Kuala Lumpur.
9. Siddiqui, B. S., Afshan, F., Sham-Sul-Arfeen and Gulzar, T. 2006. *Natural Product Research* 20: 1036–1040.
10. Siddiqui, B. S., Tariq Ali, S. and Kashif Ali, S. 2006. *Natural Product Research* 20: 241–245.
11. Siddiqui, B. S., Rasheed, M., Ilyas, F., Gulzar, T., Tariq, R. M. and Naqvi, S. N. 2004. *Zeitschrift für Naturforschung. C, Journal of Biosciences* 59: 104–112.
12. Siddiqui, B. S., Afshan, F., Gulzar, T. and Hanif, M. 2004. *Phytochemistry* 65: 2363–2367.
13. Siddiqui, B. S., Afshan, F., Gulzar, T., Sultana, R., Naqvi, S. N. and Tariq, R. M. 2003. *Chemical and Pharmaceutical Bulletin (Tokyo)* 51: 415–417.
14. Siddiqui, B. S., Afshan, F., Faizi, S., Naqvi, S. N. and Tariq, R. M. 2002. *Journal of Natural Products* 65: 1216–1218.
15. Garg, H. S. and Bhakuni, D. S. 1985. *Phytochemistry* 24: 866–867.
16. Garg, H. S. and Bhakuni, D. S. 1984. *Phytochemistry* 23: 2115–2118.
17. Garg, H. S. and Bhakuni, D. S. *Phytochemistry* 23: 2383–2385.
18. Govindachari, T. R. and Gopalakrishnan, G. 1997. *Phytochemistry* 45: 397–399.
19. Lee, S. M., Olsen, J. I., Schweizer, M. P. and Klocke, J. A. 1988. *Phytochemistry* 27: 2773–2775.
20. Luo, X. D., Wu, S. H., Ma, Y. B. and Wu, D. G. 2000. *Fitoterapia* 71: 668–672.
21. Mangalam, S. N., Sindhu, G. and Deepa, I. 1997. *Phytochemistry* 46: 1177–1178.
22. Nanduri, S., Thunuguntla, S. S. R., Nyavanandi, V. K., Kasu, S., Kumar, P. M., Ram, P. S., Rajagopal, S., Kumar, R. A., Deevi, D. S., Rajagopalan, R. and Venkateswarlu, A. 2003. *Bioorganic & Medicinal Chemistry Letters* 22: 4111–4115.
23. Ramji, N., Venkatakrishnan, K. and Madyastha, K. M. 1996. *Phytochemistry* 42: 561–562.
24. Rojatkar, S. R. and Nagasampagi, B. A. 1992. *Phytochemistry* 32: 213–214.
25. Suresh, G., Narasimhan, N. S. and Palani, N. 1997. *Phytochemistry* 45: 807–810.
26. Singh, U. P., Maurya, S. and Singh, D. P. 2005. *Journal of Herbal Pharmacotherapy* 5: 35–43.
27. Kanokmedhakul, S., Kanokmedhakul, K., Prajuabsuk, T., Panichajakul, S., Panyamee, P., Prabpai, S. and Kongsaeree, P. 2005. *Journal of Natural Products* 68: 1047–1050.
28. Sharma, V., Bali, A. and Singh, M. 1998. *Phytochemistry* 49: 2121–2123.
29. Ghazanfar, S. A. 2004. Handbook of Arabian Medicinal Plants. CRC Press: Boca Raton, FL.
30. Jain, S. K. and DeFilipps, R. A. 1991. Medicinal Plants of India, volume one and two. Reference Publications Inc.: United States of America.
31. Ajose, F. O. 2007. *International Journal of Dermatology* 46(Suppl 1): 48–55.
32. Subapriya, R. and Nagini, S. 1995. *Current Medicinal Chemistry, Anticancer Agents* 5: 149–146.
33. Baswa, M., Rath, C. C., Dash, S. K. and Mishra, R. K. 2001. *Microbios* 105: 183–189.
34. Pai, M. R., Acharya, L. D. and Udupa, N. 2004. *Journal of Ethnopharmacology* 90: 99–103.

35. SaiRam, M., Ilavazhagan, G., Sharma, S. K., Dhanraj, S. A., Suresh, B., Parida, M. M., Jana, A. M., Devendra, K. and Selvamurthy, W. 2000. *Journal of Ethnopharmacology* 71: 377–382.

36. Mahfuzul Hoque, M. D., Bari, M. L., Inatsu, Y., Juneja, V. K. and Kawamoto, S. 2007. *Foodborne Pathogens and Disease* 4: 481–488.

37. Thakurta, P., Bhowmik, P., Mukherjee, S., Hajra, T. K., Patra, A. and Bag, P. K. 2007. *Journal of Ethnopharmacology* 111: 607–612.

38. Baral, R. and Chattopadhyay, U. 2004. *International Immunopharmacology* 4: 355–366.

39. Bose, A. and Baral, R. 2007. *Human Immunology* 68: 823–831.

40. Baral, R., Mandal, I. and Chattopadhyay, U. 2005. *International Immunopharmacology* 5: 1343–1352.

41. Roy, M. K., Kobori, M., Takenaka, M., Nakahara, K., Shinmoto, H., Isobe, S. and Tsushida, T. 2007. *Phytotherapy Research* 21: 245–250.

42. Roy, M. K., Kobori, M., Takenaka, M., Nakahara, K., Shinmoto, H. and Tsushida, T. 2006. *Planta Medica* 72: 917–923.

43. Kumar, S., Suresh, P. K., Vijayababu, M. R., Arunkumar, A. and Arunakaran, J. 2006. *Journal of Ethnopharmacology* 105: 246–250.

44. Koul, A., Mukherjee, N. and Gangar, S. C. 2006. *Molecular and Cellular Biochemistry* 283: 47–55.

45. Sastry, B. S., Suresh Babu, K., Hari Babu, T., Chandrasekhar, S., Srinivas, P. V., Saxena, A. K. and Madhusudana Rao, J. 2006. *Bioorganic and Medicinal Chemistry Letters* 16: 4391–4394.

46. Subapriya, R., Kumaraguruparan, R. and Nagini, S. 2006. *Clinical Biochemistry* 39: 1080–1087.

47. Gangar, S. C. and Koul, A. 2007. *Indian Journal of Biochemistry and Biophysics* 44: 209–215.

48. Subapriya, R., Bhuvaneswari, V. and Nagini, S. 2005. *Asian Pacific Journal of Cancer Prevention* 6: 515–520.

49. Subapriya, R., Bhuvaneswari, V., Ramesh, V. and Nagini, S. 2005. *Cell Biochemistry and Function* 23: 229–238.

50. Sarkar, K., Bose, A., Laskar, S., Choudhuri, S. K., Dey, S., Roychowdhury, P. K. and Baral, R. 2007. *International Immunopharmacology* 7: 306–312.

51. Niture, S. K., Rao, U. S. and Srivenugopal, K. S. 2006. *International Journal of Oncology* 29: 1269–1278.

52. Gangar, S. C., Sandhir, R., Rai, D. V. and Koul, A. 2006. *World Journal of Gastroenterology* 12: 2749–2755.

53. Sritanaudomchai, H., Kusamran, T., Kuakulkiat, W., Bunyapraphatsara, N., Hiransalee, A., Tepsuwan, A. and Kusamran, W. R. 2005. *Asian Pacific Journal of Cancer Prevention* 6: 263–269.

54. Arakaki, J., Suzui, M., Morioka, T., Kinjo, T., Kaneshiro, T., Inamine, M., Sunagawa, N., Nishimaki, T. and Yoshimi, N. 2006. *Asian Pacific Journal of Cancer Prevention* 7: 467–471.

55. Dasgupta, T., Banerjee, S., Yadava, P. K. and Rao, A. R. 2004. *Journal of Ethnopharmacology* 92: 23–36.

56. Parshad, O., Young, L. E. and Young, R. E. 1997. *Phytotherapy Research* 11: 398–400.

57. Parshad, O., Gardner, M. T., The, T. L., Williams, L. A. D. and Fletcher, C. K. 1997. *Phytotherapy Research* 11: 168–170.

58. Khillare, B. and Shrivastav, T. G. 2003. *Contraception* 68: 225–229.

59. Amadioha, A. C. 2000. *Crop Protection* 19: 287–290.

60. Govindachari, T. R., Suresh, G. and Masilamani, S. 1999. *Fitoterapia* 70: 417–420.

61. Polaquini, S. R., Svidzinski, T. I., Kemmelmeier, C. and Gasparetto, A. 2006. *Archieves of Oral Biology* 51: 482–490.

62. Kaur, G., Alam, M. S. and Athar, M. 2004. *Phytotherapy Research* 18: 419–424.

63. El Tahir, A., Satti, G. M. and Khalid, S. A. 1999. *Journal of Ethnopharmacology* 64: 227–233.

64. RaviDhar, Zhang, K. Y., Talwar, G. P., Garg, S. and Kumar, N. 1998. *Journal of Ethnopharmacology* 61: 31–39.

65. Nathan, S. S., Kalaivani, K. and Murugan, K. 2005. *Acta Tropica* 96: 47–55.

66. Udeinya, I. J., Brown, N., Shu, E. N., Udeinya, F. I. and Quakeyie, I. 2006. *Annals of Tropical Medicine and Parasitology* 100: 17–22.

67. Kirira, P. G., Rukunga, G. M., Wanyonyi, A. W., Muregi, F. M., Gathirwa, J. W., Muthaura, C. N., Omar, S. A., Tolo, F., Mungai, G. M. and Ndiege, I. O. 2006. *Journal of Ethnopharmacology* 106: 403–407.

68. Soh, P. N. and Benoit-Vical, F. 2007. *Journal of Ethnopharmacology* 114: 130–140.

69. Gopal, M., Gupta, A., Arunachalam, V. and Magu, S. P. 2007. *Bioresource Technology* 98: 3154–3158.

70. Gupta, S., Kataria, M., Gupta, P. K., Murganandan, S. and Yashroy, R. C. 2004. *Journal of Ethnopharmacology* 90: 185–189.

71. Subapriya, R., Kumaraguruparan, R., Abraham, S. K. and Nagini, S. 2005. *Journal of Herbal Pharmacotherapy* 5: 39–50.

72. Sithisarn, P., Supabphol, R. and Gritsanapan, W. 2006. *Medical Principles and Practice* 15: 219–222.

73. Sithisarn, P., Supabphol, R. and Gritsanapan, W. 2005. *Journal of Ethnopharmacology* 99: 109–112.

74. Prakash, D., Suri, S., Upadhyay, G. and Singh, B. N. 2007. *International Journal of Food Science and Nutrition* 58: 18–28.

75. Di Ilio, V., Pasquariello, N., van der Esch, A. S., Cristofaro, M., Scarsella, G. and Risuleo, G. 2006. *Molecular and Cellular Biochemistry* 287: 69–77.

76. Chaube, S. K., Prasad, P. V., Khillare, B. and Shrivastav, T. G. 2006. *Fertility and Sterility* 85 Suppl 1: 1223–1231.

77. Ashorobi, R. B. 1998. *Phytotherapy Research* 12: 41–43.

78. Parida, M. M., Upadhyay, C., Pandya, G. and Jana, A. M. 2002. *Journal of Ethnopharmacology* 79: 273–278.

79. Joshi, S. N., Katti, U., Godbole, S., Bharucha, K., Kishore Kumar, B., Kulkarni, S., Risbud, A. and Mehendale, S. 2005. *Transactions of the Royal Society of Tropical Medicine and Hygiene* 99: 769–774.

80. Yinusa, R., Isiaka, A. O., Caleb, A. O. and Godwin, J. 2004. *Journal of Ethnopharmacology* 90: 167–170.

81. Bandyopadhyay, U., Biswas, K., Chatterjee, R., Bandyopadhyay, D., Chattopadhyay, I., Ganguly, C. K., Chakraborty, T., Bhattacharya, K. and Banerjee, R. K. 2002. *Life Sciences* 71: 2845–2865.

82. Dorababu, M., Joshi, M. C., Bhawani, G., Kumar, M. M., Chaturvedi, A. and Goel, R. K. 2006. *Indian Journal of Physiology and Pharmacology* 50: 241–249.

83. Chattopadhyay, R. R. 2003. *Journal of Ethnopharmacology* 89: 217–219.

84. Koul, A., Ghara, A. R. and Gangar, S. C. 2006. *Phytotherapy Research* 20: 169–177.

85. Koul, A., Binepal, G. and Gangar, S. C. 2007. *Indian Journal of Experimental Biology* 45: 359–366.

86. Chattopadhyay, R. R. 1999. *Journal of Ethnopharmacology* 67: 367–372.

87. Waheed, A., Miana, G. A. and Ahmad, S. I. 2006. *Pakistan Journal of Pharmaceutical Sciences* 19: 322–325.

88. Oiefuna, I. and Young, R. 2004. Novel Compounds from Natural Products in the New Millennium. Potential and Challenges. World Scientific Publishing: Singapore.

89. Chattopadhyay, R. R. 1997. *General Pharmacology: The Vascular System* 28: 449–451.

90. Obiefuna, I. and Young, R. 2005. *Phytotherapy Research* 19: 792–795.

91. Beuth, J., Schneider, H. and Ko, H. L. 2006. *In Vivo* 20: 247–251.

92. Haque, E. and Baral, R. 2006. *Immunobiology* 211: 721–731.

93. Haque, E., Mandal, I., Pal, S. and Baral, R. 2006. *Immunopharmacology and Immunotoxicology* 28: 33–50.

94. Yanpallewar, S., Rai, S., Kumar, M., Chauhan, S. and Acharya, S. B. 2005. *Life Sciences* 76: 1325–1338.

95. Hördegen, P., Cabaret, J., Hertzberg, H., Langhans, W. and Laurer, V. 2006. *Journal of Ethnopharmacology* 108: 85–89.

96. Fatima, F., Khalid, A., Nazar, N., Abdalla, M., Mohomed, H., Toum, A. M., Magzoub, M. and Alı, M. S. 2005. *Turkish Society for Parasitology* 29: 3–6.

97. Nakahara, K., Roy, M. K., Ono, H., Maeda, I., Ohnishi-Kameyama, M., Yoshida, M. and Trakoontivakorn, G. 2003. *Journal of Agriculture and Food Chemistry* 51: 6456–6460.

98. Nakahara, K., Trakoontivakorn, G., Alzoreky, N. S., Ono, H., Onishi-Kameyama, M. and Yoshida, M. 2002. *Journal of Agriculture and Food Chemistry* 50: 4796–4802.

99. Farah, M. A., Ateeq, B. and Ahmad, W. 2006. *The Science of the Total Environment* 364: 200–214.

100. Singh, K., Singh, A. and Singh, D. K. 1996. *Journal of Ethnopharmacology* 52: 35–40.

101. Abdel-Shafy, S. and Zayed, A. A. 2002. *Veterinary Parasitology* 106: 89–96.

102. Al-Rajhy, D. H., Alahmed, A. M., Hussein, H. I. and Kheir, S. M. 2003. *Pest Management Science* 59: 1250–1254.

103. Paul, P. K. and Sharma, P. D. 2002. *Physiological and Molecular Plant Pathology* 61: 3–13.

104. Singh, U. P. and Prithiviraj, B. 1997. *Physiological and Molecular Plant Pathology* 51: 181–194.

105. Thacker, J. R. M., Bryan, W. J., McGinley, C., Heritage, S. and Strang, R. H. C. 2003. *Crop Protection* 22: 753–760.

106. Nathan, S. S., Savitha, G., George, D. K., Narmadha, A., Suganya, L. and Chung, P. G. 2006. *Bioresource Technology* 97: 1316–1323.

107. Nathan, S. S., Kalaivani, K. and Murugan, K. 2006. *Ecotoxicology and Environmental Safety* 65: 102–107.

108. Nathan, S. S., Chung, P. G. and Murugan, K. 2006. *Ecotoxicology and Environmental Safety* 64: 382–389.

109. Anuradha, A., Annadurai, R. S. and Shashidhara, L. S. 2007. *Insect Biochemistry and Molecular Biology* 37: 627–634.

110. Okumu, F. O., Knols, B. G. and Fillinger, U. 2007. *Malaria Journal* 6: 63.

111. Skidmore-Roth L. 2001. Mosby's handbook of herbs and natural supplements. Mosby, Inc.: St. Louis.

112. Boeke, S. J., Boersma, M. G., Alink, G. M., van Loon, J. J., van Huis, A., Dicke, M. and Rietjens, I. M. 2004. *Journal of Ethnopharmacology* 94: 25–41.

Barringtonia asiatica L.

1. Hsuan, K. 1992. Orders and Families of Malayan Seed Plants. Singapore University Press: Singapore.

2. Hsuan, K. 1990. The Concise Flora of Singapore, Gymnosperms and Dicotyledons. Singapore University Press: National University of Singapore.

3. Wee, Y. C. 1992. A Guide to Medicinal Plants. Singapore Science Centre Publication: Singapore.

4. USDA, ARS GRIN Taxonomy for Plants 2008 http://www.ars-grin.gov/cgi-bin/npgs/html/taxon.pl?6512
Date Accessed: 30ᵗʰ June 2008

5. Herlt, A. J., Mander, L. N., Pongoh, E., Rumampuk, R. J. and Tarigan, P. 2002. *Journal of Natural Products* 65: 115–120.

6. Nozoe, T. 1934. *Nippon Kagaku Kaishi* 55: 1106–1114.

7. Burton, R. A., Wood, S. G. and Owen, N. L. 2003. *Arkivoc* Part 13: 137–146.

8. Locher, C. P., Burch, M. T., Mower, H. F., Berestecky, J., Davis, H., Van Poel, B., Lasure, A., Vanden Berghe, D. A. and Vlietinck, A. J. 1995. *Journal of Ethnopharmacology* 49: 23–32.

9. Whistler, A. W. 1992. Polynesian Herbal Medicine. Everbest Publishing Co.: Hong Kong.

10. Perry, L. M. 1980. Medicinal plants of East and Southeast Asia: Attributed properties and uses. The MIT Press: Cambridge, Massachusetts and London.

11. Steiner, R. P. 1986. Folk medicine: the art and the science. American Chemical Society: Washington D. C.

12. Woodley, E. (Ed.). 1991. Medicinal plants of Papua New Guinea, part I. Veerlag Josef Margraf Scientific Books: Weikersheim.

13. Khan, M. R. and Omoloso, A. D. 2002. *Fitoterapia* 73: 255–260.

Barringtonia racemosa (L.) K. Spreng

1. Hsuan, K. 1992. Orders and Families of Malayan Seed Plants. Singapore University Press: Singapore.

2. Hsuan, K. 1990. The Concise Flora of Singapore, Gymnosperms and Dicotyledons. Singapore University Press: National University of Singapore.

3. USDA, ARS GRIN Taxonomy for Plants 2008 http://www.ars-grin.gov/cgi-bin/npgs/html/taxon.pl?6514
Date Accessed: 30ᵗʰ June 2008

4. Anantaraman, R., Pillai, K. S. and Madhavan. 1956. *Journal of the Chemical Society, Abstracts*: 4369–4373.

5. Lin, Y. T., Lo, T. B. and Su, S. C. 1957. *Journal of the Chinese Chemical Society, Taipei, Taiwan* 4 (Ser. II): 77–81.
6. Pillai, K. S. and Madhavan. 1959. *Bull. Cent. Research Inst. University of Kerala, Trivandrum, Ser. A* 6: 15–20.
7. Hasan, C. M., Khan, S., Jabbar, A. and Rashid, M. A. 2000. *Journal of Natural Products* 63: 410–411.
8. Sun, H. Y., Long, L. J. and Wu, J. 2006. *Zhong Yao Cai* 29: 671–672.
9. Yang, Y., Deng, Z., Proksch, P. and Lin, W. 2006. *Pharmazie* 61: 365–366.
10. Deraniyagala, S. A., Ratnasooriya, W. D. and Goonasekara, C. L. 2003. *Journal of Ethnopharmacology* 86: 21–26.
11. Jain, S. K. and DeFilipps, R. A. 1991. Medicinal Plants of India, volume one and two. Reference Publications Inc.: United States of America.
12. Khan, S., Jabbar, A., Hasan, C. M. and Rashid, M. A. 2001. *Fitoterapia* 72: 162–164.
13. Gowri. P. M., Tiwari, A. K., Ali, A. Z. and Rao, J. M. 2007. *Phytotherapy Research* 2: 796–799.
14. Thomas, T. J., Panikkar, B., Subramoniam, A., Nair, M. K. and Panikkar, K. R. 2002. *Journal of Ethnopharmacology* 82: 223–227.
15. Mackeen, M. M., Ali, A. M., El-Sharkawy, S. H., Manap, M. Y., Salleh, K. M., Lajis, N. H. and Kawazu, K. 1997. *International Journal of Pharmacognosy* 35: 174–178.

7. Bhartiya, H. P., Dubey, P., Katiyar, S. B. and Gupta, P. C. 1979. *Phytochemistry* 18: 689.
8. Pettit, G. R., Numata, A., Iwamoto, C., Usami, Y., Yamada, T., Ohishi, H., Cragg, G. M. 2006. *Journal of Natural Products* 69: 323–327.
9. Boonphong, S., Puangsombat, P., Baramee, A., Mahidol, C., Ruchirawat, S. and Kittakoop, P. 2007. *Journal of Natural Products* 70: 795–801.
10. Perry, L. M. 1980. Medicinal plants of East and Southeast Asia: Attributed properties and uses. The MIT Press: Cambridge, Massachusetts and London.
11. Brennan, M. J., Cisar, J. O., Vatter, A. E. and Sandberg, A. L. 1984. *Infection and Immunity* 46: 459–464.
12. Zakaria, Z. A., Wen, L. Y., Abdul Rahman, N. I., Abdul Ayub, A. H., Sulaiman, M. R. and Gopalan, H. K. 2007. *Medical Principles and Practice* 16: 443–449.
13. Panda, S. and Kar, A. 1999. *Journal of Ethnopharmacology* 67: 233–239.

Bauhinia purpurea L.

1. Hsuan, K. 1990. The Concise Flora of Singapore, Gymnosperms and Dicotyledons. Singapore University Press: National University of Singapore.
2. Wiart, C. 2006. Medicinal Plants of the Asia-Pacific: Drugs for the future? World Scientific Publishing Co. Pte. Ltd.: Singapore.
3. Wu, A. M., Wu, J. H., Liu, J. H. and Singh, T. 2004. *Life Sciences* 74: 1763–1779.
4. Yadava, R. N. and Tripathi, P. 2000. *Fitoterapia* 71: 88–90.
5. Kuo, Y. H., Yeh, M. H. and Huang, S. L. 1998. *Phytochemistry* 49: 2529–2530.
6. Bhartiya, H. P. and Gupta, P. C. 1981. *Phytochemistry* 20: 2051.

Bixa orellana L.

1. Wiart, C. 2006. Medicinal Plants of the Asia-Pacific: Drugs for the future? World Scientific Publishing Co. Pte. Ltd.: Singapore.
2. Hsuan, K. 1990. The Concise Flora of Singapore, Gymnosperms and Dicotyledons. Singapore University Press: National University of Singapore.
3. Wiart, C. 2000. Medicinal plants of Southeast Asia. Pelanduk Publications (M) Sdn Bhd: Malaysia.
4. eFloras.org 2008 http://www.efloras.org/florataxon.aspx?flora_id=2&taxon_id=200014319 Date Accessed: 30th March 2008
5. Houssiau, L., Felicissimo, M., Bittencourt, C. and Pireaux, J. J. 2004. *Applied Surface Science* 231–232: 416–419.
6. Mercadante, A. Z., Steck, A. and Pfander, H. 1997. *Phytochemistry* 46: 1379–1383.
7. Mercadante, A. Z., Steck, A. and Pfander, H. 1999. *Phytochemistry* 52: 135–139.
8. Mercadante, A. Z., Steck, A., Rodriguez-Amaya, D., Pfander, H. and Britton, G. 1996. *Phytochemistry* 41: 1201–1203.

9. Lawrence, B. M. and Hogg, J. W. 1973. *Phytochemistry* 12: 2995.
10. Schneider, W. P., Caron, E. L. and Hinman, J. W. 1965. *Journal of Organic Chemistry* 30: 2856–2857.
11. Jain, S. K. and DeFilipps, R. A. 1991. Medicinal Plants of India, volume one and two. Reference Publications Inc.: United States of America.
12. Fleischer, T. C., Ameade, E. P. K., Mensah, M. L. K. and Sawer, I. K. 2003. *Fitoterapia* 74: 136–138.
13. Lans, C. A. 2006. *Journal of Ethnobiology and Ethnomedicine* 2: 45.
14. Morton, J. F. 1981. Atlas of Medicinal Plants of Middle America: Bahamas to Yucatan. C.C. Thomas Publishers Co.: Springfield, IL.
15. Cáceres, A., Menéndez, H., Méndez, E., Cohobón, E., Samayoa, B. E., Jauregui, E., Peralta, E. and Carrillo, G. 1995. *Journal of Ethnopharmacology* 48: 85–88.
16. Shilpi, J. A., Taufiq-Ur-Rahman, M., Uddin, S. J., Alam, M. S., Sadhu, S. K. and Seidel, V. 2006. *Journal of Ethnopharmacology* 108: 264–271.
17. Rojas, J. J., Ochoa, V. J., Ocampo, S. A. and Muñoz, J. F. 2006. *BMC Complementary and Alternative Medicine* 6: 2.
18. Galindo-Cuspinera, V. and Rankin, S. A. 2005. *Journal of Agriculture and Food Chemistry* 53: 2524–2529.
19. Reddy, M. K., Alexander-Lindo, R. L. and Nair, M. G. 2005. *Journal of Agriculture and Food Chemistry* 53: 9268–9273.
20. Nunez, V., Otero, R., Barona, J., Saldarriaga, M., Osorio, R. G., Fonnegra, R., Jimenez, S. L., Diaz, A. and Quintanam J. C. 2004. *Brazilian Journal of Medical and Biological Research* 37: 969–977.
21. Villar, R., Calleja, J. M., Morales, C. and Cáceres, A. 1997. *Phytotherapy Research* 11: 441–445.
22. Russell, K. R., Morrison, E. Y. and Ragoobirsingh, D. 2005. *Phytotherapy Research* 19: 433–436.
23. Júnior, A. C., Asad, L. M., Oliveira, E. B., Kovary, K., Asad, N. R. and Felzenszwalb, I. 2005. *Genetics and Molecular Research* 4: 94–99.
24. Braga, F. G., Bouzada, M. L., Fabri, R. L., de O Matos, M., Moreira, F. O., Scio, E. and Coimbra, E. S. 2007. *Journal of Ethnopharmacology* 111: 396–402.
25. Duke, J. A., Bogenschutz-Godwin, M. J., duCellier, J. and Duke, P. A. K. 2002. Handbook of medicinal herbs. CRC Press: Boca Raton, FL.
26. Duke, J. A. and duCellier, J. L. 1993. Handbook of Alternative Cash Crops. CRC Press: Boca Raton, FL.
27. Agner, A. R., Barbisan, L. F., Scolastici, C. and Salvadori, D. M. F. 2004. *Food and Chemical Toxicology* 42: 1687–1693.
28. Paumgartten, F. J. R., De-Carvalho, R. R., Araujo, I. B., Pinto, F. M., Borges, O. O., Souza, C. A. M. and Kuriyama, S. N. 2002. *Food and Chemical Toxicology* 40: 1595–1601.

Calophyllum inophyllum L.

1. Wiart, C. 2006. Medicinal Plants of the Asia-Pacific: Drugs for the future? World Scientific Publishing Co. Pte. Ltd.: Singapore.
2. Hsuan, K. 1990. The Concise Flora of Singapore, Gymnosperms and Dicotyledons. Singapore University Press: National University of Singapore.
3. USDA, ARS GRIN Taxonomy for Plants 2008 http://www.ars-grin.gov/cgi-bin/npgs/html/taxon.pl?8631 Date Accessed: 30th June 2008
4. Govindachari, T. R. and Viswanathan, B. R. P. N., Ramadas, R. U. and Srinivasan, M. 1967. *Tetrahedron* 23: 1901–1910.
5. Kawazu, K., Ohigashi, H. and Mitsui, T. 1968. *Tetrahedron Letters* 9: 2383–2385.
6. Jackson, B., Locksley, H. D. and Scheinmann, F. 1969. *Phytochemistry* 8: 927–929.
7. Subramanian, S. S. and Nair, A. G. R. 1971. *Phytochemistry* 10: 1679–1680.
8. Al-Jeboury, F. S. and Locksley, H. D. 1971. *Phytochemistry* 10: 603–606.
9. Bhushan, B., Rangaswami, S. and Seshadri, T. R. 1975. *Indian Journal of Chemistry* 13: 746–747.
10. Kumar, V., Ramachandran, S. and Sultanbawa, M. U. S. 1976. *Phytochemistry* 15: 2016–2017.
11. Goh, S. H. and Jantan, I. 1991. *Phytochemistry* 30: 366–367.
12. Iinuma, M., Tosa, H., Tanaka, T. and Yonemori, S. 1994. *Phytochemistry* 35: 527–532.

13. Iinuma, M., Tosa, H., Tanaka, T. and Yonemori, S. 1995. *Phytochemistry* 38: 725–728.
14. Khan, N. U. D., Parveen, N., Singh, M. P., Singh, R., Achari, B., Dastidar, P. P. G. and Dutta, P. K. 1996. *Phytochemistry* 42: 1181–1183.
15. Spino, C., Dodier, M. and Sotheeswaran, S. 1998. *Bioorganic & Medicinal Chemistry Letters* 8: 3475–3478.
16. Muhammad, S. A., Shaukat, M., Shaista, P., Viqar, U. A. and Ghazala, H. R. 1999. *Phytochemistry* 50: 1385–1389.
17. Yao, C. F. and Zeng, H. P. 2000. *Huanan Shifan Daxue Xuebao, Ziran Kexueban* 3: 62–64.
18. Joshi, S. P., Deodhar, V. B. and Phalgune, U. D. 2000. *Indian Journal of Chemistry, Section B: Organic Chemistry Including Medicinal Chemistry* 39B: 560–561.
19. Itoigawa, M., Ito, C., Tan, H. T. W., Kuchide, M., Tokuda, H., Nishino, H. and Furukawa, H. 2001. *Cancer Letters* 169: 15–19.
20. Shen, Y. C., Hung, M. C., Wang, L. T. and Chen, C. Y. 2003. *Chemical and Pharmaceutical Bulletin Tokyo* 51: 802–806.
21. Wu, Y., Zhang, P. C., Chen, R. Y., Yu, D. Q. and Liang, X. T. 2003. *Huaxue Xuebao* 61: 1047–1051.
22. Cheng, H. C., Wang, L. T., Khalil, A. T., Chang, Y. T., Lin, Y. C. and Shen, Y. C. 2004. *Journal of the Chinese Chemical Society, Taipei, Taiwan* 51: 431–435.
23. Sekino, E., Kumamoto, T., Tanaka, T., Ikeda, T. and Ishikawa, T. 2004. *Journal of Organic Chemistry* 69: 2760–2767.
24. Yimdjo, M. C., Azebaze, A. G., Nkengfack, A. E., Meyer, A. M., Bodo, B. and Fomum, Z. T. 2004. *Phytochemistry* 65: 2789–2795.
25. Laure, F., Herbette, G., Faure, R., Bianchini, J. P., Raharivelomanana, P. and Fogliani, B. 2005. *Magnetic Resonance in Chemistry: MRC* 43: 65–68.
26. Crane, S., Aurore, G., Joseph, H., Mouloungui, Z. and Bourgeois, P. 2005. *Phytochemistry* 66: 1825–1831.
27. Ee, G. C., Kua, A. S., Lim, C. K., Jong, V. and Lee, H. L. 2006. *Natural Product Research* 20: 485–491.
28. Li, Y. Z., Li, Z. L., Hua, H. M., Li, Z. G. and Liu, M. S. 2007. *Zhongguo Zhong Yao Za Zhi* 32: 692–694.
29. Jain, S. K. and DeFilipps, R. A. 1991. Medicinal Plants of India, volume one and two. Reference Publications Inc.: United States of America.
30. Woodley, E. (Ed.). 1991. Medicinal plants of Papua New Guinea, part I. Veerlag Josef Margraf Scientific Books: Weikersheim.
31. Sundaram, B. M., Gopalkrishnan, C. and Subramanian, S. 1986. *Arogya, Manipal, India* 12: 48–49.
32. Ali, M. S., Mahmud, S., Perveen, S., Rizwani, G. H. and Ahmad, V. U. 1999. *Journal of the Chemical Society of Pakistan* 21: 174–178.
33. Gopalakrishnan, C., Shankaranarayanan, D., Nazimudeen, S. K., Viswanathan, S. and Kameswaran, L. 1980. *Indian Journal of Pharmacology* 12: 181–191.
34. Jantan, I. B., Jalil, J. and Abd, W. N. M. 2001. *Pharmaceutical Biology* 39: 243–246.
35. Oku, H., Ueda, Y., Iinuma, M. and Ishiguro, K. 2005. *Planta Medica* 71: 90–92.
36. Patil, A. D., Freyer, A. J., Eggleston, D. S., Haltiwanger, R. C., Bean, M. F., Taylor, P. B., Caranfa, M. J., Breen, A. L., Bartus, H. R., Johnson, R. K., Hertzberg, R. P. and Westley, J. W. 1993. *Journal of Medicinal Chemistry* 36: 4131–4138.
37. Clercq, E. D. 2000. *Medicinal Research Reviews* 20: 323–349.
38. Ishikawa, T. 2000. *Heterocycles* 53: 453–474.
39. Said, T., Dutot, M., Martin, C., Beaudeux, J. L., Boucher, C., Enee, E., Baudouin, C., Warnet, J. M. and Rat, P. 2007. *European Journal of Pharmaceutical Sciences* 30: 203–210.
40. Ravelonjato, B., Libot, F., Ramiandrasoa, F., Kunesch, N., Gayral, P. and Poisson, J. 1992. *Planta Medica* 58: 51–55.
41. Perry, L. M. 1980. Medicinal plants of East and Southeast Asia: Attributed properties and uses. The MIT Press: Cambridge, Massachusetts and London.
42. Ajayi, I. A., Oderinde, R. A., Talwo, V. O. and Agbedana, E. O. 2008. *Food Chemistry* 106: 458–465.

Cananga odorata (Lam.) Hook. f. & Th.

1. Hsuan, K. 1992. Orders and Families of Malayan Seed Plants. Singapore University Press: Singapore.

2. Hsuan, K. 1990. The Concise Flora of Singapore, Gymnosperms and Dicotyledons. Singapore University Press: National University of Singapore.
3. Wiart, C. 2000. Medicinal plants of Southeast Asia. Pelanduk Publications (M) Sdn Bhd: Malaysia.
4. USDA, ARS GRIN Taxonomy for Plants 2008 http://www.ars-grin.gov/cgi-bin/npgs/html/taxon.pl?8805
   Date Accessed: 30th June 2008
5. Caloprisco, E., Fourneron, J. D., Faure, R. and Demarne, F. E. 2002. *Journal of Agriculture and Food Chemistry* 50: 78–80.
6. Phan, T. S., Phan, M. G. and Nguyen, D. H. 2001. *Tap Chi Duoc Hoc* 7: 9–11.
7. Hsieh, T. J., Chang, F. R., Chia, Y. C., Chen, C. Y., Chiu, H. F. and Wu, Y. C. 2001. *Journal of Natural Products* 64: 616–619.
8. Kamarudin, M. S. 1988. Proceedings: Malaysian Traditional Medicine, Kuala Lumpur, pp. 80–87. Institute of Advanced Studies, University of Malaya, Kuala Lumpur.
9. Jain, S. K. and DeFilipps, R. A. 1991. Medicinal Plants of India, volume one and two. Reference Publications Inc.: United States of America.
10. Baratta, M. T., Dorman, H. J. D., Deans, S. G., Figueiredo, A. C., Barroso, J. G. and Ruberto, G. 1998. *Flavour and Fragrance Journal* 13: 235–244.
11. Rahman, M. M., Lopa, S. S., Sadik, G., Harun-Or-Rashid, Islam, R., Khondkar, P., Alam, A. H. and Rashid, M. A. 2005. *Fitoterapia* 76: 758–761.
12. Kluza, J., Clark, A. M. and Bailly, C. 2003. *Annals of the New York Academy of Sciences* 1010: 331–334.
13. Hongratanaworakit, T. and Buchbauer, G. 2004. *Planta Medica* 70: 632–636.
14. Woo, S. H., Reynolds, M. C., Sun, N. J., Cassady, J. M. and Snapka, R. M. 1997. Inhibition of topoisomerase II by liriodenine *Biochemical Pharmacology* 54: 467–473.
15. Kluza, J., Mazinghien, R., Degardin, K., Lansiaux, A. and Bailly, C. 2005. *European Journal of Pharmacology* 525: 32–40.
16. Chu, D. M., Miles, H., Toney, D., Ngyuen, C. and Marciano-Cabral, F. 1998. *Parasitology Research* 84: 746–752.
17. Anitha, P. and Indira, M. 2006. *Indian Journal of Experimental Biology* 44: 976–980.
18. Burdock, G. A. and Carabin, I. G. 2008. *Food and Chemical Toxicology* 46: 433–445.

Capsicum annuum L.

1. Hsuan, K. 1990. The Concise Flora of Singapore, Gymnosperms and Dicotyledons. Singapore University Press: National University of Singapore.
2. Li, T. S. C. 2000. Medicinal Plants. Culture, Utilisation and Phytopharmacology. Technomic Publishing Co., Inc.: Lancaster.
3. Wee, Y. C. 1992. A Guide to Medicinal Plants. Singapore Science Centre Publication: Singapore.
4. Wee, Y. C. and Hsuan, K. 1990. An illustrated Dictionary of Chinese Medicinal Herbs. Times Edition and Eu Yan Seng Holdings Ltd: Singapore.
5. eFloras.org 2008 http://www.efloras.org/florataxon.aspx?flora_id=2&taxon_id=200020513
   Date Accessed: 30th March 2008
6. Uquiche, E., del Valle, J. M. and Ortiz, J. 2004. *Journal of Food Engineering* 65: 55–66.
7. Mazida, M. M., Salleh, M. M. and Osman, H. 2005. *Journal of Food Composition and Analysis* 18: 427–437.
8. Materska, M., Piacente, S., Stochmal, A., Pizza, C., Oleszek, W. and Perucka, I. 2003. *Phytochemistry* 63: 893–898.
9. Yahara, S., Ura, T., Sakamoto, C. and Nohara, T. 1994. *Phytochemistry* 37: 831–835.
10. Ochi, T., Takaishi, Y., Kogure, K. and Yamauti, I. 2003. *Journal of Natural Products* 66: 1094–1096.
11. Iorizzi, M., Lanzotti, V., Ranalli, G., De Marino, S. and Zollo, F. 2002. *Journal of Agricultural and Food Chemistry* 50: 4310–4316.
12. Lee, J. H., Kiyota, N., Ikeda, T. and Nohara, T. 2007. *Chemical and Pharmaceutical Bulletin* 55: 1151–1156.
13. Lee, J. H., Kiyota, N., Ikeda, T. and Nohara, T. 2006. *Chemical and Pharmaceutical Bulletin (Tokyo)* 54: 1365–1369.
14. Nagy, V., Agócs, A., Turcsi, E., Molnár, P., Szabó, Z. and Deli, J. 2007. *Tetrahedron Letters* 48: 9012–9014.

15. Jain, S. K. and DeFilipps, R. A. 1991. Medicinal Plants of India, volume one and two. Reference Publications Inc.: United States of America.

16. Sumner J. 2000. The Natural History of Medicinal Plants. Timber Press, Portland.

17. Blumenthal, M., Goldberg, A. and Gruenwald, J. (Eds.). 2000. Herbal Medicine. Expanded Commission E Monographs. Integrative Medicine Communications: Austin, Texas.

18. Careaga, M., Fernández, E., Dorantes, L., Mota, L., Jaramillo, M. E. and Hernandez-Sanchez, H. 2003. *International Journal of Food Microbiology* 83: 331–335.

19. Kuda, T., Iwai, A. and Yano, T. 2004. *Food and Chemical Toxicology* 42: 1695–1700.

20. Ribeiro, S. F. F., Carvalho, A. O., Cunha M. D., Rodrigues, R., Cruz, L. P., Melo, V. M. M., Vasconcelos, I. M., Melo, E. J. T. and Gomes, V. M. 2007. *Toxicon* 50: 600–611.

21. Diz, M. S. S., Carvalho, A. O., Rodrigues, R., Neves-Ferreira, A. G. C., Cunha, M. D., Alves, E. W., Okorokova-Façanha, A. L., Oliveira, M. A., Perales, J., Machado, O. L. T. and Gomes, V. M. 2006. *Biochimica et Biophysica Acta* 1760: 1323–1332.

22. Han, S. S., Keum, Y. S., Chun, K. S. and Surh, Y. J. 2002. *Archives of Pharmacal Research* 25: 475–479.

23. Maoka, T., Mochida, K., Kozuka, M., Ito, Y., Fujiwara, Y., Hashimoto, K., Enjo, F., Ogata, M., Nobukuni, Y., Tokuda, H. and Nishino, H. 2001. *Cancer Letters* 172: 103–109.

24. Maoka, T., Goto, Y., Isobe, K., Fujiwara, Y., Hashimoto, K. and Mochida, K. 2001. *Journal of Oleo Science* 50: 663–665.

25. Sun, T., Xu, Z., Wu, C.-T., Janes, M., Prinyawiwatkul, W. and No, H. K. 2007. *Journal of Food Science* 72: 98–102.

26. Materska, M. and Perucka, I. 2005. *Journal of Agriculture and Food Chemistry* 53: 1750–1756.

27. Frischkorn, C. G., Frischkorn, H. E. and Carrazzoni, E. 1978. *Naturwissenschaften* 65: 480–483.

28. Kwon, M. J., Song, Y. S., Choi, M. S. and Song, Y. O. 2003. *Clinica Chimica Acta* 332: 37–44.

29. Takano, F., Yamaguchi, M., Takada, S., Shoda, S., Yahagi, N., Takahashi, T. and Ohta T. 2007. *Life Sciences* 80: 1553–1563.

30. El Hamss, R., Idaomar, M., Alonso-Moraga, A. and Munoz Serrano, A. 2003. *Food and Chemical Toxicology* 41: 41–47.

31. Laohavechvanich, P., Kangsadalampai, K., Tirawanchai, N. and Ketterman, A. J. 2006. *Food and Chemical Toxicology* 44: 1348–1354.

32. Antonious, G. F., Meyer, J. E. and Snyder, J. C. 2006. *Journal of Environmental Science and Health B* 41: 1383–1391.

33. Fleming, T. (Ed.). 2000. PDR for Herbal Medicines, 2nd edition. Medical Economics Company: New Jersey.

34. Skidmore-Roth L. 2001. Mosby's handbook of herbs and natural supplements. Mosby, Inc.: St. Louis.

35. Johnson, W. Jr. 2007. *International Journal of Toxicology* 26: 3–106.

36. Lininger, S. W., Gaby, A. R., Austin, S., Batz, F., Yarnell, E., Brown, D. J. and Constantine, G. 1999. A–Z guide to drug-herb-vitamin interactions: how to improve your health and avoid problems when using common medications and natural supplements together. Prima Health: Rocklin, CA.

Cassia fistula L.

1. Hsuan, K. 1990. The Concise Flora of Singapore, Gymnosperms and Dicotyledons. Singapore University Press: National University of Singapore.

2. Palaniswamy, U. R. 2003. A guide to Medicinal Plants of Asian origin and culture. CPL Press: United States of America.

3. Wee, Y. C. 1992. A Guide to Medicinal Plants. Singapore Science Centre Publication: Singapore.

4. Duke, J. A. and Ayensu, E. S. 1985. Medicinal Plants of China, volume one and two. Reference Publications, Inc.: United States of America.

5. Sircar, P. K., Dey, B., Sanyal, T., Ganguly, S. N. and Sircar, S. M. 1970. *Phytochemistry* 9: 735–736.

6. Murty, V. K., Rao, T. V. P. and Venkateswarlu, V. 1967. *Tetrahedron* 23: 515–518.

7. Kuo, Y. H., Lee, P. H. and Wein, Y. S. 2002. *Journal of Natural Products, Technical Note* 65: 1165–1167.

8. Yadava, R. N. and Verma, V. 2003. *Journal of Asian Natural Product Research* 5: 57–61.
9. Chowdhury, S. A. C., Mustafa, A. K. M., Alam, M. N., Gafur, M. A., Ray, B. K., Ahmed, K. and Faruq, O. 1996. *Bangladesh Journal of Scientific and Industrial Research* 31: 91–97.
10. Jain, S. K. and DeFilipps, R. A. 1991. Medicinal Plants of India, volume one and two. Reference Publications Inc.: United States of America.
11. Kumar, V. P., Chauhan, N. S., Padh, H. and Rajani, M. 2006. *Journal of Ethnopharmacology* 107: 182–188.
12. Perumal Samy, R., Ignacimuthu, S. and Sen, A. 1998. *Journal of Ethnopharmacology* 62: 173–182.
13. Duraipandiyan, V. and Ignacimuthu, S. 2007. *Journal of Ethnopharmacology* 112: 590–594.
14. Ali, N. H., Kazmi, S. U. and Faizi, S. 2007. *Pakistan Journal of Pharmaceutical Sciences* 20: 140–145.
15. Ingkaninan, K., Temkitthawon, P., Chuenchom, K., Yuyaem, T. and Thongnoi, W. 2003. *Journal of Ethnopharmacology* 89: 261–264.
16. Yadav, R. and Jain, G. C. 1999. *Advances in Contraception* 15: 293–301.
17. Sunil Kumar, K. C. and Klaus, M. 1998. *Phytotherapy Research* 12: 526–528.
18. Gupta, M., Mazumder, U. K., Rath, N. and Mukhopadhyay, D. K. 2000. *Journal of Ethnopharmacology* 72: 151–156.
19. Munasinghe, T. C. J., Seneviratne, C. K., Thabrew, M. I. and Abeysekera, A. M. 2001. *Phytotherapy Research* 15: 519–523.
20. Siddhuraju, P., Mohan, P. S. and Becker, K. 2002. *Food Chemistry* 79: 61–67.
21. Prakash, D., Suri, S., Upadhyay, G. and Singh, B. N. 2007. *International Journal of Food Sciences and Nutrition* 58: 18–28.
22. Manonmani, G., Bhavapriya, V., Kalpana, S., Govindasamy, S. and Apparanantham, T. 2005. Antioxidant activity of *Cassia fistula* (Linn.) flowers in alloxan induced diabetic rats. *Journal of Ethnopharmacology* 97: 39–42.
23. Mazumder, U. K., Gupta, M. and Rath, N. 1998. *Phytotherapy Research* 12: 520–522.

24. Bhakta, T., Mukherjee, P. K., Mukherjee, K., Banerjee, S., Mandal, S. C., Maity, T. K., Pal, M. and Saha, B. P. 1999. *Journal of Ethnopharmacology* 66: 277–282.
25. Pradeep, K., Mohan, C. V., Gobianand, K. and Karthikeyan, S. 2006. *Chemico-Biological Interactions* 167: 12–18.
26. Pradeep, K., Mohan, C. V., Anand, K. G. and Karthikeyan, S. 2005. *Indian Journal of Experimental Biology* 43: 526–530.
27. el-Saadany, S. S., el-Massry, R. A., Labib, S. M. and Sitohy, M. Z. 1991. *Die Nahrung* 35: 807–815.
28. Sartorelli, P., Andrade, S. P., Melhem, M. S., Prado, F. O. and Tempone, A. G. 2007. *Phytotherapy Research* 21: 644–647.
29. Govindarajan, M., Jebanesan, A. and Pushpanathan, T. 2008. *Parasitology Research* 102: 289–292.
30. Bhakta, T., Mukherjee, P. K., Mukherjee, K., Pal, M. and Saha, B. P. 1998. *Natural Product Sciences* 4: 84–87.
31. Senthil Kumar, M., Sripriya, R., Vijaya Raghavan, H. and Sehgal, P. K. 2006. *Journal of Surgical Research* 131: 283–289.
32. Fleming, T. (Ed.). 2000. PDR for Herbal Medicines, 2nd edition. Medical Economics Company: New Jersey.
33. de Smet, P. A. G. M., Keller, K., Hasen, R. and Chandler, R. F. 1993. Adverse Effects of Herbal Drugs, Volume 2. Springer-Verlag: Berlin.

Catharanthus roseus (L.) G. Don

1. Hsuan, K., Chin, S. C. and Tan, H. T. W. 1998. The Concise Flora of Singapore volume II: Monocotyledons. Singapore University Press: Singapore.
2. Wiart, C. 2006. Medicinal Plants of the Asia-Pacific: Drugs for the future? World Scientific Publishing Co. Pte. Ltd.: Singapore.
3. eFloras.org 2008 http://www.efloras.org/florataxon.aspx?flora_id=2&taxon_id=200018366 Date Accessed: 30th March 2008
4. Chan, G. L., Chang, S., Chow, K. K. G., Goh, K. L. H. and Hoo, K. 1998. A Guide to Toxic Plants of Singapore. Singapore Science Centre Publication: Singapore.

5. Goh, S. H., Chuah, C. H., Mok, J. S. L. and Soepadmo, E. 1995. Malaysian Medicinal Plants for the Treatment of Cardiovascular Diseases. Pelanduk Publications (M) Sdn Bhd: Malaysia.

6. Li, T. S. C. 2000. Medicinal Plants. Culture, Utilisation and Phytopharmacology. Technomic Publishing Co., Inc.: Lancaster.

7. Nguyen, V. D. and Doan, T. N. 1989. Medicinal Plants in Vietnam. World Health Organisation, Regional Office for the Western Pacific, Manila, Institute of Materia Medica, Hanoi.

8. Contin, A., Heijden, R. V. D. and Verpoorte, R. 1999. *Plant Science* 147: 177–183.

9. Gilles, B., Marie-Geneviève, D., Bruno, D. and Anne-Marie, M. 1999. *Phytochemistry* 50: 167–169.

10. Atta-ur-Rahman and Fatimak, J. A. 1984. *Tetrahedron Letters* 25: 6051–6054.

11. Atta-ur-Rahman, Bashir, M., Kaleem, S. and Fatima, T. 1983. *Phytochemistry* 22: 1021–1023.

12. Sumner, J. 2000. The Natural History of Medicinal Plants. Timber Press, Portland.

13. Wiart, C. 2000. Medicinal plants of Southeast Asia. Pelanduk Publications (M) Sdn Bhd: Malaysia.

14. Jain, S. K. and DeFilipps, R. A. 1991. Medicinal Plants of India, volume one and two. Reference Publications Inc.: United States of America.

15. Duke, J. A. and Ayensu, E. S. 1985. Medicinal Plants of China, volume one and two. Reference Publications, Inc.: United States of America.

16. Goh, S. H. 1988. Proceedings: Malaysian Traditional Medicine, Kuala Lumpur, pp. 7–26. Institute of Advanced Studies, University of Malaya, Kuala Lumpur.

17. Koh, Y. K., Chang, P. and Geh, S. L. 1988. Proceedings: Malaysian Traditional Medicine, Kuala Lumpur, pp. 27–31. Institute of Advanced Studies, University of Malaya, Kuala Lumpur.

18. Ueda, J. Y., Yasuhiro, T., Hari, B. A., Quan, L. T., Kim, T. Q., Yuko, H., Ikuo, S. and Shigetoshi, K. 2002. *Biological & Pharmaceutical Bulletin* 25: 753–760.

19. Haque, N., Chowdhury, S. A. R., Nutan, M. T. H., Rahman, G. M. S., Rahman, K. M. and Rashid, M. A. 2000. *Fitoterapia* 71: 547–552.

20. Zheng, W. and Wang, S. Y. 2001. *Journal of Agricultural and Food Chemistry* 49: 5165–5170.

21. Wang, S. S., Zheng, Z. G., Weng, Y. Q., Yu, Y. J., Zhang, D. F., Fan, W. H., Dai, R. H. and Hu, Z. B. 2004. *Life Sciences* 74: 2467–2478.

22. Mohanta, B., Sudarshan, M., Boruah, M. and Chakraborty, A. 2007. *Biological Trace Element Research* 117: 139–151.

23. Singh, S. N., Vats, P., Suri, S., Shyam, R., Kumria, M. M. L., Ranganathan, S. and Sridharan, K. 2001. *Journal of Ethnopharmacology* 76: 269–277.

24. Chattopadhyay, R. R. 1999. *Journal of Ethnopharmacology* 67: 367–372.

25. Nammi, S., Boini, M. K., Lodagala, S. D. and Behara, R. B. 2003. *BMC Complementary and Alternative Medicine* 3: 4.

26. Elgorashi, E. E., Taylor, J. L. S., Maes, A., Staden, J. V., Kimpe, N. D. and Verschaeve, L. 2003. *Toxicology Letters* 143: 195–207.

27. Ahmed, A. U., Ferdous, A. H., Saha, S. K., Nahar, S., Awal, M. A. and Parvin, F. 2007. *Mymensingh Medical Journal* 16: 143–148.

28. Rau, O., Wurglics, M., Dingermann, T., Abdel-Tawab, M. and Schubert-Zsilavecz, M. 2006. *Pharmazie* 61: 952–956.

29. Nayak, B. S., Anderson, M. and Pinto Pereira, L. M. 2007. *Fitoterapia* 78: 540–544.

30. Nayak, B. S. and Pinto Pereira, L. M. 2006. *BMC Complementary and Alternative Medicine* 6: 41.

31. Duke, J. A., Bogenschutz-Godwin, M. J., duCellier, J. and Duke, P. A. K. 2002. Handbook of medicinal herbs. CRC Press: Boca Raton, FL.

32. van Wyk, B. E. and Wink, M. 2004. Medicinal plants of the world: an illustrated scientific guide to important medicinal plants and their uses. Timber Press: Portland.

33. Ghosh, D., Roy, I., Chanda, S. and Gupta-Bhattacharya, S. 2007. *Annals of Agricultural and Environmental Medicine* 14: 39–43.

34. Sweetman, S. C. (Ed). 2007. Martindale: the complete drug reference, 35th edition, volume one. Pharmaceutical Press: London.

Celosia argentea L.

1. Wiart, C. 2006. Medicinal Plants of the Asia-Pacific: Drugs for the future? World Scientific Publishing Co. Pte. Ltd.: Singapore.
2. Hsuan, K. 1990. The Concise Flora of Singapore, Gymnosperms and Dicotyledons. Singapore University Press: National University of Singapore.
3. Wee, Y. C. 1992. A Guide to Medicinal Plants. Singapore Science Centre Publication: Singapore.
4. Wee, Y. C. and Hsuan, K. 1990. An illustrated Dictionary of Chinese Medicinal Herbs. Times Edition and Eu Yan Seng Holdings Ltd: Singapore.
5. Wiart, C. 2000. Medicinal plants of Southeast Asia. Pelanduk Publications (M) Sdn Bhd: Malaysia.
6. eFloras.org 2008
   http://www.efloras.org/florataxon.aspx?flora_id=1&taxon_id=200006992
   Date Accessed: 30th March 2008
7. Suzuki, H., Morita, H., Iwasaki, S. and Kobayashi, J. 2003. *Tetrahedron* 59: 5307–5315.
8. Suzuki, H., Morita, H., Shiro, M. and Kobayashi, J. 2004. *Tetrahedron* 60: 2489–2495.
9. Morita, H., Suzuki, H. and Kobayashi, J. 2004. *Journal of Natural Products* 67: 1628–1630.
10. Morita, H., Shimbo, K., Shigemori, H. and Kobayashi, J. 2000. *Bioorganic and Medicinal Chemistry Letters* 10: 469–471.
11. Hase, K., Kadota, S., Basnet, P., Namba, T. and Takahashi, T. 1996. *Phytotherapy Research* 10: 387–392.
12. Duke, J. A. and Ayensu, E. S. 1985. Medicinal Plants of China, volume one and two. Reference Publications, Inc.: United States of America.
13. Jain, S. K. and DeFilipps, R. A. 1991. Medicinal Plants of India, volume one and two. Reference Publications Inc.: United States of America.
14. Keys, J. D. 1976. Chinese herbs: their botany, chemistry, and pharmacodynamics. Charles E. Tuttle Co. of Rutland: Vermont & Tokyo, Japan.
15. Gnanamani, A., Shanmuga Priya, K., Radhakrishnan, N. and Babu, M. 2003. *Journal of Ethnopharmacology* 86: 59–61.
16. Kobayashi, J., Suzuki, H., Shimbo, K., Takeya, K. and Morita, H. 2001. *Journal of Organic Chemistry* 66: 6626–6633.
17. Hayakawa, Y., Fujii, H., Hase, K., Ohnishi, Y., Sakukawa, R., Kadota, S., Namba, T. and Saiki, I. 1998. *Biological and Pharmaceutical Bulletin* 21: 1154–1159.
18. Vetrichelvan, T., Jegadeesan, M. and Devi, B. A. 2002. *Biological and Pharmaceutical Bulletin* 25: 526–528.
19. Hase, K., Kadota, S., Basnet, P., Takahashi, T. and Namba, T. 1996. Protective effect of celosian, an acidic polysaccharide, on chemically and immunologically induced liver injuries. *Biological and Pharmaceutical Bulletin* 19: 567–572.
20. Hase, K., Basnet, P., Kadota, S. and Namba, T. 1997. *Planta Medica* 63: 216–219.
21. Iwalewa, E. O., Adewunmi, C. O., Omisore, N. O., Adebanji, O. A., Azike, C. K., Adigun, A. O., Adesina, O. A. and Olowoyo, O. G. 2005. *Journal of Medical Food* 8: 539–544.
22. Priya, K. S., Arumugam, G., Rathinam, B., Wells, A. and Babu, M. 2004. *Wound Repair and Regeneration* 12: 618–625.
23. Perry, L. M. 1980. Medicinal plants of East and Southeast Asia: Attributed properties and uses. The MIT Press: Cambridge, Massachusetts and London.

Centella asiatica (L.) Urban

1. Hsuan, K. 1990. The Concise Flora of Singapore, Gymnosperms and Dicotyledons. Singapore University Press: National University of Singapore.
2. Wee, Y. C. 1992. A Guide to Medicinal Plants. Singapore Science Centre Publication: Singapore.
3. eFloras.org 2008
   http://www.efloras.org/florataxon.aspx?flora_id=5&taxon_id=200015478
   Date Accessed: 30th March 2008
4. Goh, S. H. 1988. Proceedings: Malaysian Traditional Medicine, Kuala Lumpur,

pp. 7–26. Institute of Advanced Studies, University of Malaya, Kuala Lumpur.

5. Goh, S. H., Chuah, C. H., Mok, J. S. L. and Soepadmo, E. 1995. Malaysian Medicinal Plants for the Treatment of Cardiovascular Diseases. Pelanduk Publications (M) Sdn Bhd: Malaysia.

6. Duke, J. A. and Ayensu, E. S. 1985. Medicinal Plants of China, volume one and two. Reference Publications, Inc.: United States of America.

7. Li, T. S. C. 2000. Medicinal Plants. Culture, Utilisation and Phytopharmacology. Technomic Publishing Co., Inc.: Lancaster.

8. Park, B. C., Bosire, K. O., Lee, E. S., Lee, Y. S. and Kim, J. A. 2005. *Cancer Letters* 218: 81–90.

9. Cheng, C. L., Guo, J. S., Luk, J. and Koo, M. W. L. 2004. *Life Sciences* 74: 2237–2249.

10. Inamdar, P. K., Yeole, R. D., Ghogare, A. B. and de Souza, N. J. 1996. *Journal of Chromatography A* 742: 127–130.

11. Yu, Q. L., Duan, H. Q., Takaishi, Y. and Gao, W. Y. 2006. *Molecules* 11: 661–665.

12. Yu, Q. L., Gao, W. Y., Zhang, Y. W., Teng, J. and Duan, H. Q. 2007. *Zhongguo Zhong Yao Za Zhi* 32: 1182–1184.

13. Govindan, G., Sambandan, T. G., Govindan, M., Sinskey, A., Vanessendelft, J., Adenan, I. and Rha, C. K. 2007. *Planta Medica* 73: 597–599.

14. Siddiqui, B. S., Aslam, H., Ali, S. T., Khan, S. and Begum, S. 2007. *Journal of Asian Natural Product Research* 9: 407–414.

15. Zheng, C. J. and Qin, L. P. 2007. *Zhong Xi Yi Jie He Xue Bao* 5: 348–351.

16. Wiart, C. 2000. Medicinal plants of Southeast Asia. Pelanduk Publications (M) Sdn Bhd: Malaysia.

17. Wee, Y. C. and Hsuan, K. 1990. An illustrated Dictionary of Chinese Medicinal Herbs. Times Edition and Eu Yan Seng Holdings Ltd: Singapore.

18. Nguyen, V. D. and Doan, T. N. 1989. Medicinal Plants in Vietnam. World Health Organisation, Regional Office for the Western Pacific, Manila, Institute of Materia Medica, Hanoi.

19. Jain, S. K. and DeFilipps, R. A. 1991. Medicinal Plants of India, volume one and

two. Reference Publications Inc.: United States of America.

20. Zaidan, M. R., Noor Rain, A., Badrul, A. R., Adlin, A., Norazah, A. and Zakiah, I. 2005. *Tropical Biomedicine* 22: 165–170.

21. Chen, Y., Han, T., Rui, Y., Yin, M., Qin, L. and Zheng, H. 2005. *Zhong Yao Cai* 28: 492–496.

22. Shobi, V. and Goel, H. C. 2001. *Physiology & Behavior* 73: 19–23.

23. Babu, T. D., Kuttan, G. and Padikkala, J. 1995. *Journal of Ethnopharmacology* 48: 53–57.

24. Bunpo, P., Kataoka, K., Arimochi, H., Nakayama, H., Kuwahara, T., Bando, Y., Izumi, K., Vinitketkumnuen, U. and Ohnishi, Y. 2004. *Food and Chemical Toxicology* 42: 1987–1997.

25. Govindan, G., Sambandan, T. G., Govindan, M., Sinskey, A., Vanessendelft, J., Adenan, I. and Rha, C. K. 2007. *Planta Medica* 73: 597–599.

26. Yoshida, M., Fuchigami, M., Nagao, T., Okabe, H., Matsunaga, K., Takata, J., Karube, Y., Tsuchihashi, R., Kinjo, J., Mihashi, K. and Fujioka, T. 2005. *Biological and Pharmaceutical Bulletin* 28: 173–175.

27. Jayashree, G., Muraleedhara, G. K., Sudarslal, S. and Jacob, V. B. 2003. *Fitoterapia* 74: 431–434.

28. Zainol, M. K., Abd-Hamid, A., Yusof, S. and Muse, R. 2003. *Food Chemistry* 81: 575–581.

29. Ramanathan, M., Sivakumar, S., Anandvijayakumar, P. R., Saravanababu, C. and Pandian, P. R. 2007. *Indian Journal of Experimental Biology* 45: 425–431.

30. Gupta, R. and Flora, S. J. 2006. *Journal of Applied Toxicology* 26: 213–222.

31. Subathra, M., Shila, S., Devi, M. A. and Panneerselvam, C. 2005. *Experimental Gerontology* 40: 707–715.

32. Bajpai, M., Pande, A., Tewari, S. K. and Prakash, D. 2005. *International Journal of Food Science and Nutrition* 56: 287–291.

33. Satake, T., Kamiya, K., An, Y., Oishi Nee Taka, T. and Yamamoto, J. 2007. *Biological and Pharmaceutical Bulletin* 30: 935–940.

34. Awad, R., Levac, D., Cybulska, P., Merali, Z., Trudeau, V. L. and Arnason, J. T. 2007.

Canadian Journal of Physiology and Pharmacology 85: 933–942.

35. Wijeweera, P., Arnason, J. T., Koszycki, D. and Merali, Z. 2006. *Phytomedicine* 13: 668–676.

36. Cheng, C. L. and Koo, M. W. L. 2000. *Life Sciences* 67: 2647–2653.

37. Punturee, K., Wild, C. P. and Vinitketkumneun, U. 2004. *Journal of Ethnopharmacology* 95: 183–189.

38. Punturee, K., Wild, C. P., Kasinrerk, W. and Vinitketkumnuen, U. 2005. *Asian Pacific Journal of Cancer Prevention* 6: 396–400.

39. Siddique, Y. H., Ara, G., Beg, T., Faisal, M., Ahmad, M. and Afzal, M. 2007. *Toxicology in vitro* 22: 10–17.

40. Soumyanath, A., Zhong, Y. P., Gold, S. A., Yu, X., Koop, D. R., Bourdette, D. and Gold, B. G. 2005. *Journal of Pharmacy and Pharmacology* 57: 1221–1229.

41. Sharma, J. and Sharma, R. 2002. *Phytotherapy Research* 16: 785–786.

42. Sharma, R. and Sharma, J. 2005. *Phytotherapy Research* 19: 605–611.

43. Shukla, A., Rasik, A. M., Jain, G. K., Shankar, R., Kulshrestha, D. K. and Dhawan, B. N. 1999. *Journal of Ethnopharmacology* 65: 1–11.

44. Shetty, B. S., Udupa, S. L., Udupa, A. L. and Somayaji, S. N. 2006. *International Journal of Lower Extremity Wounds* 5: 137–143.

45. van Wyk, B. E. and Wink, M. 2004. Medicinal plants of the world: an illustrated scientific guide to important medicinal plants and their uses. Timber Press: Portland.

46. Arpaia, M. R., Ferrone, R., Amitrano, M., Nappo, C., Leonardo, G. and del Guercio, R. 1990. *Journal of Clinical Pharmacology Research* 10: 229–233.

47. Belcaro, G. V., Grimaldi, R. and Guidi, G. 1990. *Angiology* 41: 12–18.

48. Pointel, J. P., Boccalon, H., Cloarec, M. Ledevehat, C. and Joubert, M. 1987. *Angiology* 38(1 Pt 1): 46–50.

49. Fleming, T. (Ed.). 2000. PDR for Herbal Medicines, 2nd edition. Medical Economics Company: New Jersey.

50. Skidmore-Roth L. 2001. Mosby's handbook of herbs and natural supplements. Mosby, Inc.: St. Louis.

51. Jorge, O. A. and Jorge, A. D. 2005. *Revista española de enfermedades digestivas* 97: 115–124.

52. Lininger, S. W., Gaby, A. R., Austin, S., Batz, F., Yarnell, E., Brown, D. J. and Constantine, G. 1999. A-Z guide to drug-herb-vitamin interactions: how to improve your health and avoid problems when using common medications and natural supplements together. Prima Health: Rocklin, CA.

Cerbera odollam Gaertn.

1. Hsuan, K. 1992. Orders and Families of Malayan Seed Plants. Singapore University Press: Singapore.

2. Hsuan, K. 1990. The Concise Flora of Singapore, Gymnosperms and Dicotyledons. Singapore University Press: National University of Singapore.

3. Wee, Y. C. 1992. A Guide to Medicinal Plants. Singapore Science Centre Publication: Singapore.

4. Wiart, C. 2006. Medicinal Plants of the Asia-Pacific: Drugs for the future? World Scientific Publishing Co. Pte. Ltd.: Singapore.

5. USDA, ARS GRIN Taxonomy for Plants 2008 http://www.ars-grin.gov/cgi-bin/npgs/html/taxon.pl?9932
Date Accessed: 30th June 2008

6. Abe, F., Yamauchi, T. and Wan, A. S. C. 1988. *Phytochemistry* 27: 3627–3631.

7. Chantrapromma, S., Usman, A., Fun, H. K., Laphookhieo, S., Karalai, C., Rat-a-pa, Y. and Chantrapromma, K. 2003. *Acta Crystallographica C* 59: 68–70.

8. Laphookhieo, S., Cheenpracha, S., Karalai, C., Chantrapromma, S., Rat-a-pa, Y., Ponglimanont, C. and Chantrapromma, K. 2004. *Phytochemistry* 65: 507–510.

9. Rao, E. V. and Rao, M. A. 1976. *Phytochemistry* 15: 848.

10. Shen, L. R., Jin, S. M., Yin, B. W., Du, X. F., Wang, Y. L. and Huo, C. F. 2007. *Chemistry & Biodiversity* 4: 1438–1449.

11. Wiart, C. 2000. Medicinal plants of Southeast Asia. Pelanduk Publications (M) Sdn Bhd: Malaysia.

12. Yamauchi, T., Abe, F. and Wan, A. S. C. 1987. *Chemical & Pharmaceutical Bulletin* 35: 4813–4818.

13. Chopra, R. N., Nayar, S. L. and Chopra, I. C. 1956. Glossary of Indian Medicinal Plants. Council of Scientific and Industrial Research: New Delhi.

14. Hiên, T. T. M., Navarro-Delmasure, C. and Vy, T. 1991. *Journal of Ethnopharmacology* 34: 201–206.

15. Gaillard, Y., Krishnamoorthy, A. and Bevalot, F. 2004. *Journal of Ethnopharmacology* 95: 123–126.

16. Iyer, G. V. and Narendranath, M. 1975. *Indian Journal of Medical Research* 63: 321–324.

17. Narendranathan, M., Krishna Das, K. V. and Vijayaraghavan, G. 1975. *Indian Heart Journal* 27: 283–286.

Cissus quadrangularis L.

1. Hsuan, K. 1990. The Concise Flora of Singapore, Gymnosperms and Dicotyledons. Singapore University Press: National University of Singapore.

2. eFloras.org 2008
http://www.efloras.org/florataxon.aspx?flora_id=5&taxon_id=242444305
Date Accessed: 30th March 2008

3. Adesanya, S. A., Nia, R., Marie-Therese, M., Boukamcha, N., Montagnac, A. and Paies, M. 1999. *Journal of Natural Products* 62: 1694–1695.

4. Bhutani, K. K., Kapoor, R. and Atal, C. K. 1984. *Phytochemistry* 23: 407–410.

5. Gupta, M. M. and Verma, R. K. 1991. *Phytochemistry* 30: 875–878.

6. Gupta, M. M. and Verma, R. K. 1990. *Phytochemistry* 29: 336–337.

7. Singh, G., Rawat, P. and Maurya, R. 2007. *Natural Product Research* 21: 522–528.

8. Bah, S., Diallo, D., Dembélé, S. and Paulsen, B. S. 2006. *Journal of Ethnopharmacology* 105: 387–399.

9. Chidambara Murthy, K. N., Vanitha, A., Mahadeva Swamy, M. and Ravishankar, G. A. 2003. *Journal of Medicinal Food* 6: 99–105.

10. Panthong, A., Supraditaporn, W., Kanjanapothi, D., Taesotikul, T. and Reutrakul, V. 2007. *Journal of Ethnopharmacology* 110: 264–270.

11. Bah, S., Jäger, A. K., Adsersen, A., Diallo, D. and Paulsen, B. S. 2007. *Journal of Ethnopharmacology* 110: 451–457.

12. Jainu, M. and Devi, C. S. 2004. *Journal of Medicinal Food* 7: 372–376.

13. Jainu, M., Mohan, K. V. and Devi, C. S. 2006. *Journal of Ethnopharmacology* 104: 302–305.

14. Jainu, M. and Shyamala Devi, C. S. 2005. *Journal of Herbal Pharmacotherapy* 5: 33–42.

15. Jainu, M., Vijai Mohan, K. and Shyamala Devi, C. S. 2006. *Indian Journal of Medical Research* 123: 799–806.

16. Jainu, M. and Devi, C. S. 2006. *Chemico-Biological Interactions* 161: 262–270.

17. Shirwaikar, A., Khan, S. and Malini, S. 2003. *Journal of Ethnopharmacology* 89: 245–250.

Cocos nucifera L.

1. Hsuan, K. 1992. Orders and Families of Malayan Seed Plants. Singapore University Press: Singapore.

2. Hsuan, K., Chin, S. C. and Tan, H. T. W. 1998. The Concise Flora of Singapore volume II: Monocotyledons. Singapore University Press: Singapore.

3. eFloras.org 2008
http://www.efloras.org/florataxon.aspx?flora_id=1&taxon_id=200027077
Date Accessed: 30th March 2008

4. Ge, L. Y., Yong, J. W. H., Tan, S. N., Yang, X. H. and Ong, E. S. 2004. *Journal of Chromatography A* 48: 119–126.

5. Kooiman, P. 1971. *Carbohydrate Research* 20: 329–337.

6. Duke, J. A. and Ayensu, E. S. 1985. Medicinal Plants of China, volume one and two. Reference Publications, Inc.: United States of America.

7. Jain, S. K. and DeFilipps, R. A. 1991. Medicinal Plants of India, volume one and two. Reference Publications Inc.: United States of America.

8. Woodley, E. (Ed.). 1991. Medicinal plants of Papua New Guinea, part I. Veerlag Josef Margraf Scientific Books: Weikersheim.

9. Esquenazi, D., Wigg, M. D., Miranda, M. M. F. S., Rodrigues, H. M., Tostes, J. B. F., Rozental, S., da Silva, A. J. R. and Alviano, C. S. 2002. *Research in Microbiology* 153: 647–652.

10. Pummer, S., Heil, P., Maleck, W. and Petroianu, G. 2001. *American Journal of Emergency Medicine* 19: 287–289.

11. Alviano, D. S., Rodrigues, K. F., Leitão, S. G., Rodrigues, M. L., Matheus, M. E., Fernandes, P. D., Antoniolli, A. R. and Alviano, C. S. 2004. *Journal of Ethnopharmacology* 92: 269–273.

12. Alanís, A. D., Calzada, F., Cervantes, J. A., Torres, J. and Ceballos, G. M. 2005. *Journal of Ethnopharmacology* 100: 153–157.

13. Wang, H. X. and Ng, T. B. 2005. *Peptides* 26: 2392–2396.

14. Namasivayam, N., Vaiyapuri, M. and Venugopal, P. M. 2004. *Clinica Chimica Acta* 342: 203–210.

15. Koschek, P. R., Alviano, D. S., Alviano, C. S. and Gattass, C. R. 2007. *Brazilian Journal of Medical and Biological Research* 40: 1339–1343.

16. Mantena, S. K., Jagadish, Badduri, S. R., Siripurapu, K. B. and Unnikrishnan, M. K. 2003. *Nahrung* 47: 126–131.

17. Mendonça-Filho, R. R., Rodrigues, I. A., Alviano, D. S., Santos, A. L. S., Soares, R. M. A., Alviano, C. S., Lopes, A. H. C. S. and Rosa, M. D. S. S. 2004. *Research in Microbiology* 155: 136–143.

18. Sindurani, J. A. and Rajamohan, T. 2000. *Indian Journal of Physiology and Pharmacology* 44: 97–100.

19. Salil, G. and Rajamohan, T. 2001. *Journal of Experimental Biology* 39: 1028–1034.

20. Sandhya, V. G. and Rajamohan, T. 2006. *Journal of Medicinal Food* 9: 400–407.

21. Alleyne, T., Roache, S., Thomas, C. and Shirley, A. 2005. *West Indian Medical Journal* 54: 3–8.

22. Kirszberg, C., Esquenazi, D., Alviano, C. S. and Rumjanek, V. M. 2003. *Phytotherapy Research* 17: 1054–1058.

22. Calzada, F., Yépez-Mulia, L. and Tapia-Contreras, A. 2007. *Journal of Ethnopharmacology* 113: 248–251.

23. Wittczak, T., Pas-Wyroslak, A. and Palczynski, C. 2005. *Allergy* 60: 970–971.

Coix lacryma-jobi L.

1. Hsuan, K. 1990. The Concise Flora of Singapore, Gymnosperms and Dicotyledons. Singapore University Press: National University of Singapore.

2. Wiart, C. 2006. Medicinal Plants of the Asia-Pacific: Drugs for the future? World Scientific Publishing Co. Pte. Ltd.: Singapore.

3. Wee, Y. C. and Hsuan, K. 1990. An illustrated Dictionary of Chinese Medicinal Herbs. Times Edition and Eu Yan Seng Holdings Ltd: Singapore.

4. USDA, ARS GRIN Taxonomy for Plants 2008 http://www.ars-grin.gov/cgi-bin/npgs/html/taxon.pl?11129 Date Accessed: 30th June 2008

5. Wee, Y. C. 1992. A Guide to Medicinal Plants. Singapore Science Centre Publication: Singapore.

6. Targon, M. L. N., Ottoboni, L. M. M., Leite, A., Ludevid, D., Puigdomenech, P. and Arruda, P. 1992. *Plant Science* 83: 169–180.

7. Yamada, H., Yanahira, S., Kiyohara, H., Cyong, J. C. and Otsuka, Y. 1987. *Phytochemistry* 26: 3269–3275.

8. Kuo, C. C., Chiang, W., Liu, G. P., Chien, Y. L., Chang, J. Y., Lee, C. K., Lo, J. M., Huang, S. L., Shih, M. C. and Kuo, Y. H. 2002. *Journal of Agriculture and Food Chemistry* 50: 5850–5855.

9. Takahashi, M., Konno, C. and Hikino, H. 1986. *Planta Medica* 52: 64–65.

10. Duke, J. A. and Ayensu, E. S. 1985. Medicinal Plants of China, volume one and two. Reference Publications, Inc.: United States of America.

11. Keys, J. D. 1976. Chinese herbs: their botany, chemistry, and pharmacodynamics. Charles E. Tuttle Co. of Rutland: Vermont & Tokyo, Japan.

12. Nguyen, V. D. and Doan, T. N. 1989. Medicinal Plants in Vietnam. World Health Organisation, Regional Office for the Western Pacific, Manila, Institute of Materia Medica, Hanoi.

13. Jain, S. K. and DeFilipps, R. A. 1991. Medicinal Plants of India, volume one and two. Reference Publications Inc.: United States of America.

14. Hsu, H. Y., Lin, B. F., Lin, J. Y., Kuo, C. C. and Chiang, W. 2003. *Journal of Agriculture and Food Chemistry* 51: 3763–3769.
15. Hung. W. C. and Chang, H. C. 2003. *Journal of Agriculture and Food Chemistry* 51: 7333–7337.
16. Hsu, H. Y., Lin, B. F., Lin, J. Y., Kuo, C. C. and Chiang, W. 2003. *Journal of Agriculture and Food Chemistry* 51: 3656–3660.
17. Kubota, T., Haramaki, Y., Oki, Y., Saito, F., Kuroshima, A. and Numata, M. 1988. *Nippon Nogei Kagaku Kaishi* 62: 23–28.
18. Kuo, C. C., Shih, M. C., Kuo, Y. H. and Chiang, W. 2001. *Journal of Agriculture and Food Chemistry* 49: 1564–1570.
19. Huang, T., Wu, W., Li, Y., Zhang, J and Huang, L. 2002. *Zhonghua Gan Zang Bing Za Zhi* 10: 452–454.
20. Shih, C. K., Chiang, W. C. and Kuo, M. L. 2004. *Food and Chemical Toxicology* 42: 1339–1347.
21. Bao, Y., Yuan, Y., Xia, L., Jiang, H., Wu, W. and Zhang, X. 2005. *Journal of Gastroenterology and Hepatology* 20: 1046–1053.
22. Yeh, P. H., Chiang, W. and Chiang, M. T. 2006. *International Journal for Vitamin and Nutrition Research* 76: 299–305.
23. Kim, S. O., Yun, S. J., Jung, B., Lee, E. H., Hahm, D. H., Shim, I. and Lee, H. J. 2004. *Life Sciences* 75: 1391–1404.
24. Huang, B. W., Chiang, M. T., Yao, H. T. and Chiang, W. 2005. *Phytomedicine* 12: 433–439.
25. Ko, B. J. and Lee, S. U. 1974. *Chonyonmul Hwahak Yonguso Yongu Pogo (Yongnam Taehakkyo)* 2: 19–21.
26. Tzeng, H. P., Chiang, W., Ueng, T. H. and Liu, S. H. 2005. *Journal of Toxicology and Environmental Health A* 68: 1557–1565.

Crinum asiaticum L.

1. Hsuan, K. 1992. Orders and Families of Malayan Seed Plants. Singapore University Press: Singapore.
2. Hsuan, K., Chin, S. C. and Tan, H. T. W. 1998. The Concise Flora of Singapore volume II: Monocotyledons. Singapore University Press: Singapore.
3. Wee, Y. C. 1992. A Guide to Medicinal Plants. Singapore Science Centre Publication: Singapore.
4. Wee, Y. C. and Corlett, R. 1986. The City and the Forest: Plant Life in Urban Singapore. Singapore University Press: National University of Singapore.
5. eFloras.org 2008 http://www.efloras.org/florataxon.aspx?flora_id=1&taxon_id=200028036 Date Accessed: 30th March 2008
6. Ochi, M., Otsuki, H. and Nagao, K. 1976. *Bulletin of the Chemical Society of Japan* 49: 3363–3364.
7. Duke, J. A. and Ayensu, E. S. 1985. Medicinal Plants of China, volume one and two. Reference Publications, Inc.: United States of America.
8. Ghosal, S., Saini, K. S., Razdan, S. and Kumar, Y., 1985a. *Journal of Chemical Research, Synopses* 3: 100–101.
9. Ghosal, S., Shanthy, A., Kumar, A. and Kumar, Y. 1985b. *Phytochemistry* 24: 2703–2706.
10. Ghosal, S., Shanthy, A., Mukhopadhyay, M., Sarkar, M. K. and Das, P. K. 1986. *Pharmaceutical Research* 3: 240–243.
11. Ghosal, S., Shanthy, A. and Singh, S. K. 1988. *Phytochemistry* 27: 1849–1852.
12. Kobayshi, S., Ishikawa, H., Kihara, M., Shing, T. and Uyeo, S. 1976. *Chemical and Pharmaceutical Bulletin* 24: 2553–2555.
13. Yui, S., Mikami, M., Kitahara, M. and Yamazaki, M. 1998. *Immunopharmacology* 40: 151–162.
14. Tang, R. J., Bi, N. J. and Guang, E. 1994. *Chinese Chemical Letters* 5: 855–858.
15. Min, B. S., Jiang, J., Nakamura, N., Kim, Y. H. and Hattori, M. 2001. *Chemical & Pharmaceutical Bulletin* 49: 1217–1219.
16. Phan, T. S., Tran, B. D., Phan, M. G., Nguyen, T. M., Nguyen, Q. V. and Taylor, W. C. 2002. *Tap Chi Hoa Hoc* 40: 53–58.
17. Sun, Q., Zhang, W. D., Shen, Y. H., Zhang, C. and Li H. L. 2008. *Chinese Chemical Letters* 19: 447–449.
18. Samud, A. M., Asmawi, M. Z., Sharma, J. N. and Yusof, A. P. M. 1999. *Immunopharmacology* 43: 311–316.
19. Wiart, C. 2000. Medicinal plants of Southeast Asia. Pelanduk Publications (M) Sdn Bhd: Malaysia.

20. Woodley, E. (Ed.). 1991. Medicinal plants of Papua New Guinea, part I. Veerlag Josef Margraf Scientific Books: Weikersheim.
21. Nguyen, V. D. and Doan, T. N. 1989. Medicinal Plants in Vietnam. World Health Organisation, Regional Office for the Western Pacific, Manila, Institute of Materia Medica, Hanoi.
22. Jain, S. K. and DeFilipps, R. A. 1991. Medicinal Plants of India, volume one and two. Reference Publications Inc.: United States of America.
23. Min, B. S., Kim, Y. H., Tomiyama, M., Nakamura, N., Miyashiro, H., Otake, T. and Hattori, M. 2001. *Phytotherapy Research* 15: 481–486.
24. Kim, Y. H., Park, E. J., Park, M. H., Badarch, U., Woldemichael, G. M. and Beutler, J. A. 2006. *Biological & Pharmaceutical Bulletin* 29: 2140–2142.

Cymbopogon citratus (DC.) Stapf.

1. Hsuan, K., Chin, S. C. and Tan, H. T. W. 1998. The Concise Flora of Singapore volume II: Monocotyledons. Singapore University Press: Singapore.
2. Wee, Y. C. and Hsuan, K. 1990. An illustrated Dictionary of Chinese Medicinal Herbs. Times Edition and Eu Yan Seng Holdings Ltd: Singapore.
3. Palaniswamy, U. R. 2003. A guide to Medicinal Plants of Asian origin and culture. CPL Press: United States of America.
4. Li, T. S. C. 2000. Medicinal Plants. Culture, Utilisation and Phytopharmacology. Technomic Publishing Co., Inc.: Lancaster.
5. Nguyen, V. D. and Doan, T. N. 1989. Medicinal Plants in Vietnam. World Health Organisation, Regional Office for the Western Pacific, Manila, Institute of Materia Medica, Hanoi.
6. Hanson, S. W., Crawford, M., Koker, M. E. S. and Menezes, F. A. 1976. *Phytochemistry* 15: 1074–1075.
7. Chisowa, E. H., Hall, D. R. and Farman, D. I. 1998. *Flavour and Fragrance Journal* 13: 29–30.
8. Cheel, J., Theoduloz, C., Rodríguez, J. and Schmeda-Hirschmann, G. 2005. *Journal of Agriculture and Food Chemistry* 53: 2511–2517.
9. Wee, Y. C. 1992. A Guide to Medicinal Plants. Singapore Science Centre Publication: Singapore.
10. Jain, S. K. and DeFilipps, R. A. 1991. Medicinal Plants of India, volume one and two. Reference Publications Inc.: United States of America.
11. Viana, G. S. B., Vale, T. G., Pinho, R. S. N. and Matos, F. J. A. 2000. *Journal of Ethnopharmacology* 70: 323–327.
12. Lorenzetti, B. B., Souza, G. E., Sarti, S. J., Santos Filho, D. and Ferreira, S. H. 1991. *Journal of Ethnopharmacology* 34: 43–48.
13. Muhammad, M. M., Abdul, M. A., Mohd, A. A., Rozita, M. N., Nashriyah, B. M., Abdul, R. R. and Kazuyoshi, K. 1997. *Pesticide Science* 51: 165–170.
14. Ohno, T., Kita, M., Yamaoka, Y., Imamura, S., Yamamoto, T., Mitsufuji, S., Kodama, T., Kashima, K. and Imanishi, J. 2003. *Helicobacter* 8: 207–215.
15. Ogunlana, E. O., Hoglund, S., Onawunmi, G. and Skold, O. 1987. *Microbios* 50: 43–59.
16. Adegoke, G. O. and Odesola, B. A. 1996. *International Biodeterioration & Biodegradation* 37: 81–84.
17. Betoni, J. E., Mantovani, R. P., Barbosa, L. N., Di Stasi, L. C. and Fernandes Junior, A. 2006. *Memórias do Instituto Oswaldo Cruz* 101: 387–390.
18. Wannissorn, B., Jarikasem, S., Siriwangchai, T. and Thubthimthed, S. 2005. *Fitoterapia* 76: 233–236.
19. Abe, S., Sato, Y., Inoue, S., Ishibashi, H., Maruyama, N., Takizawa, T., Oshima, H. and Yamaguchi, H. 2003. *Nippon Ishinkin Gakkai Zasshi* 44: 285–291.
20. Paranagama, P. A., Abeysekera, K. H., Abeywickrama, K. and Nugaliyadde, L. 2003. *Letters in Applied Microbiology* 37: 86–90.
21. Helal, G. A., Sarhan, M. M., Abu Shahla, A. N. and Abou El-Khair, E. K. 2007. *Journal of Basic Microbiology* 47: 5–15.
22. Helal, G. A., Sarhan, M. M., Abu Shahla, A. N. and Abou El-Khair, E. K. 2006. *Journal of Basic Microbiology* 46: 456–469.
23. Helal, G. A., Sarhan, M. M., Abu Shahla, A. N. and Abou El-Khair, E. K. 2006. *Journal of Basic Microbiology* 46: 375–386.

24. Dutta, B. K., Karmakar, S., Naglot, A., Aich, J. C. and Begam, M. 2007. *Mycoses* 50: 121–124.

25. Pawar, V. C. and Thaker, V. S. 2006. In vitro efficacy of 75 essential oils against *Aspergillus niger*. *Mycoses* 49: 316–323.

26. Kinouchi, T., Suaeyun, R., Chewonarin, T., Intiyot, Y., Arimochi, H., Kataoka, K., Akimoto, S., Vinitketkumnuen, U. and Ohnishi, Y. 1997. *Mutation Research/ Fundamental and Molecular Mechanisms of Mutagenesis* 379: S181.

27. Dudai, N., Weinstein, Y., Krup, M., Rabinski, T. and Ofir, R. 2005. *Planta Medica* 71: 484–488.

28. Tchoumbougnang, F., Zollo, P. H., Dagne, E. and Mekonnen, Y. 2005. *Planta Medica* 71: 20–23.

29. Melo, C. D. F., Soares, S. F., da Costa, R. F., da Silva, C. R. and de Oliveira, *Mutation Research/Genetic Toxicology and Environmental Mutagenesis* 496: 33–38.

30. Tapia, A., Cheel, J., Theoduloz, C., Rodríguez, J., Schmeda-Hirschmann, G., Gerth, A., Wilken, D., Jordan, M., Jiménez-González, E., Gomez-Kosky, R. and Mendoza, E. Q. 2007. *Zeitschrift für Naturforschung. C, Journal of Biosciences* 62: 447–457.

31. Tognolini, M., Barocelli, E., Ballabeni, V., Bruni, R., Bianchi, A., Chiavarini, M. and Impicciatore, M. 2006. *Life Sciences* 78: 1419–1432.

32. Puatanachokchai, R., Kishida, H., Denda, A., Murata, N., Konishi, Y., Vinitketkumnuen, U. and Nakae, D. 2002. *Cancer Letters* 183: 9–15.

33. Adeneye, A. A. and Agbaje, E. O. 2007. *Journal of Ethnopharmacology* 112: 440–444.

34. Blanco, M. M., Costa, C. A., Freire, A. O., Santos, J. G. Jr. and Costa, M. 2007. Article in Press, Available online 11th June 2007.

35. Runnie, I., Salleh, M. N., Mohamed, S., Head, R. J. and Abeywardena, M. Y. 2004. *Journal of Ethnopharmacology* 92: 311–316.

36. Vinitketkumnuen, U., Puatanachokchai, R., Kongtawelert, P., Lertprasertsuke, N. and Matsushima, T. 1994. *Mutation Research/ Genetic Toxicology* 341: 71–75.

37. Chungsamarnyart, N. and Jiwajinda, S. 1992. *Kasetsart Journal: Natural Sciences* 26: 46–51.

38. Fuentes, J. L., Alonso, A., Cuétara, E., Vernhe, M., Alvarez, N., Sánchez-Lamar, A. and Llagostera, M. 2006. *International Journal of Radiation Biology* 82: 323–329.

39. Fleming, T. (Ed.). 2000. PDR for Herbal Medicines, 2nd edition. Medical Economics Company: New Jersey.

40. Akinboro, A. and Bakare, A. A. 2007. *Journal of Ethnopharmacology* 112: 470–475.

41. Skidmore-Roth, L. 2001. Mosby's Handbook of Herbs and Natural Supplements. Mosby, Inc.: St. Louis.

Dolichos lablab L.

1. Nguyen, V. D. 1993. Medicinal Plants of Vietnam, Cambodia and Laos. Nguyen Van Duong: Vietnam.

2. Hsuan, K. 1990. The Concise Flora of Singapore, Gymnosperms and Dicotyledons. Singapore University Press: National University of Singapore.

3. Wee, Y. C. 1992. A Guide to Medicinal Plants. Singapore Science Centre Publication: Singapore.

4. Wee, Y. C. and Hsuan, K. 1990. An illustrated Dictionary of Chinese Medicinal Herbs. Times Edition and Eu Yan Seng Holdings Ltd: Singapore.

5. USDA, ARS GRIN Taxonomy for Plants 2008 http://www.ars-grin.gov/cgi-bin/npgs/html/taxon.pl?104887 Date Accessed: 30th June 2008

6. Ye, X. T., Wang, H. X. and Ng, T. B. 2000. *Biochemical and Biophysical Research Communications* 269: 155–159.

7. Yoshiki, Y., Kim, J. H., Okubo, K., Nagoya, I., Sakabe, T. and Tamura, N. 1995. *Phytochemistry* 38: 229–231.

8. Yoshikawa, M., Murakami, T., Komatsu, H. and Matsuda, H. 1998. *Chemical and Pharmaceutical Bulletin (Tokyo)* 46: 812–816.

9. Yokota, T., Kobayashi, S., Yamane, H. and Takahashi, N. 1978. *Agricultural and Biological Chemistry* 42: 1811–1812.

10. Yokota, T., Ueda, J. and Takahashi, N. 1981. *Phytochemistry* 20: 683–686.

11. Salimath, P. V. and Tharanathan, R. N. 1982. *Carbohydrate Research* 104: 341–347.

12. Salimath, P. V. and Tharanathan, R. N. 1982. *Carbohydrate Research* 106: 251–257.
13. Hamana, K., Niitsu, M., Samejima, K. and Matsuzaki, S. 1992. *Phytochemistry* 31: 893–894.
14. Ghosh, R. and Das, A. 1984. *Carbohydrate Research* 126: 287–296.
15. Duke, J. A. and Ayensu, E. S. 1985. Medicinal Plants of China, volume one and two. Reference Publications, Inc.: United States of America.
16. Jain, S. K. and DeFilipps, R. A. 1991. Medicinal Plants of India, volume one and two. Reference Publications Inc.: United States of America.
17. Kim, Y. H., Woloshuk, C. P., Cho, E. H., Bae, J. M., Song, Y. S. and Huh, G. H. 2007. *Plant Cell Reports* 26: 395–405.
18. Favero, J., Miquel, F., Dornand, J. and Mani, J. C. 1988. *Cellular Immunology* 112: 302–314.
19. Silva-Lima, M., Pusztai, A., Nunes, D. C. and Farias, M. E. 1988. *Brazilian Journal of Medical and Biological Research* 21: 219–222.
20. Guran, A., Ticha, M., Filka, K. and Kocourek, J. 1983. *Biochemical Journal* 209: 653–657.
21. Nguyen, V. D. and Doan, T. N. 1989. Medicinal Plants in Vietnam. World Health Organisation, Regional Office for the Western Pacific, Manila, Institute of Materia Medica, Hanoi.

Elephantopus scaber L.

1. Wee, Y. C. 1992. A Guide to Medicinal Plants. Singapore Science Centre Publication: Singapore.
2. Hsuan, K. 1990. The Concise Flora of Singapore, Gymnosperms and Dicotyledons. Singapore University Press: National University of Singapore.
3. Wee, Y. C. and Corlett, R. 1986. The City and the Forest: Plant Life in Urban Singapore. Singapore University Press: National University of Singapore.
4. USDA, ARS GRIN Taxonomy for Plants 2008 http://www.ars-grin.gov/cgi-bin/npgs/html/

taxon.pl?401632
Date Accessed: 30th June 2008
5. Goh, S. H. 1988. Proceedings: Malaysian Traditional Medicine, Kuala Lumpur, pp. 7–26. Institute of Advanced Studies, University of Malaya, Kuala Lumpur.
6. Goh, S. H., Chuah, C. H., Mok, J. S. L. and Soepadmo, E. 1995. Malaysian Medicinal Plants for the Treatment of Cardiovascular Diseases. Pelanduk Publications (M) Sdn Bhd: Malaysia.
7. But, P. P. H., Hon, P. M., Cao, H., Chan, T. W. D., Wu, B. M., Mak, T. C. W. and Che, C. T. 1997. *Phytochemistry* 44: 113–116.
8. DeSilva, L. B., Herath, W. H. M. W., Jennings, R. C., Mahendran, M. and Wannigama, G. E. 1982. *Phytochemistry* 21: 1173–1175.
9. Sim, K. Y. and Lee, H. T. 1969. *Phytochemistry* 8: 933–934.
10. Wiart, C. 2000. Medicinal Plants of Southeast Asia. Pelanduk Publications (M) Sdn Bhd: Malaysia.
11. Duke, J. A. and Ayensu, E. S. 1985. Medicinal Plants of China, volume one and two. Reference Publications, Inc.: United States of America.
12. Jain, S. K. and DeFilipps, R. A. 1991. Medicinal Plants of India, volume one and two. Reference Publications Inc.: United States of America.
13. Chen, C. P., Lin, C. C. and Namba, T. 1989. *Journal of Ethnopharmacology* 27: 285–295.
14. Ichikawa, H., Nair, M. S., Takada, Y., Sheeja, D. B., Kumar, M. A., Oommen, O. V. and Aggarwal, B. B. 2006. *Clinical Cancer Research* 12: 5910–5918.
15. Xu, G., Liang, Q., Gong, Z., Yu, W., He, S. and Xi, L. 2006. *Experimental Oncology* 28: 106–109.
16. Tsai, C. C. and Lin, C. C. 1998. *Journal of Ethnopharmacology* 64: 85–89.
17. Li, Y., Ooi, L. S., Wang, H., But, P. P. and Ooi, V. E. 2004. *Phytotherapy Research* 18: 718–722.
18. Lin, C. C., Tsai, C. C. and Yen, M. H. 1995. *Journal of Ethnopharmacology* 45: 113–123.
19. Rajesh, M. G. and Latha, M. S. 2001. Hepatoprotection by *Elephantopus scaber* Linn. in CCl4. *Indian Journal of Physiology and Pharmacology* 45: 481–486.

Euphorbia hirta L.

1. Hsuan, K. 1990. The Concise Flora of Singapore, Gymnosperms and Dicotyledons. Singapore University Press: National University of Singapore.
2. USDA, ARS GRIN Taxonomy for Plants 2008 http://www.ars-grin.gov/cgi-bin/npgs/html/taxon.pl?400049 Date Accessed: 30th June 2008
3. Duke, J. A. and Ayensu, E. S. 1985. Medicinal Plants of China, volume one and two. Reference Publications, Inc.: United States of America.
4. Chen, L. 1991. *Zhongguo Zhong Yao Za Zhi* 16: 38–39.
5. Gupta, D. R. and Garg, S. K. 1966. *Bulletin of the Chemical Society of Japan* 39: 2532–2534.
6. Galvez, J., Zarzuelo, A., Crespo, M. E., Lorente, M. D., Ocete, M. A. and Jiménez, J. 1993. *Planta Medica* 59: 333–336.
7. Jain, S. K. and DeFilipps, R. A. 1991. Medicinal Plants of India, volume one and two. Reference Publications Inc.: United States of America.
8. Woodley, E. (Ed.). 1991. Medicinal plants of Papua New Guinea, part I. Veerlag Josef Margraf Scientific Books: Weikersheim.
9. Lanhers, M. C., Fleurentin, J., Dorfman, P., Mortier, F. and Pelt, J. M. 1991. *Planta Medica* 57: 225–231.
10. Vijaya, K., Ananthan, S. and Nalini, R. 1995. *Journal of Ethnopharmacology* 49: 115–118.
11. Oyewale, A. O., Mika, A. and Peters, F. A. 2002. *Global Journal of Pure and Applied Sciences* 8: 49–55.
12. Wang, Y. C. and Huang, T. L. 2005. *FEMS Immunology and Medical Microbiology* 43: 295–300.
13. Ndip, R. N., Malange Tarkang, A. E., Mbullah, S. M., Luma, H. N., Malongue, A., Ndip, L. M., Nyongbela, K., Wirmum, C. and Efange, S. M. 2007. *Journal of Ethnopharmacology* 114: 452–457.
14. Sudhakar, M., Rao, Ch. V., Rao, P. M., Raju, D. B. and Venkateswarlu, Y. 2006. *Fitoterapia* 77: 378–380.
15. Hore, S. K., Ahuja, V., Mehta, G., Kumar, P., Pandey, S. K. and Ahmad, A. H. 2006. *Fitoterapia.* 77: 35–38.
16. Martinez-Vazquez, M., Apan, T. O. R., Lazcano, M. E. and Bye, R. 1999. *Revista de la Sociedad Quimica de Mexico* 43: 103–105.
17. Singh, G. D., Kaiser, P., Youssouf, M. S., Singh, S., Khajuria, A., Koul, A., Bani, S., Kapahi, B. K., Satti, N. K., Suri, K. A. and Johri, R. K. 2006. *Phytotherapy Research* 20: 316–321.
18. Pretorius, E., Ekpo, O. E. and Smit, E. 2007. *Experimental and Toxicologic Pathology* 59: 105–114.
19. Tona, L., Cimanga, R. K., Mesia, K., Musuamba, C. T., De Bruyne, T., Apers, S., Hernans, N., Van Miert, S., Pieters, L., Totté, J. and Vlietinck, A. J. 2004. *Journal of Ethnopharmacology* 93: 27–32.
20. Tona, L., Ngimbi, N. P., Tsakala, M., Mesia, K., Cimanga, K., Apers, S., De Bruyne, T., Pieters, L., Totté, J. and Vlietinck, A. J. 1999. *Journal of Ethnopharmacology* 68: 193–203.
21. Anuradha, H., Srikumar, B. N., Shankaranarayana Rao, B. S. and Lakshmana, M. 2008. *Journal of Neural Transmission* 115: 35–42.
22. Johnson, P. B., Abdurahman, E. M., Tiam, E. A., Abdu-Aguye, I. and Hussaini, I. M. 1999. *Journal of Ethnopharmacology* 65: 63–69.
23. Lanhers, M. C., Fleurentin, J., Cabalion, P., Rolland, A., Dorfman, P., Misslin, R. and Pelt, J. M. 1990. *Journal of Ethnopharmacology* 29: 189–198.
24. Youssouf, M. S., Kaiser, P., Tahir, M., Singh, G. D., Singh, S., Sharma, V. K., Satti, N. K., Haque, S. E. and Johri, R. K. 2007. *Fitoterapia* 78: 535–539.
25. Singh, S. K., Yadav, R. P., Tiwari, S. and Singh, A. 2005. *Chemosphere* 59: 263–270.
26. Singh, S. K., Yadav, R. P., Singh, D. and Singh, A. 2004. *Environmental Toxicology and Pharmacology* 15: 87–93.

Eurycoma longifolia Jack.

1. Hsuan, K. 1992. Orders and Families of Malayan Seed Plants. Singapore University Press: Singapore.
2. Hsuan, K. 1990. The Concise Flora of Singapore, Gymnosperms and Dicotyledons.

Singapore University Press: National University of Singapore.

3. Goh, S. H., Chuah, C. H., Mok, J. S. L. and Soepadmo, E. 1995. Malaysian Medicinal Plants for the Treatment of Cardiovascular Diseases. Pelanduk Publications (M) Sdn Bhd: Malaysia.

4. USDA, ARS GRIN Taxonomy for Plants 2008 http://www.ars-grin.gov/cgi-bin/npgs/html/taxon.pl?417514 Date Accessed: 30th June 2008

5. Chan, K. L., O'neill, M. J., Phillipson, J. D. and Warhurst, D. C. 1986. *Planta Medica* 52: 105–107.

6. Kardono, L. B., Angerhofer, C. K., Tsauri, S., Padmawinata, K., Pezzuto, J. M. and Kinghorn, A. D. 1991. *Journal of Natural Products* 54: 1360–1367.

7. Chan, K. L., Lee, S. P., Sam, T. W., Tan, S. C., Noguchi, H. and Sankawa, U. 1991. *Phytochemistry* 30: 3138–3141.

8. Itokawa, H., Kishi, E., Morita, H. and Takeya, K. 1992. *Chemical and Pharmaceutical Bulletin* 40: 1053–1055.

9. Morita, H., Kishi, E., Takeya, K. and Itokawa, H. 1992. *Phytochemistry* 31: 3993–3995.

10. Morita, H., Kishi, E., Takeya, K., Itokawa, H. and Iitaka, Y. 1993a. *Phytochemistry* 33: 691–696.

11. Morita, H., Kishi, E., Takeya, K., Itokawa, H. and Iitaka, Y. 1993b. *Phytochemistry* 34: 765–771.

12. Chan, K. L., Iitaka, Y., Noguchi, H., Sugiyama, H., Saito, I. and Sankawa, U. 1992. *Phytochemistry* 31: 4295–4298.

13. Ang, H. H., Chan, K. L. and Mak, J. W. 1995. *Planta Medica* 61: 177–178.

14. Mitsunaga, K., Koike, K., Tanaka, T., Ohkawa, Y., Kobayashi, Y., Sawaguchi, T. and Ohmoto, T. 1994. *Phytochemistry* 35: 799–802.

15. Ang, H. H., Cheang, H. S. and Yusof, A. P. 2000. *Experimental Animals* 49: 35–38.

16. Jiwajinda, S., Santisopasri, V., Murakami, A., Hirai, N. and Ohigashi, H. 2001. *Phytochemistry* 58: 959–962.

17. Kuo, P. C., Shi, L. S., Damu, A. G., Su, C. R., Huang, C. H., Ke, C. H., Wu, J. B., Lin, A. J., Bastow, K. F., Lee, K. H. and Wu, T. S. 2003. *Journal of Natural Products* 66: 1324–1327.

18. Kuo, P. C., Damu, A. G., Lee, K. H. and Wu, T. S. 2004. *Bioorganic and Medicinal Chemistry* 12: 537–544.

19. Bedir, E., Abou-Gazar, H., Ngwendson, J. N. and Khan, I. A. 2004. *Chemical and Pharmaceutical Bulletin* 51: 1301–1303.

20. Latiff, A. 1988. Proceedings: Malaysian Traditional Medicine, Kuala Lumpur, pp. 1–6. Institute of Advanced Studies, University of Malaya, Kuala Lumpur.

21. Ang, H.H. and Cheang, H.S. 1999. *Japanese Journal of Pharmacology* 79: 497–500.

22. Farouk A. and Benafri A. 2007. *Saudi Medical Journal* 28: 477–479.

23. Jiwajinda, S., Santisopasri, V., Murakami, A., Kawanaka, M., Kawanaka, H., Gasquet, M., Eilas, R., Balansard, G. and Ohigashi, H. 2002. *Journal of Ethnopharmacology* 82: 55–58.

24. Mohd-Fuat, A. R., Aidoo, K. E. and Candlish, A. A. G. 2007. *Tropical Biomedicine* 24: 49–59.

25. Tee, T. T. and Azimahtol, H. L. 2005. *Anticancer Research* 25: 2205–2213.

26. Nurhanan, M. Y., Azimahtol, Hawariah, L. P., Mohd Ilham, A. and Mohd Shukri, M. A. 2005. *Phytotherapy Research.* 19: 994–996.

27. Li, Y., Liang, F., Jiang, W., Yu, F., Cao, R., Ma, Q., Dai, X., Jiang, J., Wang, Y. and Si, S. 2007. *Cancer Biology and Therapy* 6: 1193–1199.

28. Chan, K. L., Choo, C. Y., Abdullah, N. R. and Ismail, Z. 2004. *Journal of Ethnopharmacology* 92: 223–227.

29. Chan, K. L., Choo, C. Y. and Abdullah, N. R. 2005. *Planta Medica* 71: 967–969.

30. Mohd Ridzuan, M. A., Rain, A. N., Zhari, I. and Zakiah, I. 2005. *Tropical Biomedicine* 22: 155–163.

31. Hout, S., Chea, A., Bun, S. S., Elias, R., Gasquet, M., Timon-David, P., Balansard, G. and Azas, N. 2006. *Journal of Ethnopharmacology* 107: 12–18.

32. Mohd Ridzuan, M. A., Sow, A., Noor Rain, A., Mohd Ilham, A. and Zakiah, I. 2007. *Tropical Biomedicine* 24: 111–118.

33. Tada, H., Yasuda, F., Otani, K., Doteuchi, M., Ishara, Y. and Shiro, M. 1991. *European Journal of Medical Chemistry* 26: 345–349.

34. Widjajakusama, L. E. 1987. Androgenic Activity in Root and Stem of Pasak Bumi (*Eurycoma longifolia*) as Determined by a Modified Chick-comb Growth Method of Dorfman. XVI Pacific Science Congress, Seoul, Korea.

35. Ang, H. H. and Sim, M. K. 1998. *Archives of Pharmacal Research* 21: 779–781.
36. Ang, H. H. and Ngai, T. H. 2001a. *Fundamental and Clinical Pharmacology* 15: 265–268.
37. Ang, H. H., Ikeda, S. and Gan, E. K. 2001b. *Phytotherapy Research* 15: 435–436.
38. Ang, H. H. and Cheang, H. S. 2001c. *Archives of Pharmacal Research* 24: 437–440.
39. Satayavivad, J., Soonthornchareonnon, N., Somanabandhu, A. and Thebtaranonth, Y. 1998. *Thai Journal of Phytopharmacy* 52: 14–27.
40. Chan, K. L. and Choo, C. Y. 2002. *Planta Medica* 68: 662–664.

Hibiscus mutabilis L.

1. Hsuan, K. 1992. Orders and Families of Malayan Seed Plants. Singapore University Press: Singapore.
2. Hsuan, K. 1990. The Concise Flora of Singapore, Gymnosperms and Dicotyledons. Singapore University Press: National University of Singapore.
3. Wee, Y. C. and Hsuan, K. 1990. An illustrated Dictionary of Chinese Medicinal Herbs. Times Edition and Eu Yan Seng Holdings Ltd: Singapore.
4. eFloras.org 2008 http://www.efloras.org/florataxon.aspx?flora_id=2&taxon_id=200013711 Date Accessed: 30th March 2008
5. Duke, J. A. and Ayensu, E. S. 1985. Medicinal Plants of China, volume one and two. Reference Publications, Inc.: United States of America.
6. Chauhan, J. S., Vidyapati, T. J. and Gupta, A. K. A. 1979. *Phytochemistry* 18: 1766–1767.
7. Chen, R. T., Chen, L., Wei, K. X. and Fang, S. D. 1993. *Zhongcaoyao* 24: 227–229.
8. Lowry, J. B. 1971. *Phytochemistry* 10: 673–674.
9. Vidyapati, T. J., Gupta, A. K. and Chauhan, J. S. 1979. *Indian Journal of Chemistry, Section B: Organic Chemistry Including Medicinal Chemistry* 17B: 536.
10. Yao, L. Y., Lu, Y. and Chen, Z. N. 2003. *Zhongcaoyao* 34: 201–203.
11. Duke, J. A. and Ayensu, E. S. 1985. Medicinal Plants of China, volume one and two.

Reference Publications, Inc.: United States of America.
12. Nguyen, V. D. and Doan, T. N. 1989. Medicinal Plants in Vietnam. World Health Organisation, Regional Office for the Western Pacific, Manila, Institute of Materia Medica, Hanoi.
13. Fu, S. C., Zhang, F. H., Shi, W. B., Du, N., Yao, L. Y. and Lu, Y. 2001. *Shanghai Dier Yike Daxue Xuebao* 21: 14–16.

Hibiscus rosa-sinensis L.

1. Hsuan, K. 1992. Orders and Families of Malayan Seed Plants. Singapore University Press: Singapore.
2. Palaniswamy, U. R. 2003. A guide to Medicinal Plants of Asian origin and culture. CPL Press: United States of America.
3. USDA, ARS GRIN Taxonomy for Plants 2008 http://www.ars-grin.gov/cgi-bin/npgs/html/taxon.pl?19075 Date Accessed: 30th June 2008
4. Duke, J. A. and Ayensu, E. S. 1985. Medicinal Plants of China, volume one and two. Reference Publications, Inc.: United States of America.
5. Li, T. S. C. 2000. Medicinal Plants. Culture, Utilisation and Phytopharmacology. Technomic Publishing Co., Inc.: Lancaster.
6. Nakatani, M., Matsuoka, K., Uchio, Y. and Hase, T. 1994. *Phytochemistry* 35: 1245–1247.
7. Woodley, E. (Ed.). 1991. Medicinal plants of Papua New Guinea, part I. Veerlag Josef Margraf Scientific Books: Weikersheim.
8. Jain, S. K. and DeFilipps, R. A. 1991. Medicinal Plants of India, volume one and two. Reference Publications Inc.: United States of America.
9. Kasture, V. S., Chopde, C. T. and Deshmukh, V. K. 2000. *Journal of Ethnopharmacology* 71: 65–75.
10. Sharma, S., Khan, N. and Sultana, S. 2004. *European Journal of Cancer Prevention* 13: 53–63.
11. Kholkute, S. D. 1977. *Planta Medica* 31: 127–135.
12. Reddy, C. M., Murthy, D. R. and Patil, S. B. 1997. *Indian Journal of Experimental Biology* 35: 1170–1174.

13. Murthy, D. R., Reddy, C. M. and Patil, S. B. 1997. *Biological and Pharmaceutical Bulletin* 20: 756–758.
14. Nivsarkar, M., Patel, M., Padh, H., Bapu, C. and Shrivastava, N. 2005. *Contraception* 71: 227–230.
15. Gauthaman, K. K., Saleem, M. T., Thanislas, P. T., Prabhu, V. V., Krishnamoorthy, K. K., Devaraj, N. S. and Somasundaram, J. S. 2006. *BMC Complementary and Alternative Medicine* 6: 32.
16. Obi, F. O., Ichide, I. I., Okororo, A. A., Esiri, O. and Oghre, S. 2002. *Global Journal of Pure and Applied Sciences* 8: 259–264.
17. Sachdewa, A. and Khemani, L. D. 1999. *Biomedical and Environmental Science* 12: 222–226.
18. Sachdewa, A. and Khemani, L. D. 2003. *Journal of Ethnopharmacology* 89: 61–66.
19. Prakash, A. O. 1979. *Experientia* 35: 1122–1123.
20. Singwi, M. S. and Lall, S. B. 1981. *Indian Journal of Experimental Biology* 19: 359–362.
21. Shivananda Nayak, B., Sivachandra Raju, S., Orette, F. A. and Chalapathi Rao, A. V. 2007. *International Journal of Lower Extremity Wounds* 6: 76–81.
22. Duke, J. A., Bogenschutz-Godwin, M. J., duCellier, J. and Duke, P. A. K. 2002. Handbook of medicinal herbs. CRC Press: Boca Raton, FL.

Hibiscus tiliaceus L.

1. Hsuan, K. 1990. The Concise Flora of Singapore, Gymnosperms and Dicotyledons. Singapore University Press: National University of Singapore.
2. Wee, Y. C. 1992. A Guide to Medicinal Plants. Singapore Science Centre Publication: Singapore.
3. Wee, Y. C. and Corlett, R. 1986. The City and the Forest: Plant Life in Urban Singapore. Singapore University Press: National University of Singapore.
4. eFloras.org 2008 http://www.efloras.org/florataxon.aspx?flora_id=2&taxon_id=200013733 Date Accessed: 30th March 2008

5. Melecchi, M. I. S., Martinez, M. M., Abad, F. C., Zini, P. P., Filho, I. D. N. and Caramão, E. B. 2002. *Journal of Separation Science* 25: 86–90.
6. Chen, J. J., Huang, S. Y., Duh, C. Y., Chen, I. S., Wang, T. C. and Fang, H. Y. 2006. *Planta Medica* 72: 935–938.
7. Li, L., Huang, X., Sattler, I., Fu, H., Grabley, S. and Lin, W. 2006. *Magnetic Resonance Chemistry* 44: 624–628.
8. Jain, S. K. and DeFilipps, R. A. 1991. Medicinal Plants of India, volume one and two. Reference Publications Inc.: United States of America.
9. Masuda, T., Yamashita, D., Takeda, Y. and Yonemori, S. 2005. *Bioscience, Biotechnology, and Biochemistry* 69: 197–201.
10. Rosa, R. M., Melecchi, M. I., da Costa Halmenschlager, R., Abad, F. C., Simoni, C. R., Caramão, E. B., Henriques, J. A., Saffi, J. and de Paula Ramos, A. L. 2006. *Journal of Agricultural and Food Chemistry* 54: 7324–7330.
11. Rosa, R. M., Moura, D. J., Melecchi, M. I., dos Santos, R. S., Richter, M. F., Camarão, E. B., Henriques, J. A., de Paula Ramos, A. L. and Saffi, J. 2007. *Toxicology in vitro* 21: 1442–1452.

Impatiens balsamina L.

1. Hsuan, K. 1978. Orders and Families of Malayan Seed Plants. Singapore University Press: Singapore.
2. Hsuan, K. 1990. The Concise Flora of Singapore, Gymnosperms and Dicotyledons. Singapore University Press: National University of Singapore.
3. Wee, Y. C. 1992. A Guide to Medicinal Plants. Singapore Science Centre Publication: Singapore.
4. Wiart, C. 2000. Medicinal plants of Southeast Asia. Pelanduk Publications (M) Sdn Bhd: Malaysia.
5. Wiart, C. 2006. Medicinal Plants of the Asia-Pacific: Drugs for the future? World Scientific Publishing Co. Pte. Ltd.: Singapore.
6. Ishiguro, K., Ohira, Y. and Oku, H. 1998. *Journal of Natural Products* 61: 1126–1129.
7. Ishiguro, K., Oku, H. and Kato, T. 2000. *Phytotherapy Research* 14: 54–56.

8. Oku, H. and Ishiguro, K. 2002. *Biological and Pharmaceutical Bulletin* 25: 658–660.
9. Shoji, N., Umeyama, A., Saitou, N., Yoshikawa, K., Nagai, M. and Arihara, S. 1994. *Chemical and Pharmaceutical Bulletin* 42: 1422–1426.
10. Shoji, N., Umeyama, A., Yoshikawa, K., Nagai, M. and Arihara, S. 1994. *Phytochemistry* 37: 1437–1441.
11. Duke, J. A. and Ayensu, E. S. 1985. Medicinal Plants of China, volume one and two. Reference Publications, Inc.: United States of America.
12. Yang, X., Summerhurst, D. K., Koval, S. F., Ficker, C., Smith, M. L. and Bernards, M. A. 2001. *Phytotherapy Research* 15: 676–680.
13. Choi, S. C. and Jung, J. S. 1999. *Journal of the Korean Fiber Society* 36: 338–343.
14. Lim, Y. H., Kim, I. H. and Seo, J. J. 2007. *Journal of Microbiology* 45: 473–477.
15. Tailor, R. H., Acland, D. P., Attenborough, S., Cammue, B. P., Evans, I. J., Osborn, R. W., Ray, J. A., Rees, S. B. and Broekaert, W. F. 1997. *Journal of Biological Chemistry* 272: 24480–24487.
16. Ishiguro, K., Ohira, Y. and Oku, H. 2002. *Biological and Pharmaceutical Bulletin* 25: 505–508.
17. Oku, H. and Ishiguro, K. 1999. *Phytotherapy Research* 13: 521–525.
18. Thevissen, K., François, I. E., Sijtsma, L., van Amerongen, A., Schaaper, W. M., Meloen, R., Posthuma-Trumpie, T., Broekaert, W. F. and Cammue, B. P. 2005. *Peptides* 26: 1113–1119.
19. Oku, H. and Ishiguro, K. 2001. *Phytotherapy Research* 15: 506–510.

Imperata cylindrica (L.) P. Beauv.

1. Nguyen, V. D. 1993. Medicinal Plants of Vietnam, Cambodia and Laos. Nguyen Van Duong: Vietnam.
2. eFloras.org 2008 http://www.efloras.org/florataxon.aspx?flora_id=2&taxon_id=200025558 Date Accessed: 30th March 2008
3. Duke, J. A. and Ayensu, E. S. 1985. Medicinal Plants of China, volume one and two. Reference Publications, Inc.: United States of America.

4. Goh, S. H., Chuah, C. H., Mok, J. S. L. and Soepadmo, E. 1995. Malaysian Medicinal Plants for the Treatment of Cardiovascular Diseases. Pelanduk Publications (M) Sdn Bhd: Malaysia.
5. Matsunaga, K., Shibuya, M. and Ohizumi, Y. 1994. *Journal of Natural Products* 57: 1183–1184.
6. Matsunaga, K., Shibuya, M. and Ohizumi, Y. 1995. *Journal of Natural Products* 58: 138–139.
7. Matsunaga, K., Ikeda, M., Shibuya, M. and Ohizumi, Y. 1994. *Journal of Natural Products* 57: 1290–1293.
8. Matsunaga, K., Shibuya, M. and Ohizumi, Y. 1994. *Journal of Natural Products* 57: 1734–1736.
9. Yoon, J. S., Lee, M. K., Sung, S. H. and Kim, Y. C. 2006. *Journal of Natural Products* 69: 290–291.
10. Lee, D. Y., Han, K. M., Song, M. C., Lee, D. G., Rho, Y. D. and Baek, N. I. 2008. *Journal of Asian Natural Products Research* 10: 337–341.
11. Wee, Y. C. and Hsuan, K. 1990. An illustrated Dictionary of Chinese Medicinal Herbs. Times Edition and Eu Yan Seng Holdings Ltd: Singapore.
12. Sripanidkulchai, B., Wongpanich, V., Laupattarakasem, P., Suwansaksri, J. and Jirakulsomchok, D. 2001. *Journal of Ethnopharmacology* 75: 185–190.
13. De Leon, R. M., Panlilio, B. G., Bamba, E. M., Bognot, F. L. and Dizon, D. S. 2001. Antibacterial properties of local medicinal plants against selected test organisms. Abstracts of Papers, 22nd ACS National Meeting, Chicago, IL, United States.
14. Koh, D., Goh, C. L., Tan, H. T., Ng, S. K. and Wong, W. K. 1997. *Contact Dermatitis* 37: 32–34.
15. Kumar, L., Sridhara, S., Singh, B. P., Gangal, S. V. 1998. *International Archieves of Allergy and Immunology* 117: 174–179.

Ipomoea pes-caprae (L.) Sweet

1. Hsuan, K. 1992. Orders and Families of Malayan Seed Plants. Singapore University Press: Singapore.
2. Hsuan, K. 1990. The Concise Flora of Singapore, Gymnosperms and Dicotyledons.

Singapore University Press: National University of Singapore.
3. Wee, Y. C. 1992. A Guide to Medicinal Plants. Singapore Science Centre Publication: Singapore.
4. Wiart, C. 2000. Medicinal plants of Southeast Asia. Pelanduk Publications (M) Sdn Bhd: Malaysia.
5. eFloras.org 2008
http://www.efloras.org/florataxon.aspx?flora_id=2&taxon_id=200018852
Date Accessed: 16th April 2008
6. Amor-Prats, D. and Harborne, J. B. 1993. *Biochemical Systematics and Ecology* 21: 455–461.
7. Escobedo-Martinez, C. and Pereda-Miranda, Rogelio. 2007. *Journal of Natural Products* 70: 974–978.
8. Pereda-Miranda, R., Escalante-Sanchez, E. and Escobedo-Martinez, C. 2005. *Journal of Natural Products* 68: 226–230.
9. Pongprayoon, U., Baeckstrom, P., Jacobsson, U., Lindstrom, M. and Bohlin, L. 1991. *Planta Medica* 57: 515–518.
10. Pongprayoon, U., Baeckström, P., Jacobsson, U., Lindström, M. and Bohlin, L. 1992. *Phytotherapy Research* 6: 104–107.
11. Pongprayoon, U., Bohlin, L., Baeckström, P., Jacobsson, U. and Lindström, M. 1992. *Planta Medica* 58: 19–21.
12. Teramachi, F., Koyano, T., Kowithayakorn, T., Hayashi, M., Komiyama, K. and Ishibashi, M. 2005. *Journal of Natural Products* 68: 794–796.
13. Jain, S. K. and DeFilipps, R. A. 1991. Medicinal Plants of India, volume one and two. Reference Publications Inc.: United States of America.
14. Rogers, K. L., Grice, I. D. and Griffiths, L. R. 2000. *European Journal of Pharmaceutical Sciences* 9: 355–363.
15. Pongprayoon, U., Bohlin, L., Sandberg, F. and Wasuwat, S. 1989. *Acta Pharmaceutica Nordica* 1: 41–44.
16. Maria de Souza, M., Madeira, A., Berti, C., Krogh, R., Yunes, R. A. and Cechinel-Filho, V. 2000. *Journal of Ethnopharmacology* 69: 85–90.
17. Khan, M. M., Ahmad, F., Rastogi, A. K., Kidwai, J. R., Lakshmi, V. and Bhakuni, D. S. 1994. *Fitoterapia* 65: 231–234.

18. Pongprayoon, U., Bohlin, L., Soonthornsaratune, P. and Wasuwat, S. 1991. *Phytotherapy Research* 5: 63–66.
19. Pongprayoon, U., Bohlin, L. and Wasuwat, S. 1991. *Journal of Ethnopharmacology* 35: 65–69.

Ixora chinensis Lam.

1. Hsuan, K. 1990. The Concise Flora of Singapore, Gymnosperms and Dicotyledons. Singapore University Press: National University of Singapore.
2. Takeda, Y., Nishimura, H. and Inouye, H. 1975. *Phytochemistry* 14: 2647–2650.
3. Huang, M. 1990. *Phytochemistry* 29: 1317–1319.
4. Hui, W. H. and Ho, C. T. 1968. *Australian Journal of Chemistry* 21: 547–549.
5. Goh, S. H., Liew, S. L. and Soepadmo, E. 1988. Proceedings: Malaysian Traditional Medicine, Kuala Lumpur, pp. 104–110, Institute of Advanced Studies, University of Malaya, Kuala Lumpur.
6. Wee, Y. C. and Hsuan, K. 1990. An illustrated Dictionary of Chinese Medicinal Herbs. Times Edition and Eu Yan Seng Holdings Ltd: Singapore.
7. Duke, J. A. and Ayensu, E. S. 1985. Medicinal Plants of China, volume one and two. Reference Publications, Inc.: United States of America.

Jatropha curcas L.

1. eFloras.org 2008
http://www.efloras.org/florataxon.aspx?flora_id=2&taxon_id=200012585
Date Accessed: 30th June 2008
2. Wiart, C. 2006. Medicinal Plants of the Asia-Pacific: Drugs for the future? World Scientific Publishing Co. Pte. Ltd.: Singapore.
3. USDA, ARS GRIN Taxonomy for Plants 2008
http://www.ars-grin.gov/cgi-bin/npgs/html/taxon.pl?20692
Date Accessed: 30th June 2008
4. Duke, J. A. and Ayensu, E. S. 1985. Medicinal Plants of China, volume one and two. Reference Publications, Inc.: United States of America.

5. Ravindranath, N., Ramesh, C. and Das, B. 2003. *Biochemical Systematics and Ecology* 31: 431–432.

6. Ravindranath, N., Reddy, M. R., Mahender, G., Ramu, R., Kumar, K. R. and Das, B. 2004. *Phytochemistry* 65: 2387–2390.

7. Naengchomnong, W., Thebtaranonth, Y., Wiriyachitra, P., Okamoto, K. T. and Clardy, J. 1986. *Tetrahedron Letters* 27: 2439–2442.

8. Van den Berg, A. J., Horsten, S. F, Kettenes-van den Bosch, J. J., Kroes, B. H., Beukelman, C. J., Leeflang, B. R. and Labadie, R. P. 1995. *FEBS Letters* 358: 215–218.

9. Staubmann, R., Schubert-Zsilavecz, M., Hiermann, A. and Kartnig, T. 1998. *Phytochemistry* 50: 337–338.

10. Kong, L., Min, Z. and Shi, J. 1996. *Zhiwu Xuebao* 38: 161–166.

11. Talapatra, S. K., Mandal, K. and Talapatra, B. 1993. *Journal of the Indian Chemical Society* 70: 543–548.

12. Abigor, R. D. and Uadia, P. O. 2001. *Rivista Italiana delle Sostanze Grasse* 78: 163–165.

13. Lin, J., Yan, F., Tang, L. and Chen, F. 2003. *Acta Pharmacologica Sinica* 24: 241–246.

14. Muangman, S., Thippornwong, M. and Tohtong, R. 2005. *In vivo* 19: 265–268.

15. Perry, L. M. 1980. Medicinal plants of East and Southeast Asia: Attributed properties and uses. The MIT Press: Cambridge, Massachusetts and London.

16. Jain, S. K. and DeFilipps, R. A. 1991. Medicinal Plants of India, volume one and two. Reference Publications Inc.: United States of America.

17. Rau, O., Wurglics, M., Dingermann, T., Abdel-Tawab, M. and Schubert-Zsilavecz, M. 2006. *Pharmazie* 61: 952–956.

18. Mujumdar, A. M. and Misar, A. V. 2004. *Journal of Ethnopharmacology* 90: 11–15.

19. Luo, M. J., Yang, X. Y., Liu, W. X., Xu, Y., Huang, P., Yan, F. and Chen, F. 2006. *Acta Biochimica et Biophysica Sinica (Shanghai)* 38: 663–668.

20. Ankrah, N. A., Nyarko, A. K., Addo, P. G., Ofosuhene, M., Dzokoto, C., Marley, E., Addae, M. M. and Ekuban, F. A. 2003. *Phytotherapy Research* 17: 697–701.

21. Matsuse, I. T., Lim, Y. A., Hattori, M., Correa, M. and Gupta, M. P. 1999. *Journal of Ethnopharmacology* 64: 15–22.

22. Muanza, D. N., Euler, K. L., Williams, L. and Newman, D. J. 1995. *International Journal of Pharmacognosy* 33: 98–106.

23. Osoniyi, O. and Onajobi, F. 2003. *Journal of Ethnopharmacology* 89: 101–105.

24. Goonasekera, M. M., Gunawardana, V. K., Jayasena, K., Mohammed, S. G. and Balasubramaniam, S. 1995. *Journal of Ethnopharmacology* 47: 117–123.

25. Gandhi, V. M., Cherian, K. M. and Mulky, M. J. 1995. *Food and Chemical Toxicology* 33: 39–42.

26. Staubmann, R., Ncube, I., Gubitz, G. M., Steiner, W. and Read, J. S. 1999. *Journal of Biotechnology* 75: 117–126.

27. Karmegam, N., Sakthivadivel, M., Anuradha, V. and Daniel, T. 1997. *Bioresource Technology* 59: 137–140.

28. Meshram, P. B., Kulkarni, N. and Joshi, K. C. 1996. *Journal of Environmental Biology* 17: 295–298.

29. Georges, K., Jayaprakasam, B., Dalavoy, S. S. and Nair, M. G. 2008. *Bioresource Technology* 99: 2037–2045.

30. Liu, S. Y., Sporer, F., Wink, M., Jourdane, J., Henning, R., Li, Y. L. and Ruppel, A. 1997. *Tropical Medicine ad International Health* 2: 179–188.

31. Shetty, S., Udupa, S. L., Udupa, A. L. and Vollala, V. R. 2006. *Saudi Medical Journal* 27: 1473–1476.

32. Duke, J. A. 1985. CRC Handbook of Medicinal Herbs. CRC Press: Boca Raton, FL.

33. Felke, J. 1914. *Landw. Vers. Stat.* 82: 427–463.

34. Liberalino, A. A. A., Bambirra, E. A., Moraes-Santos, T. and Vieira, E. C. 1988. *Arquivos de Biologia e Tecnologia* 31: 539–550.

35. Abdu-Aguye, I., Sannusi, A., Alafiya-Tayo, R. A. and Bhusnurmath, S. R. 1986. *Human Toxicology* 5: 269–274.

36. el Badwi, S. M. and Adam, S. E. 1992. *Veterinary and Human Toxicology* 34: 112–115.

37. Joubert, P. H., Brown, J. M., Hay, I. T. and Sebata, P. D. 1984. *South African Medical Journal* 65: 729–730.

38. Ahmed, O. M. and Adam, S. E. 1979. *Research in Veterinary Science* 27: 89–96.

39. Gandhi, V. M., Cherian, K. M. and Mulky, M. J. 1995. *Food and Chemical Toxicology* 33: 39–42.

40. Kulkarni, M. L., Sreekar, H., Keshavamurthy, K. S. and Shenoy, N. 2005. *Indian Journal of Pediatrics* 72: 75–76.

Juniperus chinensis L.

1. eFloras.org 2008
   http://www.efloras.org/florataxon.aspx?flora_
   id=2&taxon_id=210000896
   Date Accessed: 30th March 2008
2. USDA, ARS GRIN Taxonomy for Plants 2008
   http://www.ars-grin.gov/cgi-bin/npgs/html/
   taxon.pl?300290
   Date Accessed: 30th June 2008
3. Duke, J. A. and Ayensu, E. S. 1985. Medicinal Plants of China, volume one and two. Reference Publications, Inc.: United States of America.
4. Lee, C. K. and Cheng, Y. S. 2001. *Journal of the Chinese Chemical Society* 48: 1077–1080.
5. Lee, C. K., Fang, J. M. and Cheng, Y. S. 1995. *Phytochemistry* 39: 391–394.
6. Lim, J. P., Song, Y. C., Kim, J. W., Ku, C. H., Eun, J. S., Leem, K. H. and Kim, D. K. 2002. *Archieves of Pharmacal Research* 25: 449–452.
7. Miyata, M., Itoh, K. and Tachibana, S. 1998. Extractives of *Juniperus chinensis* L. *Journal of Wood Science* 44: 397–400.
8. Kim, Y. K. and Kim, W. K. 2001. *Han'guk Sikp'um Yongyang Kwahak Hoechi* 30: 1204–1214.
9. Kuo, Y. H. and Chen, W. C. 1994. *Chemical & Pharmaceutical Bulletin* 42: 2187–2189.
10. Kuo, Y. H. and Shiu, L. L. 1996. *Chemical & Pharmaceutical Bulletin* 44: 1758–1760.
11. Shiu, L. L., Chen, W. C. and Kuo, Y. H. 1999. *Chemical & Pharmaceutical Bulletin* 47: 557–560.
12. Wee, Y. C. and Hsuan, K. 1990. An illustrated Dictionary of Chinese Medicinal Herbs. Times Edition and Eu Yan Seng Holdings Ltd: Singapore.
13. Bagci, E. and Digrak, M. 1996. *Turkish Journal of Biology* 20: 191–198.
14. Mohamed, S., Saka, S., El-Sharkawy, S. H., Ali, A. M. and Muid, S. 1996. *Pesticide Science* 47: 259–264.
15. Ali, A. M., Mackeen, M. M., Intan-Safinar, I., Hamid, M., Lajis, N. H., el-Sharkawy, S. H.

and Murakoshi, M. 1996. *Journal of Ethnopharmacology* 53: 165–169.
16. Ju, J. B., Kim, J. S., Choi, C. W., Lee, H. K., Oh, T. K. and Kim, S. C. 2008. *Journal of Ethnopharmacology* 115: 110–115.

Kaempferia galanga L.

1. Hsuan, K., Chin, S. C. and Tan, H. T. W. 1998. The Concise Flora of Singapore volume II: Monocotyledons. Singapore University Press: Singapore.
2. USDA, ARS GRIN Taxonomy for Plants 2008
   http://www.ars-grin.gov/cgi-bin/npgs/html/
   taxon.pl?20916
   Date Accessed: 30th June 2008
3. Duke, J. A. and Ayensu, E. S. 1985. Medicinal Plants of China, volume one and two. Reference Publications, Inc.: United States of America.
4. Duke, J. A. 1985. CRC Handbook of Medicinal Herbs. CRC Press: Boca Raton, FL.
5. Jenie, U. A. and Sudibyo, R. S. 2000. *Majalah Farmasi Indonesia* 11: 168–173.
6. Jirovetz, L., Buchbauer, G., Shafi, P. M. and Abraham, G. T. 2001. *Acta Pharmaceutica Turcica* 43: 107–110.
7. Kiuchi, F., Nakamura, N. and Tsuda, Y. 1987. *Phytochemistry* 26: 3350–3351.
8. bin Din, L., Zakaria, Z., Hj. Abd Malek, S. N. and Samsudin. M. W. 1988. Proceedings: Malaysian Traditional Medicine, Kuala Lumpur, pp. 118–124. Institute of Advanced Studies, University of Malaya, Kuala Lumpur.
9. Wee, Y. C. and Hsuan, K. 1990. An illustrated Dictionary of Chinese Medicinal Herbs. Times Edition and Eu Yan Seng Holdings Ltd: Singapore.
10. Jain, S. K. and DeFilipps, R. A. 1991. Medicinal Plants of India, volume one and two. Reference Publications Inc.: United States of America.
11. Arambewela, L., Perera, A., Wijesundera, R. T. R. L. C. and Gunatileke, J. 2000. *Journal of the National Science Foundation of Sri Lanka* 28: 225–230.
12. Bhamarapravati, S., Pendland, S. L., Mahady, G. B. 2003. *In Vivo* 17: 541–544.
13. bin Jantan, I., Yassin, M. S. M., Chin, C. B., Chen, L. L. and Sim, N. L. 2003. *Pharmaceutical Biology (Lisse, Netherlands)* 41: 392–397.

14. Othman, R., Ibrahim, H., Mohd, M. A., Mustafa, M. R. and Awang, K. 2006. *Phytomedicine* 13: 61–66.

15. Xue, Y. and Chen, H. 2002. *Wei Sheng Yan Jiu* 31: 247–248.

16. Xue, Y., Murakami, A., Koizumi, K. and Chen, H. 2002. *Zhongguo Zhongyao Zazhi* 27: 522–524.

17. Vimala, S., Norhanom, A. W. and Yadav, M. 1999. *British Journal of Cancer* 80: 110–116.

18. Kosuge, T., Yokota, M., Sugiyama, K., Saito, M., Iwata, Y., Nakura, M. and Yamamoto, T. 1985. *Chemical & Pharmaceutical Bulletin* 33: 5565–5567.

19. Chirangini, P., Sharma, G. J. and Sinha, S. K. 2004. *Journal of Environmental Pathology, Toxicology and Oncology* 23: 227–236.

20. Chu, D. M., Miles, H., Toney, D., Ngyuen, C. and Marciano-Cabral, F. 1998. *Parasitology Research* 84: 746–752.

21. Kanjanapothi, D., Panthong, A., Lertprasertsuke, N., Taesotikul, T., Rujjanawate, C., Kaewpinit, D., Sudthayakorn, R., Choochote, W., Chaithong, U., Jitpakdi, A. and Pitasawat, B. 2004. *Journal of Ethnopharmacology* 90: 359–365.

22. Othman, R., Ibrahim, H., Mohd, M. A., Awang, K., Gilani, A. U. and Mustafa, M. R. 2002. *Planta Medica* 68: 655–657.

23. Tewtrakul, S. and Subhadhirasakul, S. 2007. *Journal of Ethnopharmacology* 109: 535–538.

24. Choochote, W., Chaithong, U., Kamsuk, K., Jitpakdi, A., Tippawangkosol, P., Tuetun, B., Champakaew, D. and Pitasawat, B. 2007. *Fitoterapia* 78: 359–364.

25. Choochote, W., Kanjanapothi, D., Panthong, A., Taesotikul, T., Jitpakdi, A., Chaithong, U. and Pitasawat, B. 1999. *Southeast Asian Journal of Tropical Medicine and Public Health* 30: 470–476.

26. Pandji, C., Grimm, C., Wray, V., Witte, L. and Proksch, P. 1993. *Phytochemistry* 34: 415–419.

27. Kiuchi, F., Nakamura, N., Tsuda, Y., Kondo, K. and Yoshimura, H. 1988. *Chemical & Pharmaceutical Bulletin* 36: 412–415.

28. Suzuki, J., Yasuda, I., Murata, I., Murata, R. and Nishitani, K. 2003. *Tokyo-toritsu Eisei Kenkyusho Kenkyu Nenpo* 53: 35–39.

29. Tara Shanbhag, V., Chandrakala, S., Sachidananda, A., Kurady, B. L., Smita, S. and Ganesh, S. 2006. *Indian Journal of Physiology and Pharmacology* 50: 384–390.

30. Nguyen, V. D. and Doan, T. N. 1989. Medicinal Plants in Vietnam. World Health Organisation, Regional Office for the Western Pacific, Manila, Institute of Materia Medica, Hanoi.

Lantana camara L.

1. Hsuan, K. 1990. The Concise Flora of Singapore, Gymnosperms and Dicotyledons. Singapore University Press: National University of Singapore.

2. USDA, ARS GRIN Taxonomy for Plants 2008 http://www.ars-grin.gov/cgi-bin/npgs/html/taxon.pl?310628 Date Accessed: 30th June 2008

3. Duke, J. A. and Ayensu, E. S. 1985. Medicinal Plants of China, volume one and two. Reference Publications, Inc.: United States of America.

4. Pan, W. D., Li, Y. J., Mai, L. T., Ohtani, K., Kasai, R. and Tanaka, O. 1992. *Yao Xue Xue Bao* 27: 515–521.

5. Pan, W. D., Li, Y. J., Mai, L. T., Ohtani, K. H., Kasai, R. T., Tanaka, O. and Yu, D. Q. 1993. *Yao Xue Xue Bao* 28: 40–44.

6. Pan, W. D., Mai, L. T., Li, Y. J., Xu, X. L. and Yu, D. Q. 1993. *Yao Xue Xue Bao* 28: 35–39.

7. Begum, S., Wahab, A., Siddiqui, B. S. and Qamar, F. 2000. *Journal of Natural Products* 63: 765–767.

8. Begum, S., Wahab, A. and Siddiqui, B. S. 2003. *Chemical and Pharmaceutical Bulletin (Tokyo)* 51: 134–137.

9. Begum, S., Wahab, A and Siddiqui, B. S. 2002. *Natural Product Letters* 16: 235–238.

10. Refahy, L. A. 2003. *Egyptian Journal of Chemistry* 46: 537–543.

11. Yadav, S. B. and Tripathi, V. 2003. *Fitoterapia* 74: 320–351.

12. Misra, L. and Laatsch, H. 2000. *Phytochemistry* 54: 969–974.

13. Barre, J. T., Bowden, B. F., Coll, J. C., DeJesus, J., De La Fuente, V. E., Janairo, G. C. and Ragasa, C. Y. 1997. *Phytochemistry* 45: 321–324.

14. Alitonou, G., Avlessi, F., Bokossa, I., Ahoussi, E., Dangou, J. and Sohounhloue, D. C. K.

2004. *Comptes Rendus Chimie* 7: 1101–1105.

15. Wee, Y. C. and Hsuan, K. 1990. An illustrated Dictionary of Chinese Medicinal Herbs. Times Edition and Eu Yan Seng Holdings Ltd: Singapore.

16. Tabuti, J. R. 2007. *Journal of Ethnopharmacology* 116: 33–42.

17. Avadhoot, Y., Dixit, V. K. and Varma, K. C. 1980. *Indian Drugs & Pharmaceuticals Industry* 15: 19–20.

18. Kasali, A, A., Ekundayo, O., Oyedeji, A. O., Adeniyi, B. A. and Adeolu, E. O. 2002. *Journal of Essential Oil-Bearing Plants* 5: 108–110.

19. Sukul, S. and Chaudhuri, S. 1999. *Journal of Phytological Research* 12: 119–121.

20. Hernandez, T., Canales, M., Avila, J. G., Duran, A., Caballero, J., Romo de Vivar, A. and Lira, R. 2003. *Journal of Ethnopharmacology* 88: 181–188.

21. Basu, S., Ghosh, A. and Hazra, B. 2005. *Phytotherapy Research* 19: 888–894.

22. Inada, A., Nakanishi, T., Tokuda, H., Nishino, H. and Sharma, O. P. 1997. *Planta Medica* 63: 272–274.

23. Sharma, M., Sharma, P. D., Bansal, M. P. and Singh J. 2007. *Chemistry and Biodiversity* 4: 932–939.

24. Fatope, M. O., Salihu, L., Asante, S. K. and Takeda, Y. 2002. *Pharmaceutical Biology (Lisse, Netherlands)* 40: 564–567.

25. Basu, S. and Hazra, B. 2006. *Phytotherapy Research* 20: 896–900.

26. Kothari, M. and Chaudhary, B. L. 2001. *Journal of Experimental Biology* 39: 1194–1198.

27. Kumar, V. P., Chauhan, N. S., Padh, H. and Rajani, M. 2006. *Journal of Ethnopharmacology* 107: 182–188.

28. Begum, S., Wahab, A. and Siddiqui, B. S. 2008. *Natural Product Research* 22: 467–470.

29. Sagar, L., Sehgal, R. and Ojha, S. 2005. *BMC Complementary and Alternative Medicine* 5: 18.

30. Saini, N., Singh, J., Sehgal, R., Ojha, S. 2007. *Cellular and Molecular Biology (Noisy-le-grand)* 53: 79–83.

31. O'Neill, M. J., Lewis, J. A., Noble, H. M., Holland, S., Mansat, C., Farthing, J. E., Foster, G., Noble, D., Lane, S. J., Sidebottom, P. J., Lynn, S. M., Hayes, M. V. and Dix, C. J. 1998. *Journal of Natural Products* 61: 1328–1331.

32. Inada, A., Nakanishi, T., Tokuda, H., Nishino, H., Iwashima A. and Sharma, O. P. 1995. *Planta Medica* 61: 558–559.

33. Misra, N., Sharma, M., Raj, K., Dangi, A., Srivastava, S. and Misra-Bhattacharya, S. 2007. *Parasitology Research* 100: 439–448.

34. Dua, V. K., Gupta, N. C., Pandey, A. C. and Sharma, V. P. 1996. *Journal of the American Mosquito Control Association* 12: 406–408.

35. Seyoum, A., Killeen, G. F., Kabiru, E. W., Knols, B. G. and Hassanali, A. 2003. *Tropical Medicine and International Health* 8: 1005–1011.

36. Bouda, H., Tapondjou, L. A., Fontem, D. A. and Gumedzoe, M. Y. 2001. *Journal of Stored Products Research* 37: 103–109.

37. Morallo-Rejesus, B. and Tantengco, G. B. 1986. *NSTA Technology Journal* 11: 37–46.

38. Pandey, U. K., Verma, G. S., Lekha, C. and Singh, A. K. 1982. *Indian Journal of Agricultural Sciences* 52: 205–206.

39. Verma, R. K. and Verma, S. K. 2006. *Fitoterapia* 77: 466–468.

40. Abdel-Hady, N. M., Abdei-Halim, A. S. and Al-Ghadban, A. M. 2005. *Journal of the Egyptian Society of Parasitology* 35: 687–698.

41. Garg, S. K., Shah, M. A., Garg, K. M., Farooqui, M. M. and Sabir, M. 1997. *Indian Journal of Experimental Biology* 35: 1315–1318.

42. de Mello, F. B., Jacobus, D., de Carvalho, K. C. and de Mello, J. R. 2003. *Veterinary and Human Toxicology* 45: 20–23.

43. Duke, J. A., Bogenschutz-Godwin, M. J., duCellier, J. and Duke, P. A. K. 2002. Handbook of medicinal herbs. CRC Press: Boca Raton, FL.

44. Johnson, J. H. and Jensen, J. M. 1998. *Journal of Zoo and Wildlife Medicine* 29: 203–207.

45. Ide, A. and Tutt, C. L. 1998. *Journal of the South African Veterinary Association* 69: 30–32.

46. Sharma, O. P., Dawra, R. K. and Makkar, H. P. 1989. *Veterinary and Human Toxicology* 31: 10–13.

47. Sharma, O. P., Dawra, R. K. and Makkar, H. P. S. 1982. *Research Communications in Chemical Pathology and Pharmacology* 38: 153–156.

48. Sharma, O. P., Vaid, J., Pattabhi, V. and Bhutani, K. K. 1992. *Journal of Biochemistry and Toxicology* 7: 73–79.

49. Sharma, O. P. and Dawra, R. K. 1984. *Chemico-Biological Interactions* 49: 369–374.

50. Sharma, O. P., Dawra, R. K., Krishna, L. and Makkar, H. P. S. 1988. *Veterinary and Human Toxicology* 30: 214–218.

51. Pass, M. A. and Goosem, M. W. 1983. *Toxicon* 3: 337–340.

52. Pass, M. A. and Heath, T. 1977. *Journal of Comparative Pathology* 87: 301–306.

53. Pass, M. A. 1986. *Australian Veterinary Journal* 63: 169–171.

54. Pass, M. A., Seawright, A. A., Lamberton, J. A. and Heath, T. J. 1979. *Pathology* 11: 89–94.

55. Ganai, G. N. and Jha, G. J. 1991. *Indian Journal of Experimental Biology* 29: 762–766.

56. Akhter, M. H., Mathur, M. and Bhide, N. K. 1990. *Indian Journal of Physiology and Pharmacology* 34: 13–16.

57. Uppal, R. P. and Paul, B. S. 1982. *Indian Veterinary Journal* 59: 18–24.

58. Sharma, O. P., Sharma, S., Pattabhi, V., Mahato, S. B. and Sharma, P. D. 2007. *Critical Reviews in Toxicology* 37: 313–352.

59. Mello, F. B., Jacobus, D., Carvalho, K. and Mello, J. R. 2005. *Toxicon* 45: 459–466.

Lonicera japonica Thunb.

1. Hsuan, K. 1990. The Concise Flora of Singapore, Gymnosperms and Dicotyledons. Singapore University Press: National University of Singapore.

2. eFloras.org 2008 http://www.efloras.org/florataxon.aspx?flora_id=5&taxon_id=200022324 Date Accessed: 30th March 2008

3. Wiart, C. 2000. Medicinal plants of Southeast Asia. Pelanduk Publications (M) Sdn Bhd: Malaysia.

4. Machida, K., Sasaki, H., Iijima, T. and Kikuchi, M. 2002. *Chemical and Pharmaceutical Bulletin (Tokyo)* 50: 1041–1044.

5. Kwak, W. J., Han, C. K., Chang, H. W., Kim, H. P., Kang, S. S. and Son, K. H. 2003. *Chemical and Pharmaceutical Bulletin (Tokyo)* 51: 333–335.

6. Son, K. H., Jung, K. Y., Chang, H. W., Kim, H. P. and Kang, S. S. 1994. *Phytochemistry* 35: 1005–1058.

7. Choi, C. W., Jung, H. A., Kang, S. S. and Choi, J. S. 2007. *Archives of Pharmacal Research* 30: 1–7.

8. Kumar, N., Singh, B., Bhandari, P., Gupta, A. P., Uniyal, S. K. and Kaul, V. K. 2005. *Phytochemistry* 66: 2740–2744.

9. Foster, S. and Hobbs, C. 2002. A Field Guide to Western Medicinal Plants and Herbs. Houghton Mifflin Company: New York.

10. Wee, Y. C. and Hsuan, K. 1990. An illustrated Dictionary of Chinese Medicinal Herbs. Times Edition and Eu Yan Seng Holdings Ltd: Singapore.

11. Wong-Leung, Y. L. 1981. *Xianggang Qinhui Xueyuan Xue Bao* 8: 115–124.

12. Phan, M. G., Nguyen, T. M. and Phan, T. S. 2002. *Tap Chi Hoa Hoc* 40: 103–107.

13. Yip, E. C., Chan, A. S., Pang, H., Tam, Y. K. and Wong, Y. H. 2006. *Cell Biology and Toxicology* 22: 293–302. Erratum in: *Cell Biology and Toxicology* 2007, 23: 139.

14. Leung, H. W., Kuo, C. L., Yang, W. H., Lin, C. H. and Lee, H. Z. 2006. *European Journal of Pharmacology* 534: 12–18.

15. Leung, H. W., Wu, C. H., Lin, C. H. and Lee, H. Z. 2005. *European Journal of Pharmacology* 508: 77–83.

16. Lee, H. K. and Chung, Y. S. 1963. *Kisul Yon'guso Pogo* 2: 76–78.

17. Cheng, J. T., Lee, Y. Y., Hsu, F. L., Chang, W. and Niu, C. S. 1994. *Chinese Pharmaceutical Journal (Taipei, Taiwan)* 46: 575–582.

18. Tae, J., Han, S. W., Yoo, J. Y., Kim, J. A., Kang, O. H., Baek, O. S., Lim, J. P., Kim, D. K., Kim, Y. H., Bae, K. H. and Lee, Y. M. 2003. *Clinica Chimica Acta* 330: 165–171.

19. Kwak, W. J., Han, C. K., Chang, H. W., Kim, H. P., Kang, S. S. and Son, K. H. 2003. *Chemical and Pharmaceutical Bulletin (Tokyo)* 51: 333–335.

20. Lee, S. J., Son, K. H., Chang, H. W., Kang, S. S. and Kim, H. P. 1998. *Phytotherapy Research* 12: 445–447.

21. Xu, Y., Oliverson, B. G. and Simmons, D. L. 2007. *Journal of Ethnopharmacology* 111: 667–670.
22. Son, M. J., Moon, T. C., Lee, E. K., Son, K. H., Kim, H. P., Kang, S. S., Son, J. K., Lee, S. H. and Chang, H. W. 2006. *Archives of Pharmacal Research* 29: 282–286.
23. Suh, S. J., Chung, T. W., Son, M. J., Kim, S. H., Moon, T. C., Son, K. H., Kim, H. P., Chang, H. W. and Kim, C. H. 2006. *Archives of Biochemistry and Biophysics* 447: 136–146.
24. Park, E., Kum, S., Wang, C., Park, S. Y., Kim, B. S. and Schuller-Levis, G. 2005. *American Journal of Chinese Medicine* 33: 415–424.
25. Kim, J. A., Kim, D. K., Kang, O. H., Choi, Y. A., Park, H. J., Choi, S. C., Kim, T. H., Yun, K. J., Nah, Y. H. and Lee, Y. M. 2005. *International Immunopharmacology* 5: 209–217.
26. Chang, W. C. and Hsu, F. L. 1992. *Prosta-glandins, Leukotrienes and Essential Fatty Acids* 45: 307–312.
27. Chang, C. W., Lin, M. T., Lee, S. S., Liu, K. C., Hsu, F. L. and Lin, J. Y. 1995. *Antiviral Research* 27: 367–374.
28. Jeong, C. S., Suh, I. O., Hyun, J. E. and Lee, E. B. 2003. *Natural Product Sciences* 9: 87–90.
29. Suh, S. J., Jin, U. H., Kim, S. H., Chang, H. W., Son, J. K., Lee, S. H., Son, K. H. and Kim, C. H. 2006. *Journal of Cellular Biochemistry* 99: 1298–1307.
30. Duke, J. A. and Ayensu, E. S. 1985. Medicinal Plants of China, volume one and two. Reference Publications, Inc.: United States of America.
31. Duke, J. A., Bogenschutz-Godwin, M. J., duCellier, J. and Duke, P. A. K. 2002. Handbook of medicinal herbs. CRC Press: Boca Raton, FL.
32. Nguyen, V. D. and Doan, T. N. 1989. Medicinal Plants in Vietnam. World Health Organisation, Regional Office for the Western Pacific, Manila, Institute of Materia Medica, Hanoi.
33. Pharmacopoeia of the People's Republic of China, volume 1. English Edition, 2000. Compiled by the State Pharmacopoeia Commission of People's Republic of China. Chemical Industry Press: Beijing.
34. Webster, R. M. 1993. *Cutis* 51: 424.
35. Thanabhorn, S., Jaijoy, K., Thamaree, S., Ingkaninan, K. and Panthong, A. 2006. *Journal of Ethnopharmacology* 107: 370–373.

Mangifera indica L.

1. Wiart, C. 2006. Medicinal Plants of the Asia-Pacific: Drugs for the future? World Scientific Publishing Co. Pte. Ltd.: Singapore.
2. USDA, ARS GRIN Taxonomy for Plants 2008 http://www.ars-grin.gov/cgi-bin/npgs/html/taxon.pl?23351 Date Accessed: 30th June 2008
3. Goh, S. H., Chuah, C. H., Mok, J. S. L. and Soepadmo, E. 1995. Malaysian Medicinal Plants for the Treatment of Cardiovascular Diseases. Pelanduk Publications (M) Sdn Bhd: Malaysia.
4. Duke, J. A. and Ayensu, E. S. 1985. Medicinal Plants of China, volume one and two. Reference Publications, Inc.: United States of America.
5. Nunez Selles, A. J., Velez Castro, H. T., Aguero-Aguero, J., Gonzalez-Gonzalez, J., Naddeo, F., De Simone, F. and Rastrelli, L. 2002. *Journal of Agriculture and Food Chemistry* 50: 762–766.
6. Rocha Ribeiro, S. M., Queiroz, J. H., Lopes Ribeiro de Queiroz, M. E., Campos, F. M. and Pinheiro Sant'ana, H. M. 2007. *Plant Foods for Human Nutrition* 62: 13–17.
7. Berardini, N., Fezer, R., Conrad, J., Beifuss, U., Carle, R. and Schieber, A. 2005. *Journal of Agriculture and Food Chemistry* 53: 1563–1570.
8. Tabuti, J. R. 2008. *Journal of Ethnopharma-cology* 116: 33–42.
9. Jain, S. K. and DeFilipps, R. A. 1991. Medicinal Plants of India, volume one and two. Reference Publications Inc.: United States of America.
10. Ojewole, J. A. 2005. *Methods and Findings in Experimental and Clinical Pharmacology* 27: 547–554.
11. Garcia, D., Escalante, M., Delgado, R., Ubeira, F. M. and Leiro, J. 2003. *Phytotherapy Research* 17: 1203–1208.
12. Nkuo-Akenji, T., Ndip, R., McThomas, A. and Fru, E. C. 2001. *Central African Journal of Medicine* 47: 155–158.
13. Sairam, K., Hemalatha, S., Kumar, A., Srinivasan, T., Ganesh, J., Shankar, M. and Venkataraman, S. 2003. *Journal of Ethnopharmacology* 84: 11–15.
14. Aqil, F. and Ahmad, I. 2007. *Methods and Findings in Experimental and Clinical Pharmacology* 29: 79–92.

15. Tona, L., Kambu, K., Ngimbi, N., Mesia, K., Penge, O., Lusakibanza, M., Cimanga, K., De Bruyne, T., Apers, S., Totte, J., Pieters, L. and Vlietinck, A. J. 2000. *Phytomedicine* 7: 31–38.

16. Cojocaru, M., Droby, S., Glotter, E., Goldman, A., Gottlieb, H. E., Jacoby, B. and Prusky, D. 1986. *Phytochemistry* 25: 1093–1095.

17. Garrido, G., Gonzalez, D., Lemus, Y., Garcia, D., Lodeiro, L., Quintero, G., Delporte, C., Nunez-Selles, A. J. and Delgado, R. 2004. *Pharmacology Research* 50: 143–149.

18. Sá-Nunes, A., Rogerio, A. P., Medeiros, A. I., Fabris, V. E., Andreu, G. P., Rivera, D. G., Delgado, R. and Faccioli, L. H. 2006. *International Immunopharmacology* 6: 1515–1523.

19. Garrido, G., González, D., Lemus, Y., Delporte, C. and Delgado, R. 2006. *Phytomedicine* 13: 412–418.

20. Muanza, D. N., Euler, K. L., Williams, L. and Newman, D. J. 1995. *International Journal of Pharmacognosy* 33: 98–106.

21. Ali, A. M., Mooi, L. Y., Yih, K. Y., Norhanom, A. W., Saleh, K. M., Lajis, N. H., Yazid, A. M., Ahmad, F. B. H. and Prasad, U. 2000. *Natural Product Sciences* 6: 147–150.

22. Mooi, L. Y., Ali, A. M., Norhanom, A. B., Salleh, K. M., Murakami, A. and Koshimizu, K. 1999. *Natural Product Sciences* 5: 33–38.

23. Percival, S. S., Talcott, S. T., Chin, S. T., Mallak, A. C., Lounds-Singleton, A. and Pettit-Moore, J. 2006. *Journal of Nutrition* 136: 1300–1304.

24. Martinez, G., Delgado, R., Perez, G., Garrido, G., Nunez Selles, A. J. and Leon, O. S. 2000. *Phytotherapy Research* 14: 424–427.

25. Martinez Sanchez, G., Candelario-Jalil, E., Giuliani, A., Leon, O. S., Sam, S., Delgado, R. and Nunez Selles, A. J. 2001. *Free Radical Research* 35: 465–473.

26. Sanchez, G. M., Re, L., Giuliani, A., Nunez-Selles, A. J., Davison, G. P. and Leon-Fernandez, O. S. 2000. *Pharmacological Research* 42: 565–573.

27. Pardo-Andreu, G. L., Sánchez-Baldoquín, C., Avila-González, R., Yamamoto, E. T., Revilla, A., Uyemura, S. A., Naal, Z., Delgado, R. and Curti, C. 2006. *Pharmacology Research* 54: 389–395.

28. Pardo-Andreu, G. L., Delgado, R., Núñez-Sellés, A. J., Vercesi, A. E. 2006. *Phytotherapy Research* 20: 120–124.

29. Pardo-Andreu, G. L., Philip, S. J., Riaño, A., Sánchez, C., Viada, C., Núñez-Sellés, A. J. and Delgado, R. 2006. *Archives of Medical Research* 37: 158–164.

30. Pardo-Andreu, G., Delgado, R., Velho, J., Inada, N. M., Curti, C. and Vercesi, A. E. 2005. *Pharmacology Research* 51: 427–435.

31. Pardo-Andreu, G. L., Paim, B. A., Castilho, R. F., Velho, J. A., Delgado, R., Vercesi, A. E. and Oliveira, H. C. 2008. *Pharmacology Research* 57: 332–338.

32. Pardo-Andreu, G. L., Barrios, M. F., Curti, C., Hernández, I., Merino, N., Lemus, Y., Martínez, I., Riaño, A. and Delgado, R. 2008. *Pharmacology Research* 57: 79–86.

33. Larrauri, J. A., Ruperez, P. and Saura-Calixto, F. 1997. *Zeitschrift fuer Lebensmittel-Untersuchung und — Forschung A: Food Research and Technology* 205: 39–42.

34. Ajila, C. M. and Prasada Rao, U. J. 2007. *Food and Chemical Toxicology* 46: 303–309.

35. Rodríguez, J., Di Pierro, D., Gioia, M., Monaco, S., Delgado, R., Coletta, M. and Marini, S. 2006. *Biochimica et Biophysica Acta* 1760: 1333–1342.

36. Prabhu, S., Jainu, M., Sabitha, K. E. and Devi, C. S. 2006. *Journal of Ethnopharmacology* 107: 126–133.

37. Remirez, D., Tafazoli, S., Delgado, R., Harandi, A. A. and O'Brien, P. J. 2005. *Drug Metabolism and Drug Interactions* 21: 19–29.

38. Bafna, P. A. and Balaraman, R. 2005. *Phytotherapy Research* 19: 216–221.

39. Yoosook, C., Bunyapraphatsara, N., Boonyakiat, Y. and Kantasuk, C. 2000. *Phytomedicine* 6: 411–419.

40. Zhu, X. M., Song, J. X., Huang, Z. Z., Wu, Y. M. and Yu, M. J. 1993. *Zhongguo Yao Li Xue Bao* 14: 452–454.

41. Carvalho, A. C., Guedes, M. M., de Souza, A. L., Trevisan, M. T., Lima, A. F., Santos, F. A. and Rao, V. S. 2007. *Planta Medica* 73: 1372–1376; Erratum in: *Planta Medica* 73: 1522.

42. Lima, Z. P., Severi, J. A., Pellizzon, C. H., Brito, A. R., Solis, P. N., Cáceres, A., Girón, L. M., Vilegas, W. and Hiruma-Lima, C. A.

2006. *Journal of Ethnopharmacology* 106: 29–37.

43. Prasad, S., Kalra, N. and Shukla, Y. 2007. *Molecular Nutrition and Food Research* 51: 352–359.
44. Sanchez, G. M., Rodriguez, H., MA, Giuliani, A., Nunez Selles, A. J., Rodriguez, N. P., Leon Fernandez, O. S. and Re, L. 2003. *Phytotherapy Research* 17: 197–201.
45. Prashanth, D., Amit, A., Samiulla, D. S., Asha, M. K. and Padmaja, R. 2001. *Fitoterapia* 72: 686–688.
46. Muruganandan, S., Srinivasan, K., Gupta, S., Gupta, P. K. and Lal, J. 2005. *Journal of Ethnopharmacology* 97: 497–501.
47. Aderibigbe, A. O., Emudianughe, T. S. and Lawal, B. A. 2001. *Phytotherapy Research* 15: 456–458.
48. Anila, L. and Vijayalakshmi, N. R. 2002. *Journal of Ethnopharmacology* 79: 81–87.
49. Garcia, D., Delgado, R., Ubeira, F. M. and Leiro. J. 2002. *International Immunopharmacology* 2: 797–806.
50. Mohammed, A. and Chadee, D. D. 2007. *Journal of the American Mosquito Control Association* 23: 172–176.
51. Jagetia, G. C. and Baliga, M. S. 2005. *Phytomedicine* 12: 209–215.
52. Jagetia, G. C. and Venkatesha, V. A. 2005. *Environmental and Molecular Mutagenesis* 46: 12–21.
53. Watt, G. 1972. Dictionary of the Economic Products of India, volume 1–6. Cosmo Publications: New Delhi.
54. Hegde, V. L. and Venkatesh, Y. P. 2007. *Journal of Investigational Allergology and Clinical Immunology* 17: 341–344.
55. Duke, J. A., Bogenschutz-Godwin, M. J., duCellier, J. and Duke, P. A. K. 2002. Handbook of Medicinal Herbs. CRC Press: Boca Raton, FL.
56. Rodeiro, I., Donato, M. T., Lahoz, A., González-Lavaut, J. A., Laguna, A., Castell, J. V., Delgado, R. and Gómez-Lechón, M. J. 2008. *Chemico-Biological Interactions* 172: 1–10.
57. Rodeiro, I., Donato, M. T., Jiménez, N., Garrido, G., Delgado, R. and Gómez-Lechón, M. J. 2007. *Food and Chemical Toxicology* 45: 2506–2512.

Manihot esculenta Crantz.

1. Hsuan, K. 1990. The Concise Flora of Singapore, Gymnosperms and Dicotyledons. Singapore University Press: National University of Singapore.
2. eFloras.org 2008 http://www.efloras.org/florataxon.aspx?flora_id=2&taxon_id=200012596 Date Accessed: 30th March 2008
3. Duke, J. A. 1985. CRC Handbook of Medicinal Herbs. CRC Press: Boca Raton, FL.
4. Sumner J. 2000. The Natural History of Medicinal Plants. Timber Press, Portland.
5. Jain, S. K. and DeFilipps, R. A. 1991. Medicinal Plants of India, volume one and two. Reference Publications Inc.: United States of America.
6. Gomez-Vasquez, R., Day, R., Buschmann, H., Randles, S., Beeching, J. R. and Cooper, R. M. 2004. *Annals of Botany (London)* 94: 87–97.
7. Chaturvedula, V. S., Schilling, J. K., Malone, S., Wisse, J. H., Werkhoven, M. C. and Kingston, D. G. 2003. *Planta Medica* 69: 271–274.
8. Padmaja, G. and Panikkar, K. R. 1989. *Indian Journal of Physiology and Pharmacology* 27: 635–639.
9. Boby, R. G. and Indira, M. 2004. *Indian Journal of Physiology and Pharmacology* 48: 41–50.
10. Gaitan, E., Cooksey, R. C., Legan, J., Lindsay, R. H., Ingbar, S. H. and Medeiros-Neto, G. 1994. *European Journal of Endocrinology* 131: 138–144.
11. Sreeja, V. G., Nagahara, N., Li, Q. and Minami, M. 2003. *British Journal of Nutrition* 90: 467–472.
12. Izokun-Etiobhio, B. O. and Ugochukwu, E. N. 1995. *Nigerian Journal of Physiological Sciences* 11: 46–49.
13. Akintonwa, A., Tunwashe, O. and Onifade, A. 1994. *Acta Horticulturae* 375: 285–288.
14. Kamalu, B. P. 1991. *British Journal of Nutrition* 65: 365–372.
15. Llorens, J. 2004. *Endocrinologia y Nutricion* 51: 418–425.

Melaleuca cajuputi Roxb.

1. Hsuan, K. 1990. The Concise Flora of Singapore, Gymnosperms and Dicotyledons.

Singapore University Press: National University of Singapore.

2. Wee, Y. C. 1992. A Guide to Medicinal Plants. Singapore Science Centre Publication: Singapore.

3. Wee, Y. C. and Corlett, R. 1986. The City and the Forest: Plant Life in Urban Singapore. Singapore University Press: National University of Singapore.

4. USDA, ARS GRIN Taxonomy for Plants 2008 http://www.ars-grin.gov/cgi-bin/npgs/html/taxon.pl?23778
Date Accessed: 30th June 2008

5. Li, T. S. C. 2000. Medicinal Plants. Culture, Utilisation and Phytopharmacology. Technomic Publishing Co., Inc.: Lancaster.

6. bin Din, L., Zakaria, Z., Hj. Abd Malek, S. N. and Samsudin, M. W. 1988. Proceedings: Malaysian Traditional Medicine, Kuala Lumpur, pp. 118–124. Institute of Advanced Studies, University of Malaya, Kuala Lumpur.

7. Nguyen, V. D. and Doan, T. N. 1989. Medicinal Plants in Vietnam. World Health Organisation, Regional Office for the Western Pacific, Manila, Institute of Materia Medica, Hanoi.

8. Fleming, T. (Ed.). 2000. PDR for Herbal Medicines, 2nd edition. Medical Economics Company: New Jersey.

Melastoma malabathricum L.

1. Hsuan, K. 1990. The Concise Flora of Singapore, Gymnosperms and Dicotyledons. Singapore University Press: National University of Singapore.

2. eFloras.org 2008 http://www.efloras.org/florataxon.aspx?flora_id=2&taxon_id=242413697
Date Accessed: 30th March 2008

3. Das, K. K. and Kotoky, J. 1988. *Journal of the Indian Chemical Society* 65: 385–386.

4. Dinda, B. and Saha, M. K. 1988. *Journal of the Indian Chemical Society* 65: 209–211.

5. Yoshida, T., Nakata, F., Hosotani, K., Nitta, A. and Okuda, T. 1992a. *Chemical and Pharmaceutical Bulletin* 40: 1727–1732.

6. Yoshida, T., Nakata, F., Hosotani, K., Nitta, A. and Okuda, T. 1992b. *Phytochemistry* 31: 2829–2833.

7. Yoshida, T., Ito, H. and Hipolito, I. J., 2005. *Phytochemistry* 66: 1972–1983.

8. Mohandoss, S. and Ravindran, P. 1993. *Fitoterapia* 64: 277–278.

9. Malek, S. N. H. A., Baek, S. H. and Asari, A. 2003. *ACGC Chemical Research Communications* 16: 28–33.

10. Wee, Y. C. 1992. A Guide to Medicinal Plants. Singapore Science Centre Publication: Singapore.

11. Jain, S. K. and DeFilipps, R. A. 1991. Medicinal Plants of India, volume one and two. Reference Publications Inc.: United States of America.

12. Sulaiman, M. R., Somchit, M. N., Israf, D. A., Ahmad, Z. and Moin, S. 2004. *Fitoterapia* 75: 667–672.

13. Zakaria, Z. A., Nor, R. N. S. R. M., Kumar, G. H., Ghani, Z. D. F. A., Sulaiman, M. R., Devi, G. R., Jais, A. M. M., Somchit, M. N. and Fatimah, C. A. 2006. *Canadian Journal of Physiology and Pharmacology* 84: 1291–1299.

14. Lohezic-Le Devehat, F., Bakhtiar, A., Bezivin, C., Amoros, M. and Boustie, J. 2002. *Fitoterapia* 73: 400–405.

15. Susanti, D., Sirat, H. M., Ahmad, F., Mat Ali, R., Aimi, N. and Kitajima, M. 2007. *Food Chemistry* 103: 710–716.

16. Mazura, M. P., Susanti, D. and Rasadah, M. A. 2007. *Pharmaceutical Biology* 45: 372–375.

17. Rengel, Z. 2004. *Biometals* 17: 669–689.

Mimosa pudica L.

1. Wee, Y. C. and Corlett, R. 1986. The City and the Forest: Plant Life in Urban Singapore. Singapore University Press: National University of Singapore.

2. Hsuan, K. 1990. The Concise Flora of Singapore, Gymnosperms and Dicotyledons. Singapore University Press: National University of Singapore.

3. Nguyen, V. D. 1993. Medicinal Plants of Vietnam, Cambodia and Laos. Nguyen Van Duong: Vietnam.

4. Wee, Y. C. 1992. A Guide to Medicinal Plants. Singapore Science Centre Publication: Singapore.

5. Kang, J. G., Shin, S. Y., Kim, M. J., Bajpai, V., Maheshwari, D. K. and Kang, S. C. 2004. *Journal of Antibiotics* 57: 726–731.
6. Yuan, K., Lü, J. L. and Yin, M. W. 2006. *Yao Xue Xue Bao* 41: 435–438.
7. Jain, S. K. and DeFilipps, R. A. 1991. Medicinal Plants of India, volume one and two. Reference Publications Inc.: United States of America.
8. Lans, C. 2007. *Journal of Ethnobiology and Ethnomedicine* 3: 13.
9. Molina, M., Contreras, C. M. and Tellez-Alcantara, P. 1999. *Phytomedicine* 6: 319–323.
10. Robinson, R. D., Williams, L. A., Lindo, J. F., Terry, S. I. and Mansingh, A. 1990. *West Indian Medical Journal* 39: 213–217.
11. Akinsinde, K. A. and Olukoya, D. K. 1995. *Journal of Diarrhoeal Diseases Research* 13: 127–129.
12. Ngo Bum, E., Dawack, D. L., Schmutz, M., Rakotonirina, A., Rakotonirina, S. V., Portet, C., Jeker, A., Olpe, H. R. and Herrling, P. 2004. *Fitoterapia* 75: 309–314.
13. Ganguly, M., Devi, N., Mahanta, R. and Borthakur, M. K. 2007. *Contraception* 76: 482–485.
14. Amalraj, T. and Ignacimuthu, S. 2002. *Fitoterapia* 73: 351–352.
15. Alsala, S. V. 2000. *Journal of Ecotoxicology & Environmental Monitoring* 10: 25–29.
16. Girish, K. S., Mohanakumari, H. P., Nagaraju, S., Vishwanath, B. S. and Kemparaju, K. 2004. *Fitoterapia* 75: 378–380.
17. Mahanta, M. and Mukherjee, A. K. 2001. *Journal of Ethnopharmacology* 75: 55–60.

Mirabilis jalapa L.

1. Hsuan, K. 1990. The Concise Flora of Singapore, Gymnosperms and Dicotyledons. Singapore University Press: National University of Singapore.
2. eFloras.org 2008 http://www.efloras.org/florataxon.aspx?flora_id=2&taxon_id=200007009 Date Accessed: 30th March 2008
3. Wee, Y. C. 1992. A Guide to Medicinal Plants. Singapore Science Centre Publication: Singapore.
4. Yang, S. W., Ubillas, R., McAlpine, J., Stafford, A., Ecker, D. M., Talbot, M. K. and

Rogers, B. 2001. *Journal of Natural Products* 64: 313–317.
5. Duke, J. A. and Ayensu, E. S. 1985. Medicinal Plants of China, volume one and two. Reference Publications, Inc.: United States of America.
6. Jain, S. K. and DeFilipps, R. A. 1991. Medicinal Plants of India, volume one and two. Reference Publications Inc.: United States of America.
7. Watt, J. M. and Breyer-Brandwijk, M. G. 1962. The Medicinal and Poisonous Plants of South Africa. E. and S. Livingstone: London.
8. Kusamba, C., Byamana, K. and Mbuyi, W. M. 1991. *Journal of Ethnopharmacology* 35: 197–199.
9. Michalet, S., Cartier, G., David, B., Mariotte, A. M., Dijoux-franca, M. G., Kaatz, G. W., Stavri, M. and Gibbons, S. 2007. *Bioorganic and Medicinal Chemistry Letters* 17: 1755–1758.
10. De Bolle, M. F. C., Terras, F. R. G., Cammue, B. P. A., Rees, S. B. and Broekaert, W. F. 1993. *Developments in Plant Pathology* 2: 433–436.
11. Wong, R. N., Ng, T. B., Chan, S. H., Dong, T. X. and Yeung, H. W. 1992. *Biochemistry International* 28: 585–593.
12. Morton, J. F. 1981. Atlas of Medicinal Plants of Middle America: Bahamas to Yucatan. C.C. Thomas Publishers Co.: Springfield, IL.

Morinda citrifolia L.

1. Wiart, C. 2006. Medicinal Plants of the Asia-Pacific: Drugs for the future? World Scientific Publishing Co. Pte. Ltd.: Singapore.
2. Hsuan, K. 1990. The Concise Flora of Singapore, Gymnosperms and Dicotyledons. Singapore University Press: National University of Singapore.
3. Goh, S. H. 1988. Proceedings: Malaysian Traditional Medicine, Kuala Lumpur, pp. 7–26. Institute of Advanced Studies, University of Malaya, Kuala Lumpur.
4. Goh, S. H., Chuah, C. H., Mok, J. S. L. and Soepadmo, E. 1995. Malaysian Medicinal Plants for the Treatment of Cardiovascular Diseases. Pelanduk Publications (M) Sdn Bhd: Malaysia.

5. Wee, Y. C. and Hsuan, K. 1990. An illustrated Dictionary of Chinese Medicinal Herbs. Times Edition and Eu Yan Seng Holdings Ltd: Singapore.

6. USDA, ARS GRIN Taxonomy for Plants 2008 http://www.ars-grin.gov/cgi-bin/npgs/html/taxon.pl?318237 Date Accessed: 30th June 2008

7. Nguyen, V. D. and Doan, T. N. 1989. Medicinal Plants in Vietnam. World Health Organisation, Regional Office for the Western Pacific, Manila, Institute of Materia Medica, Hanoi.

8. Sang, S. M., Liu, G. M., He, K., Zhu, N. Q., Dong, Z. G., Zheng, Q. Y., Rosen, R. T. and Ho, C. T. 2003. *Bioorganic & Medicinal Chemistry* 11: 2499–2502.

9. Farine, J. P., Legal, L., Moreteau, B. and Le Quere, J. 1996. *Phytochemistry* 41: 433–438.

10. Dyas, L., Threlfall, D. R. and Goad, L. J. 1994. *Phytochemistry* 35: 655–660.

11. Saludes, J. P., Garson, M. J., Franzblau, S. G. and Aguinaldo, A. M. 2002. *Phytotherapy Research* 16: 683–685.

12. Inoue, K., Nayeshiro, H., Inouyet, H. and Zenk, M. 1981. *Phytochemistry* 20: 1693–1700.

13. Liu, G., Bode, A., Ma, W. Y., Sang, S., Ho, C. T. and Dong, Z. 2001. *Cancer Research* 61: 5749–5756.

14. Su, B. N., Pawlus, A. D., Jung, H. A., Keller, W. J., McLaughlin, J. L. and Kinghorn, A. D. 2005. *Journal of Natural Products* 68: 592–595.

15. Kamiya, K., Tanaka, Y., Endang, H., Umar, M. and Satake, T. 2005. *Chemical & Pharmaceutical Bulletin* 53: 1597–1599.

16. Pawlus, A. D., Su, B. N., Keller, W. J. and Kinghorn, A. D. 2005. *Journal of Natural Products* 68: 1720–1722.

17. Siddiqui, B. S., Sattar, F. A., Begum, S., Gulzar, T. and Ahmad F. 2006. *Natural Product Research* 20: 1136–1144.

18. Dalsgaard, P. W., Potterat, O., Dieterle, F., Paululat, T., Kühn, T. and Hamburger, M. 2006. *Planta Medica* 72: 1322–1327.

19. Bui, A. K. T., Bacic, A. and Pettolino, F. 2006. *Phytochemistry* 67: 1271–1275.

20. Samoylenko, V., Zhao, J., Dunbar, D. C., Khan, I, A., Rushing, J. W. and Muhammad, I. 2006. *Journal of Agriculture and Food Chemistry* 54: 6398–6402.

21. Takashima, J., Ikeda, A. Y., Komiyama, A. K., Hayashi, M., Kishida, A., Ohsaki, A. 2007. *Chemical & Pharmaceutical Bulletin* 55: 343–345.

22. Siddiqui, B. S., Sattar, F. A., Ahmad, F. and Begum, S. 2007. *Archieves of Pharmacal Research* 30: 919–923.

23. Siddiqui, B. S., Sattar, F. A., Begum, S., Gulzar, T. and Ahmad, F. 2007. *Archieves of Pharmacal Research* 30: 793–798.

24. Akihisa, T., Matsumoto, K., Tokuda, H. and Yasukawa, K., Seino, K., Nakamoto, K., Kuninaga, H., Suzuki, T. and Kimura, Y. 2007. *Journal of Natural Products* 70: 754–757.

25. Deng, S., West, B. J., Palu, A.K., Zhou, B. N. and Jensen, C. J. 2007. *Phytomedicine* 14: 517–522.

26. Kamiya, K., Hamabe, W., Harada, S., Murakami, R., Tokuyama, S. and Satake, T. 2008. *Biological and Pharmaceutical Bulletin* 31: 935–938.

27. Deng, Y., Chin, Y. W., Chai, H., Keller, W. J. and Kinghorn, A. D. 2007. *Journal of Natural Products* 70: 2049–2052.

28. Lin, C. F., Ni, C. L., Huang, Y. L., Sheu, S. J. and Chen, C. C. 2007. *Natural Product Research* 21: 1199–1204.

29. Jain, S. K. and DeFilipps, R. A. 1991. Medicinal Plants of India, volume one and two. Reference Publications Inc.: United States of America.

30. Wang, M.Y., West, B. J., Jensen, J. C, Nowicki, D., Su, C., Palu, A. K. and Anderson, G. 2002. *Acta Pharmacologica Sinica* 23: 1127–1141.

31. Muhammad, M. M., Abdul, M. A., Mohd, A. A., Rozita, M. N., Nashriyah, B. M., Abdul, R. R. and Kazuyoshi, K. 1997. *Pesticide Science* 51: 165–170.

32. Hornick, C. A., Myers, A., Sadowska-Krowicka H., Anthony, C. T. and Woltering, E. A. 2003. *Angiogenesis* 6: 143–149.

33. Zaidan, M. R. S., Noor Rain, A., Badrul, A. R., Adlin, A., Norazah, A. and Zakiah, I. 2005. *Tropical Biomedicine* 22: 165–170.

34. Hirazumi, A., Furusawa, E., Chou, S. C. and Hokama, Y. 1994. *Proceedings of the Western Pharmacology Society* 37: 145–146.

35. Furusawa, E., Hirazumi, A., Story, S. and Jensen, J. 2003. *Phytotherapy Research* 17: 1158–1164.

36. Arpornsuwan, T. and Punjanon, T. 2006. *Phytotherapy Research* 20: 515–517.
37. Hirazumi, A. and Furusawa, E. 1999. *Phytotherapy Research* 13: 380–387.
38. Li, R. W., Myers, S. P., Leach, D. N., Lin, G. D. and Leach, G. 2003. *Journal of Ethnopharmacology* 85: 25–32.
39. Mohd Zin, Z., Abdul Hamid, A. and Osman, A. 2002. *Food Chemistry* 78: 227–231.
40. Mohd Zin, Z., Abdul Hamid, A., Osman, A. and Saari, N. 2006. *Food Chemistry* 94: 169–178.
41. Calzuola, I., Gianfranceschi, G. L. and Marsili, V. 2006. *International Journal of Food Sciences and Nutrition* 57: 168–177.
42. Wang, M. Y. and Su, C. 2001. *Annals of the New York Academy of Sciences* 952: 161–168.
43. Yang, J., Paulino, R., Janke-Stedronsky, S. and Abawi, F. 2007. *Food Chemistry* 102: 302–308.
44. Owen, P. L., Matainaho, T., Sirois, M. and Johns, T. 2007. *Journal of Biochemical and Molecular Toxicology* 21: 231–242.
45. Ancolio, C., Azas, N., Mahiou, V., Ollivier, E., Di Giorgio, C., Keita, A., Timon-David, P. and Balansard, G. 2002. *Phytotherapy Research* 16: 646–649.
46. Nayak, B. S., Isitor, G. N., Maxwell, A., Bhogadi, V. and Ramdath, D. D. 2007. *Journal of Wound Care* 16: 83–86.
47. Wang, M. Y., Nowicki, D., Anderson, G., Jensen, J. and West, B. 2008. *Plant Foods for Human Nutrition* 63: 59–63.
48. Skidmore-Roth, L. 2001. Mosby's Handbook of Herbs and Natural Supplements. Mosby, Inc.: St. Louis.
49. Mueller, B. A., Scott, M. K., Sowinski, K. M. and Prag, K. A. 2000. *American Journal of Kidney Diseases* 35: 310–312.
50. Younos, C., Rolland, A., Fleurentin, J., Lanhers, M. C., Misslin, R. and Mortier, F. 1990. *Planta Medica* 56: 430–434.
51. Westendorf, J., Effenberger, K., Iznaguen, H. and Basar, S. 2007. *Journal of Agriculture and Food Chemistry* 55: 529–537.
52. Stadlbauer, V., Fickert, P., Lackner, C., Schmerlaib, J., Krisper, P., Trauner, M. and Stauber, R. E. 2005. *World Journal of Gastroenterology* 11: 4758–4760.
53. Millonig, G., Stadlmann, S. and Vogel, W. 2005. *European Journal of Gastroenterology and Hepatology* 17: 445–447.
54. Yuce, B., Gulberg, V., Diebold, J. and Gerbes, A. L. 2006. *Digestion* 73: 167–170.

Nelumbo nucifera Gaertn.

1. Hsuan, K. 1990. The Concise Flora of Singapore, Gymnosperms and Dicotyledons. Singapore University Press: National University of Singapore.
2. USDA, ARS GRIN Taxonomy for Plants 2008 http://www.ars-grin.gov/cgi-bin/npgs/html/taxon.pl?25110 Date Accessed: 30th June 2008
3. Fleming, T. (Ed.). 2000. PDR for Herbal Medicines, 2nd edition. Medical Economics Company: New Jersey.
4. Saeed, A., Omer, E. and Hashem, A. 1993. *Bulletin of the Faculty of Pharmacy (Cairo University)* 31: 347–351.
5. Ohkoshi, E., Miyazaki, H., Shindo, K., Watanabe, H., Yoshida, A. and Yajima, H. 2007. *Planta Medica* 73: 1255–1259.
6. Xu, H. D. and Jian, Q. J. 2008. *Chinese Chemical Letters* 19: 308–310.
7. Xiao, H., Chen, A. H., Ji, A. M., Li, Z. L. and Song, X. D. 2007. *Zhong Yao Cai* 30: 289–291.
8. Hyun, S. K., Jung, Y. J., Chung, H. Y., Jung, H. A. and Choi, J. S. 2006. *Archives of Pharmacal Research* 29: 287–292.
9. Wee, Y. C. and Hsuan, K. 1990. An illustrated Dictionary of Chinese Medicinal Herbs. Times Edition and Eu Yan Seng Holdings Ltd: Singapore.
10. Duke, J. A. and Ayensu, E. S. 1985. Medicinal Plants of China, volume one and two. Reference Publications, Inc.: United States of America.
11. Jain, S. K. and DeFilipps, R. A. 1991. Medicinal Plants of India, volume one and two. Reference Publications Inc.: United States of America.
12. Mukherjee, P. K., Saha, K., Balasubramanian, R., Pal, M. and Saha, B. P. 1996. *Journal of Ethnopharmacology* 54: 63–67.
13. Li, G. R., Qian, J. Q. and Lu, F. H. 1990. *Zhongguo Yao Li Xue Bao* 11: 158–161.
14. Shoji, N., Umeyama, A., Saito, N., Iuchi, A., Takemoto, T., Kajiwara, A. and Ohizumi, Y. 1987. *Journal of Natural Products* 50: 773–774.

15. Talukder, M. J. and Nessa, J. 1998. *Bangladesh Medical Research Council Bulletin* 24: 6–9.

16. Mukherjee, P. K., Saha, K., Das, J., Pal, M. and Saha, B. P. 1997. *Planta Medica* 63: 367–369.

17. Wu, M. J., Wang, L., Weng, C. Y. and Yen, J. H. 2003. *American Journal of Chinese Medicine* 31: 687–698.

18. Sohn, D. H., Kim, Y. C., Oh, S. H., Park, E. J., Li, X. and Lee, B. H. 2003. *Phytomedicine* 10: 165–169.

19. Rai, S., Wahile, A., Mukherjee, K., Saha, B. P. and Mukherjee, P. K. 2006. *Journal of Ethnopharmacology* 104: 322–327.

20. Ling, Z. Q., Xie, B. J. and Yang, E. L. 2005. *Journal of Agriculture and Food Chemistry* 53: 2441–2445.

21. Yu, J. and Hu, W. S. 1997. *Yao Xue Xue Bao* 32: 1–4.

22. Xiao, J. H., Zhang, J. H., Chen, H. L., Feng, X. L. and Wang, J. L. 2005. *Planta Medica* 71: 225–230.

23. Mukherjee, P. K., Das, J., Saha, K., Giri, S. N., Pal, M. and Saha, B. P. 1996. *Indian Journal of Experimental Biology* 34: 275–276.

24. Sinha, S., Mukherjee, P. K., Mukherjee, K., Pal, M., Mandal, S. C. and Saha, B. P. 2000. *Phytotherapy Research* 14: 272–274.

25 Kashiwada, Y., Aoshima, A., Ikeshiro, Y., Chen, Y. P., Furukawa, H., Itoigawa, M., Fujioka, T., Mihashi, K., Cosentino, L. M., Morris-Natschke, S. L. and Lee, K. H. 2005. *Bioorganic & Medicinal Chemistry* 13: 443–448.

26. Kuo, Y. C., Lin, Y. L., Liu, C. P. and Tsai, W. J. 2005. *Journal of Biomedical Sciences* 12: 1021–1034.

27. Mukherjee, P. K., Saha, K., Pal, M. and Saha, B. P. 1997. *Journal of Ethnopharmacology* 58: 207–213.

28. Lee, M. W., Kim, J. S., Cho, S. M., Kim, J. H. and Lee, J. S. 2001. *Natural Product Sciences* 7: 107–109.

29. la Cour, B., Molgaard, P. and Yi, Z. 1995. *Journal of Ethnopharmacology* 46: 125–129.

30. Liu, C. P., Tsai, W. J., Shen, C. C., Lin, Y. L., Liao, J. F., Chen, C. F. and Kuo, Y. C. 2006. *European Journal of Pharmacology* 531: 270–279.

31. Nguyen, V. D. and Doan, T. N. 1989. *Medicinal Plants in Vietnam*. World Health Organisation, Regional Office for the Western Pacific, Manila, Institute of Materia Medica, Hanoi.

32. van Wyk, B. E. and Wink, M. 2004. *Medicinal plants of the world: an illustrated scientific guide to important medicinal plants and their uses*. Timber Press: Portland.

33. McGuffin, M., Hobbs, C., Upton, R. and Goldberg, A. (Eds). 1997. *American Herbal Products Association's Botanical Safety Handbook*. CRC Press: Boca Raton, FL.

Nephelium lappaceum L.

1. Hsuan, K. 1992. *Orders and Families of Malayan Seed Plants*. Singapore University Press: Singapore.

2. Wee, Y. C. 1992. *A Guide to Medicinal Plants*. Singapore Science Centre Publication: Singapore.

3. eFloras.org 2008 http://www.efloras.org/florataxon.aspx?flora_id=2&taxon_id=200013210 Date Accessed: 30th March 2008

4. Nishizawa, M., Adachi, K., Sastrapradja, S. and Hayashi, Y. 1983. *Phytochemistry* 22: 2853–2855.

5. Wong, K. C., Wong, S. N., Loi, H. K. and Lim, C. L. 1996. *Flavour and Fragrance Journal* 11: 223–229.

6. Ragasa, C. Y., de Luna, R. D., Cruz, W. C. Jr. and Rideout, J. A. 2005. *Journal of Natural Products* 68: 1394–1396.

7. Avato, P., Rosito, I., Papadia, P. and Fanizzi, F. P. 2006. *Natural Product Communications* 1: 751–755.

8. Duke, J. A. and Ayensu, E. S. 1985. *Medicinal Plants of China, volume one and two*. Reference Publications, Inc.: United States of America.

9. Nawawi, A., Nakamura, N., Hattori, M., Kurokawa, M. and Shiraki, K. 1999. *Phytotherapy Research* 13: 37–41.

10. Thitilertdecha, N., Teerawutgulrag, A. and Rakariyatham N. 2008. *LWT-Food Science and Technology* 41: 2029–2035.

11. Wall, M. M. 2006. *Journal of Food Composition and Analysis* 19: 655–663.

12. Palanisamy, U., Cheng, H. M., Masilamani, T., Subramaniam, T., Ling, L. T. and

Radhakrishnan, A. K. 2008. *Food Chemistry* 109: 54–63.
13. Okonogi, S., Duangrat, C., Anuchpreeda, S., Tachakittirungrod, S. and Chowwanapoonpohn, S. 2007. *Food Chemistry* 103: 839–846.

Nerium oleander L.

1. eFloras.org 2008
   http://www.efloras.org/florataxon.aspx?flora_id=2&taxon_id=200018424
   Date Accessed: 30ᵗʰ March 2008
2. Wiart, C. 2006. Medicinal Plants of the Asia-Pacific: Drugs for the future? World Scientific Publishing Co. Pte. Ltd.: Singapore.
3. Fleming, T. (Ed.). 2000. PDR for Herbal Medicines, 2ⁿᵈ edition. Medical Economics Company: New Jersey.
4. Siddiqui, B. S., Begum, S., Siddiqui, S. and Lichter, W. 1995. *Phytochemistry* 39: 171–174.
5. Begum, S., Siddiqui, B. S., Sultana, R., Zia, A. and Suria, A. 1999. *Phytochemistry* 50: 435–438.
6. Huq, M. M., Jabbar, A., Rashid, M. A., Hasan, C. M., Ito, C. and Furukawa, H. 1999. *Journal of Natural Products* 62: 1065–1067.
7. Mueller, B. M., Rosskopf, F., Paper, D. H., Kraus, J. and Franz, G. 1991. *Pharmazie* 46: 657–663.
8. Zhao, M., Bai, L., Wang, L., Toki, A., Hasegawa, T., Kikuchi, M., Abe, M., Sakai, J., Hasegawa, R., Bai, Y., Mitsui, T., Ogura, H., Kataoka, T., Oka, S., Tsushima, H., Kiuchi, M., Hirose, K., Tomida, A., Tsuruo, T. and Ando, M. 2007. *Journal of Natural Products* 70: 1098–1103.
9. Bai, L., Wang, L., Zhao, M., Toki, A., Hasegawa, T., Ogura, H., Kataoka, T., Hirose, K., Sakai, J., Bai, J. and Ando, M. 2007. *Journal of Natural Products* 70: 14–18.
10. Zhao, M., Zhang, S., Fu, L., Li, N., Bai, J., Sakai, J., Wang, L., Tang, W., Hasegawa, T., Ogura, H., Kataoka, T., Oka, S., Kiuch, M., Hirose, K. and Ando, M. 2006. *Journal of Natural Products* 69: 1164–1167.
11. Fu, L., Zhang, S., Li, N., Wang, J., Zhao, M., Sakai, J., Hasegawa, T., Mitsui, T., Kataoka, T., Oka, S., Kiuchi, M., Hirose, K. and Ando. M. 2005. *Journal of Natural Products* 68: 198–206.

12. Passalacqua, N. G., De Fine, G. and Guarrera, P. M. 2006. *Journal of Ethnobiology and Ethnomedicine* 11: 2:52.
13. Tahraoui, A., El-Hilaly, J., Israili, Z. H. and Lyoussi, B. 2007. *Journal of Ethnopharmacology* 110: 105–117.
14. Jain, S. K. and DeFilipps, R. A. 1991. Medicinal Plants of India, volume one and two. Reference Publications Inc.: United States of America.
15. Erdemoglu, N., Kupeli, E. and Yesilada, E. 2003. *Journal of Ethnopharmacology* 89: 123–129.
16. Khafagi, I. K. 2001. *Egyptian Journal of Microbiology* 34: 613–627.
17. Afaq, F., Saleem, M., Aziz, M. H. and Mukhtar, H. 2004. *Toxicology and Applied Pharmacology* 195: 361–369.
18. Smith, J. A., Madden, T., Vijjeswarapu, M. and Newman, R. A. 2001. *Biochemical Pharmacology* 62: 469–472.
19. Pathak, S., Multani, A. S., Narayan, S., Kumar, V. and Newman, R. A. 2000. *Anticancer Drugs* 11: 455–463.
20. Newman, R. A., Kondo, Y., Yokoyama, T., Dixon, S., Cartwright, C., Chan, D., Johansen, M. and Yang, P. 2007. *Integrative Cancer Therapies* 6: 354–364.
21. Turan, N., Akgün-Dar, K., Kuruca, S. E., Kiliçaslan-Ayna, T., Seyhan, V. G., Atasever, B., Meriçli, F. and Carin, M. 2006. *Journal of Experimental Therapeutics and Oncology* 6: 31–38.
22. Mekhail, T., Kaur, H., Ganapathi, R., Budd, G. T., Elson, P. and Bukowski, R. M. 2006. *Investigational New Drugs* 24: 423–427.
23. Male, O. and Holubar, K. 1967. *Proceedings of the 5ᵗʰ International Congress of Chemotherapy* 2: 73–78.
24. Siddiqui, B. S., Sultana, R., Begum, S., Zia, A. and Suria, A. 1997. *Journal of Natural Product* 60: 540–544.
25. Zia, A., Siddiqui, B. S., Begum, S., Siddiqui, S. and Suria, A. 1995. *Journal of Ethnopharmacology* 49: 33–39.
26. Tarkowska, J. A. 1971. *Hereditas (Lund, Sweden)* 67: 205–211.
27. el-Shazly, M. M., Nassar, M. I. and el-Sherief, H. 1996. *Journal of the Egyptian Society of Parasitology* 26: 461–473.

28. Pushpalatha, E. and Muthukrishnan, J. 1995. *Indian Journal of Malariology* 32: 14–23.
29. Adome, R. O., Gachihi, J. W., Onegi, B., Tamale, J. and Apio, S. O. 2003. *African Health Sciences* 3: b77–86.
30. Murad, T. M. C., Gazzinelli, N. M., Sarsur Neto, J. M. and Murad, J. E. 1980. *Ciencia e Cultura (Sao Paulo)* 32: 172–179.
31. Sreenivasan, Y., Sarkar, A. and Manna, S. K. 2003. *Biochemical Pharmacology* 66: 2223–2239.
32. Skidmore-Roth L. 2001. Mosby's handbook of herbs and natural supplements. Mosby, Inc.: St. Louis.
33. Gupta, A., Joshi, P., Jortani, S. A., Valdes, R. Jr., Thorkelsson, T., Verjee, Z. and Shemie, S. 1997. *Therapeutic Drug Monitoring* 19: 711–714.
34. Nishioka, S. A. and Resende, E. S. 1995. *Revista da Associação Médica Brasileira* 41: 60–62.
35. Haynes, B. E., Bessen, H. A. and Wightman, W. D. 1985. *Annals of Emergency Medicines* 14: 350–353.
36. Arriola Martínez, P., Montero Aparicio, E., Martínez Odriozola, P. and Miguel de la Villa, F. 2006. *Medicina Clinica* 127: 759.
37. Bourgeois, B., Incagnoli, P., Hanna, J. and Tirard, V. 2005. *Annales françaises d'anesthèsie et de reanimation* 24: 640–642.
38. Pietsch, J., Oertel, R., Trautmann, S., Schulz, K., Kopp, B. and Dressler, J. A. 2005. *International Journal of Legal Medicine* 119: 236–240.
39. Wojtyna, W. and Enseleit, F. 2004. *Pacing and Clinical Electrophysiology* 27: 1686–1688.
40. Kirsch, M. 1997. *Tierarztl Prax* 25: 398–400.
41. Mahin, L., Marzou, A. and Huart, A. 1984. *Veterinary and Human Toxicology* 26: 303–304.
42. De Pinto, F., Palermo, D., Milillo, M. A. and Iaffaldano, D. 1981. *Clinica Veterinaria* 104: 15–18.
43. Haiba, M. H., Mohamed, A. I., Mehdi, A. W. R. and Nair, G. A. 2003. *Pollution Research* 22: 157–161.
44. Aslani, M. R., Movassaghi, A. R., Mohri, M., Abbasian, A. and Zarehpour, M. 2004. *Veterinary Research Communications* 28: 609–616.
45. Adam, S. E., Al-Yahya, M. A. and Al-Farhan, A. H. 2001. *American Journal of Chinese Medicine* 29: 525–532.
46. Al-Yahya, M. A., Al-Farhan, A. H. and Adam, S. E. 2000. *Fitoterapia* 71: 385–391.
47. Al-Yahya, M. A., Al-Farhan, A. H. and Adam, S. E. 2002. *American Journal of Chinese Medicine* 30: 579–587.
48. Barbosa, R. R., Fontenele-Neto, J. D. and Soto-Blanco, B. 2008. *Research in Veterinary Science* 85: 279–281.
49. Fetrow, C. W. and Avila, J. R. 2000. The Complete Guide to Herbal Medicines. Springhouse Corp: Springhouse, PA.

Ophiopogon japonicus Ker-Gawl.

1. Hsuan, K. 1992. Orders and Families of Malayan Seed Plants. Singapore University Press: Singapore.
2. eFloras.org 2008 http://www.efloras.org/florataxon.aspx?flora_id=2&taxon_id=200027794 Date Accessed: 30th March 2008
3. Duke, J. A. and Ayensu, E. S. 1985. Medicinal Plants of China, volume one and two. Reference Publications, Inc.: United States of America.
4. Chang, J. M., Shen, C. C., Huang, Y. L., Chien, M. Y., Ou, J. C., Shieh, B. J. and Chen, C. C. 2002. *Journal of Natural Products* 65: 1731–1733.
5. Nguyen, T. H. A., Tran, V. S. and Wessjohann, L. 2003. *Tap Chi Hoa Hoc* 41: 117–121.
6. Nguyen, T. H. A., Tran, V. S. and Wessjohann, L. 2003. *Tap Chi Hoa Hoc* 41: 136–142.
7. Tada, A., Saitoh, T. and Shoji, J. 1980. *Chemical & Pharmaceutical Bulletin* 28: 2487–2493.
8. Cheng, Z. H., Wu, T. and Yu, B. Y. 2006. *Journal of Asian Natural Product Research* 8: 555–559.
9. Dai, H. F. and Mei, W. L. 2005. *Archieves of Pharmacal Research* 28: 1236–1238.
10. Wee, Y. C. and Hsuan, K. 1990. An illustrated Dictionary of Chinese Medicinal Herbs. Times Edition and Eu Yan Seng Holdings Ltd: Singapore.
11. Chen, M., Yang, Z. W., Zhu, J. T., Xiao, Z. Y. and Xiao, R. 1990. *Zhongguo Yao Li Xue Bao* 11: 161–165.
12. Shibata, M., Noguchi, R., Suzuki, M., Iwase, H., Soeda, K., Niwayama, K., Kataoka, E.,

Kusuda, K. and Hamano, M. 1971. *Hoshi Yakka Daigaku Kiyo* 13: 66–76.

13. Kou, J., Sun, Y., Lin, Y., Cheng, Z., Zheng, W., Yu, B. and Xu, Q. 2005. *Biological and Pharmaceutical Bulletin* 28: 1234–1238.

14. Kou, J., Tian, Y., Tang, Y., Yan, J. and Yu, B. 2006. *Biological and Pharmaceutical Bulletin* 29: 1267–1270.

15. Kou, J., Yu, B. and Xu, Q. 2005. *Vascular Pharmacology* 43: 157–163.

16. Wang, Y., Yan, T., Shen, J., Guo, H. and Xiang, X. 2007. *Journal of Ethnopharmacology* 114: 246–253.

17. Yu, B., Yin, X., Zhang, C. and Xu, G. 1991. *Zhongguo Yaoke Daxue Xuebao* 22: 286–288.

18. Wu, X., Dai, H., Huang, L., Gao, X., Tsim, K. W. and Tu, P. 2006. *Journal of Natural Products* 69: 1257–1260.

19. Tao, J., Wang, H., Chen, J., Xu, H. and Li, S. 2005. *American Journal of Chinese Medicine* 33: 797–806.

20. Tao, J., Wang, H., Zhou, H. and Li, S. 2005. *Life Sciences* 77: 3021–3030.

21. Liu, J. Q. and Wu, D. W. 1993. *Zhongguo Zhong Xi Yi Jie He Za Zhi* 13: 150–152, 132.

22. Nguyen, V. D. and Doan, T. N. 1989. Medicinal Plants in Vietnam. World Health Organisation, Regional Office for the Western Pacific, Manila, Institute of Materia Medica, Hanoi.

Peltophorum pterocarpum Backer ex K. Heyne.

1. Hsuan, K. 1990. The Concise Flora of Singapore, Gymnosperms and Dicotyledons. Singapore University Press: National University of Singapore.

2. Tan, H. T. W. and Morgany, T. 2001. A Guide to Growing the Native Plants of Singapore. Singapore Science Centre Publication: Singapore.

3. eFloras.org 2008
http://www.efloras.org/florataxon.aspx?flora_id=5&taxon_id=200012271
Date Accessed: 30th March 2008

4. Menon, P. S., Gangabai, G., Swarnalakshmi, T., Sulochana, N. and Amala, B. 1982. *Indian Drugs* 19: 345–347.

5. Voravuthikunchai, S., Lortheeranuwat, A., Jeeju, W., Sririrak, T., Phongpaichit, S. and Supawita, T. 2004. *Journal of Ethnopharmacology* 94: 49–54.

6. Voravuthikunchai, S. P. and Limsuwan, S. 2006. *Journal of Food Protection* 69: 2336–2341.

7. Duraipandiyan, V., Ayyanar, M. and Ignacimuthu, S. 2006. *BMC Complementary and Alternative Medicine* 6: 35.

Persicaria hydropiper L.

1. Fleming, T. (Ed.). 2000. PDR for Herbal Medicines, 2nd edition. Medical Economics Company: New Jersey.

2. USDA, ARS GRIN Taxonomy for Plants 2008
http://www.ars-grin.gov/cgi-bin/npgs/html/taxon.pl?400981
Date Accessed: 30th June 2008

3. Fukuyama, Y., Sato, T., Asakawa, Y. and Takemoto, T. 1980. *Phytochemistry* 21: 2895–2898.

4. Fukuyama, Y., Sato, T., Miura, I., Asakawa, Y. and Takemoto, T. 1983. *Phytochemistry* 22: 549–552.

5. Miyazawa, M. and Tamura, N. 2007. *Biological and Pharmaceutical Bulletin* 30: 595–597.

6. Jain, S. K. and DeFilipps, R. A. 1991. Medicinal Plants of India, volume one and two. Reference Publications Inc.: United States of America.

7. Duke, J. A. and Ayensu, E. S. 1985. Medicinal Plants of China, volume one and two. Reference Publications, Inc.: United States of America.

8. Rahman, E., Goni, S. A., Rahman, M. T. and Ahmed, M. 2002. *Fitoterapia* 73: 704–706.

9. Chaudhuri, P. S., Chaudhuri, D., Nanda, D. K., Achari, B., Bhattacharya, D. and Saha, C. 1996. *Indian Journal of Experimental Biology* 34: 277–278.

10. Kapoor, M., Garg, S. K. and Mathur, V. S. 1974. *Indian Journal of Medical Research* 62: 1225–1227.

11. Vohora, S. B., Garg, S. K. and Chaudhury, R. R. 1969. *Indian Journal of Medical Research* 57: 893–899.

12. Haraguchi, H., Hashimoto, K. and Yagi, A. 1992. *Journal of Agricultural and Food Chemistry* 40: 1349–1351.

13. Fujita, K., Fujita, T. and Kubo, I. 2000. Molecular design of anti-*Saccharomyces cerevisiae* agents and their modes of action. Book of Abstracts, 219th ACS National Meeting, San Francisco, CA, March 26–30.
14. Fujita, K. and Kubo, I. 2005. *Journal of Agricultural and Food Chemistry* 53: 5187–5191.
15. Matsumoto, T. and Tokuda, H. 1990. *Basic Life Sciences* 52: 423–427.
16. Fujimoto, T., Ose, Y., Sato, T., Matsuda, H., Nagase, H. and Kito, H. 1987. *Mutation Research* 178: 211–216.
17. Zhang, Z. and Fang, Y. 2001. *Kunchong Zhishi* 38: 207–210.
18. Zhang, Z., Liu, X., Lou, Z., Li, H., Zhu, S. and Zou, F. 1993. *Kunchong Xuebao* 36: 172–176.
19. Cribb, J. W. and Cribb, A. B. 1981. Wild Plants in Australia. Collins: Sydney.
20. Kuroiwa, K., Shibutani, M., Inoue, K., Lee, K. Y., Woo, G. H. and Hirose, M. 2006. *Food and Chemical Toxicology* 44: 1236–1244.
21. Rahman, I., Gogoi, I., Dolui, A. K. and Handique, R. 2005. *Journal of Environmental Biology* 26: 239–241.

Phyllanthus amarus Schum. & Thonn.

1. eFloras.org 2008 http://www.efloras.org/florataxon.aspx?flora_id=2&taxon_id=242337368 Date Accessed: 30th March 2008
2. Palaniswamy, U. R. 2003. A guide to Medicinal Plants of Asian origin and culture. CPL Press: United States of America.
3. USDA, ARS GRIN Taxonomy for Plants 2008 http://www.ars-grin.gov/cgi-bin/npgs/html/taxon.pl?401814 Date Accessed: 30th June 2008
4. Ahmad, B. and Alam, T. 2003. *Indian Journal of Chemistry, Section B: Organic Chemistry Including Medicinal Chemistry* 42B: 1786–1790.
5. Wee, Y. C. 1992. A Guide to Medicinal Plants. Singapore Science Centre Publication: Singapore.
6. Foo, L. Y. 1993. *Phytochemistry* 33: 487–491.
7. Foo, L. Y. 1995. *Phytochemistry* 39: 217–224.
8. Foo, L. Y. and Wong, H. 1992. *Phytochemistry* 31: 711–713.
9. Houghton, P. J., Woldemariam, T. Z., O'Shea, S. and Thyagarajan, S. P. 1996. *Phytochemistry* 43: 715–717.
10. Rajakanan, V., Sripathi, M. S., Selvanayagam, S., Velmurugan, D., Murthy, U. M., Vishwas, M., Thyagarajan, S. P., Raj, S. S. S. and Fun, H. K. 2003. *Acta Crystallographica* 59: 203–205.
11. Ali, H., Houghton, P. J. and Soumyanath, A. 2006. *Journal of Ethnopharmacology* 107: 449–455.
12. Fleming, T. (Ed.). 2000. PDR for Herbal Medicines, 2nd edition. Medical Economics Company: New Jersey.
13. Adjobimey, T., Edaye, I., Lagnika, L., Gbenou, J., Moudachirou, M. and Sanni, A. 2004. *Comptes Rendus Chimie* 7: 1023–1027.
14. Santos, A. R., De Campos, R. O., Miguel, O. G., Filho, V. C., Siani, A. C., Yunes, R. A. and Calixto, J. B. 2000. *Journal of Ethnopharmacology* 72: 229–238.
15. Okigbo, R. N. and Igwe, D. I. 2007. *Acta microbiologica et immunologica Hungarica* 54: 353–366.
16. Mazumder, A., Mahato, A. and Mazumder, R. 2006. *Natural Product Research* 20: 323–326.
17. Odetola, A. A. and Akojenu, S. M. 2000. *African Journal of Medicine and Medical Sciences* 29: 119–122.
18. Rao, M. V. and Alice, K. M. 2001. *Phytotherapy Research* 15: 265–267.
19. Agrawal, A., Srivastava, S., Srivastava, J. N. and Srivasa, M. M. 2004. *Biomedical and Environmental Sciences* 17: 359–365.
20. Kassuya, C. A. L., Silvestre, A. A., Rehder, V. L. G. and Calixto, J. B. 2003. *European Journal of Pharmacology* 478: 145–153.
21. Kiemer, A. K., Hartung, T., Huber, C. and Vollmar, A. M. 2003. *Journal of Hepatology* 38: 289–297.
22. Raphael, K. R. and Kuttan, R. 2003. *Journal of Ethnoparmacology* 87: 193–197.
23. Kassuya, C. A., Silvestre, A., Menezes-de-Lima, O. Jr., Marotta, D. M., Rehder, V. L. and Calixto, J. B. 2006. *European Journal of Pharmacology* 546: 182–188.
24. Kassuya, C. A., Leite, D. F., de Melo, L. V., Rehder, V. L. and Calixto, J. B. 2005. *Planta Medica* 71: 721–726.

25. Gowrishanker, B. and Vivekanandan, O. S. 1994. *Mutation Research* 322: 185–192.

26. Rajeshkumar, N. V., Joy, K. L., Kuttan, G., Ramsewak, R. S., Nair, M. G. and Kuttan, R. 2002. *Journal of Ethnoparmacology* 81: 17–22.

27. Sripanidkulchai, B., Tattawasart, U., Laupatarakasem, P., Vinitketkumneun, U., Sripanidkulchai, K., Furihata, C. and Matsushima, T. 2002. *Phytomedicine* 9: 26–32.

28. Joy, K. L. and Kuttan, R. 1998. *Journal of Clinical Biochemistry and Nutrition* 24: 133–139.

29. Leite, D. F., Kassuya, C. A., Mazzuco, T. L., Silvestre, A., de Melo, L. V., Rehder, V. L., Rumjanek, V. M. and Calixto, J. B. 2006. *Planta Medica* 72: 1353–1358.

30. Raphael, K. R., Sabu, M. C. and Kuttan, R. 2002. *Indian Journal of Experimental Biology* 40: 905–909.

31. Kumaran, A. and Karunakaran, R. J. 2005. *LWT — Food Science and Technology* 40: 344–352.

32. Lim, Y. Y. and Murtijaya, J. 2007. *LWT — Food Science and Technology* 40: 1664–1669.

33. Traoré, M., Diallo, A., Nikièma, J. B., Tinto, H., Dakuyo, Z. P., Ouédraogo, J. B., Guissou, I. P. and Guiguemdé, T. R. 2008. *Phytotherapy Research* 22: 550–551.

34. Notka, F., Meier, G. R. and Wagner, R. 2003. *Antiviral Research* 58: 175–186.

35. Notka, F., Meier, G. R. and Wagner, R. 2004. *Antiviral Research* 64: 93–102.

36. Yeh, S. F., Hong, C. Y., Huang, Y. L., Liu, T. Y., Choo, K. B. and Chou, C. K. 1993. *Antiviral Research* 20: 185–192.

37. Dhiman, R. K. and Chawla, Y. K. 2005. *Digestive Diseases and Sciences* 50: 1807–1812.

38. Jayaram, S. and Thyagarajan, S. P. 1996. *Indian Journal of Pathology and Microbiology* 39: 211–215.

39. Huang, R. L., Huang, Y. L., Ou, J. C., Chen, C. C., Hsu, F. L. and Chang, C. 2003. *Phytotherapy Research* 17: 449–453.

40. Wright, C. I., Van-Buren, L., Kroner, C. I. and Koning, M. M. 2007. *Journal of Ethnoparmacology* 114: 1–31.

41. Pramyothin, P., Ngamtin, C., Poungshompoo, S. and Chaichantipyuth, C. 2007. *Journal of Ethnoparmacology* 114: 169–173.

42. Naaz, F., Javed, S. and Abdin, M. Z. 2007. *Journal of Ethnoparmacology* 113: 503–509.

43. Lee, C. Y., Peng, W. H., Cheng, H. Y., Chen, F. N., Lai, M. T., Chiu, T. H. 2006. *American Journal of Chinese Medicine* 34: 471–482.

44. Adeneye, A. A., Amole, O. O. and Adeneye, A. K. 2006. *Fitoterapia* 77: 511–514.

45. Raphael, K. R., Ajith, T. A., Joseph, S. and Kuttan, R. 2002. *Teratogenesis, Carcinogenesis, and Mutagenesis* 22: 285–291.

46. Khanna, S., Srivastava, C. N., Srivasa, M. M. and Srivastava, S. 2003. *Journal of Environmental Biology* 24: 391–394.

47. Harikumar, K. B. and Kuttan, R. 2007. *Journal of Radiation Research (Tokyo)* 48: 469–476.

48. Adedapo, A. A., Adegbayibi, A. Y. and Emikpe, B. O. 2005. *Phytotherapy Research* 19: 971–976.

49. Hari Kumar, K. B. and Kuttan, R. 2006. *Biological and Pharmaceutical Bulletin* 29: 1310–1313.

Piper nigrum L.

1. Palaniswamy, U. R. 2003. A guide to Medicinal Plants of Asian origin and culture. CPL Press: United States of America.

2. eFloras.org 2008 http://www.efloras.org/florataxon.aspx?flora_id=2&taxon_id=200005581 Date Accessed: 30th March 2008

3. Fleming, T. (Ed.). 2000. PDR for Herbal Medicines, 2nd edition. Medical Economics Company: New Jersey.

4. Duke, J. A. and Ayensu, E. S. 1985. Medicinal Plants of China, volume one and two. Reference Publications, Inc.: United States of America.

5. Siddiqui, B. S., Gulzar, T. and Begum, S. 2002. *Heterocycles* 57: 1653–1658.

6. Siddiqui, B. S., Gulzar, T., Mahmood, A., Begum, S., Khan, B. and Afshan, F. 2004. *Chemical and Pharmaceutical Bulletin* 52: 1349–1352.

7. Li, T. S. C. 2000. Medicinal Plants. Culture, Utilisation and Phytopharmacology. Technomic Publishing Co., Inc.: Lancaster.

8. Tsukamoto, S., Tomise, K., Miyakawa, K., Cha, B. C., Abe, T., Hamada, T., Hirota, H.

and Ohta, T. 2002. *Bioorganic and Medicinal Chemistry* 10: 2981–2985.
9. Rasheed, M., Afshan, F., Tariq, R. M., Siddiqui, B. S., Gulzar, T., Mahmood, A., Begum, S. and Khan, B. 2005. *Natural Product Research* 19: 703–712.
10. Wei, K., Li, W., Koike, K., Chen, Y. and Nikaido, T. 2005. *Journal of Organic Chemistry* 70: 1164–1176.
11. Wee, Y. C. 1992. A Guide to Medicinal Plants. Singapore Science Centre Publication: Singapore.
12. Tiwari, K. C., Majumder, R. and Bhattacharjee, S. 1982. *International Journal of Crude Drug Research* 20: 133–137.
13. Wee, Y. C. and Hsuan, K. 1990. An illustrated Dictionary of Chinese Medicinal Herbs. Times Edition and Eu Yan Seng Holdings Ltd: Singapore.
14. Perez, C. and Anesini, C. 1994. *The American Journal of Chinese Medicine* 22: 169–174.
15. Pradhan, K. J., Variyar, P. S. and Bandekar, J. R. 1999. *Lebensmittel-Wissenschaft and — Technologie* 32: 121–123.
16. Chaudhry, N. M. and Tariq, P. 2006. *Pakistan Journal of Pharmaceutical Sciences* 19: 214–218.
17. Kaleem, M., Sheema, Sarmad, H. and Bano, B. 2005. *Indian Journal of Physiology and Pharmacology* 49: 65–71.
18. Hector, R. J., Simon, J. E., Ramboatiana, M. M. R., Behra, O., Garvey, A. S. and Raskin, I. 2004. *Acta Horticulturae* 629: 77–81.
19. Mujumdar, A. M., Dhuley, J. N., Deshmukh, V. K., Raman, P. H. and Naik, S. R. 1990. *Journal of Medical Science and Biology* 43: 95–100.
20. Pradeep, C. R. and Kuttan, G. 2002. *Clinical and Experimental Metastasis* 19: 703–708.
21. Vijayakumar, R. S. and Nalini, N. 2006. *Cell Biochemistry and Function* 24: 491–498.
22. Vijayakumar, R. S., Surya, D. and Nalini, N. 2004. *Redox Report: Communications in Free Radical Research* 9: 105–110.
23. Natarajan, K. S., Narasimhan, M., Shanmugasundaram, K. R. and Shanmugasundaram, E. R. 2006. *Journal of Ethnopharmacology* 105: 76–83.
24. Gulcin, I. 2005. *International Journal of Food Sciences and Nutrition.* 56: 491–499.
25. Topal, U., Sasaki, M., Goto, M. and Otles, S. 2007. *International Journal of Food Science and Nutrition* 18: 1–16.
26. Chatterjee, S., Niaz, Z., Gautam, S., Adhikari, S., Variyar, P. S. and Sharma, A. 2007. *Food Chemistry* 101: 515–523.
27. Nalini, N., Sabitha, K., Viswanathan, P. and Menon, V. P. 1998. *Journal of Ethnopharmacology* 62: 15–24.
28. Koul, I. B. and Kapil, A. 1993. *Planta Medica* 59: 413–417.
29. Singh, A. and Rao, A. R. 1993. *Cancer Letters* 72: 5–9.
30. Vijayakumar, R. S. and Nalini, N. 2006. *Journal of Basic and Clinical Physiology and Pharmacology* 17: 71–86.
31. Vijayakumar, R. S., Surya, D., Senthilkumar, R. and Nalini, N. 2002. *Journal of Clinical Biochemistry and Nutrition* 32: 31–42.
32. El Hamss, R., Idaomar, M., Alonso-Moraga, A. and Munoz Serrano, A. 2003. *Food and Chemical Toxicology* 41: 41–47.
33. Panda, S. and Kar, A. 2003. *Hormone and Metabolic Research* 35: 523–526.
34. Park, I. K., Lee, S. G., Shin, S. C., Park, J. D. and Ahn, Y. J. 2002. *Journal of Agricultural and Food Chemistry* 50: 1866–1870.
35. Simas, N. K., da Costa Lima, E., Kuster, R. M. and Salgueir, C. L. 2007. *Revista da Sociedade Brasileira de Medicina Tropical* 40: 405–407.
36. Siddiqui, B. S., Gulzar, T., Begum, S. and Afshan, F. 2004. *Natural Product Research* 18: 473–477.
37. Siddiqui, B. S., Gulzar, T., Begum, S., Afshan, F. and Sattar, F. A. 2005. *Natural Product Research* 19: 143–150.
38. Scott, I. M., Gagnon, N., Lesage, L., Philogène, B. J. and Arnason, J. T. 2005. *Journal of Economic Entomology* 98: 845–855.
39. Duke, J. A., Bogenschutz-Godwin, M. J., duCellier, J. and Duke, P. A. K. 2002. Handbook of medicinal herbs. CRC Press: Boca Raton, FL.
40. Rinzler, C. A. 1990. The complete book of herbs, spices and condiments. Facts on File: New York.
41. Nguyen, V. D. and Doan, T. N. 1989. Medicinal Plants in Vietnam. World Health Organisation, Regional Office for the Western

Pacific, Manila, Institute of Materia Medica, Hanoi.

42. McGuffin, M., Hobbs, C., Upton, R. and Goldberg, A. (Eds). 1997. *American Herbal Products Association's Botanical Safety Handbook.* CRC Press: Boca Raton, FL.

43. Skidmore-Roth L. 2001. Mosby's handbook of herbs and natural supplements. Mosby, Inc.: St. Louis.

44. Bhat, B. G. and Chandrasekhara, N. 1986. *Toxicology* 40: 83–92.

45. Cha, B. C. 2003. *Saengyak Hakhoechi* 34: 86–90.

46. Tsukamoto, S., Cha, B. C. and Ohta, T. 2002. *Tetrahedron* 58: 1667–1671.

47. Subehan, Usia, T., Iwata, H., Kadota, S. and Tezuka, Y. 2006. *Journal of Ethnopharmacology* 105: 449–455.

48. Usia, T., Iwata, H., Hiratsuka, A., Watabe, T., Kadota, S. and Tezuka, Y. 2006. *Phytomedicine* 13: 67–73.

49. Subehan, Usia, T., Kadota, S. and Tezuka, Y. 2006. *Planta Medica* 72: 527–532.

50. Hiwale, A. R., Dhuley, J. N. and Naik, S. R. 2002. *Indian Journal of Experimental Biology* 40: 277–281.

51. Pattanaik, S., Hota, D., Prabhakar, S., Kharbanda, P. and Pandhi, P. 2006. *Phytotherapy Research* 20: 683–686.

52. Hu, Z., Yang, X., Ho, P. C., Chan, S. Y., Heng, P. W., Chan, E., Duan, W., Koh, H. L. and Zhou, S. 2005. *Drugs* 65: 1239–1282.

53. Fetrow, C. W. and Avila, J. R. 2000. The complete guide to herbal medicines. Springhouse Corp: Springhouse, PA.

Piper sarmentosum Roxb.

1. Hsuan, K. 1992. Orders and Families of Malayan Seed Plants. Singapore University Press: Singapore.

2. Hsuan, K. 1990. The Concise Flora of Singapore, Gymnosperms and Dicotyledons. Singapore University Press: National University of Singapore.

3. Wee, Y. C. 1992. A Guide to Medicinal Plants. Singapore Science Centre Publication: Singapore.

4. Wee, Y. C. and Hsuan, K. 1990. An illustrated Dictionary of Chinese Medicinal Herbs.

Times Edition and Eu Yan Seng Holdings Ltd: Singapore.

5. eFloras.org 2008 http://www.efloras.org/florataxon.aspx?flora_id=2&taxon_id=200005594 Date Accessed: 30th March 2008

6. Toong, Y. V. and Wong, B. L. 1988. Proceedings: Malaysian Traditional Medicine, Kuala Lumpur, pp. 280. Institute of Advanced Studies, University of Malaya, Kuala Lumpur.

7. Strunz, G. M. and Finlay, H. 1995. *Phytochemistry* 39: 731–733.

8. Masuda, T., Inazumi, A., Yamada, Y., Padolina, W. G., Kikuzaki, H. and Nakatani, N. 1991. *Phytochemistry* 30: 3227–3228.

9. Stohr, J. R., Xiao, P. G. and Bauer, R. 1999. *Planta Medica* 65: 175–177.

10. Tuntiwachwuttikul, P., Phansa, P., Pootaeng-On, Y. and Taylor, W. C. 2006. *Chemical and Pharmaceutical Bulletin* 54: 149–151.

11. Rukachaisirikul, T., Siriwattanakit, P., Sukcharoenphol, K., Wongvein, C., Ruttanaweang, P., Wongwattanavuch, P. and Suksamrarn, A. 2004. *Journal of Ethnopharmacology* 93: 173–176.

12. Duke, J. A. and Ayensu, E. S. 1985. Medicinal Plants of China, volume one and two. Reference Publications, Inc.: United States of America.

13. Zaidan, M. R. S., Noor Rain, A., Badrul, A. R., Adlin, A., Norazah, A. and Zakiah, I. 2005. *Tropical Biomedicine* 22: 165–170.

14. Rahman, N. N. N. A., Furuta, T., Kojima, S., Takane, K. and Ali Mohd, M. 1999. *Journal of Ethnopharmacology* 64: 249–254.

15. Sawangjaroen, N., Sawangjaroen, K. and Poonpanang, P. 2004. *Journal of Ethnopharmacology* 91: 357–360.

16. Ridtitid, W., Rattanaprom, W., Thaina, P., Chittrakarn, S. and Sunbhanich, M. 1998. *Journal of Ethnopharmacology* 61: 135–142.

17. Peungvicha, P., Thirawarapan, S. S., Temsiririrkkul, R., Watanabe, H., Kumar Prasain, J. and Kadota, S. 1998. *Journal of Ethnopharmacology* 60: 27–32.

18. Chaithong, U., Choochote, W., Kamsuk, K., Jitpakdi, A., Tippawangkosol, P., Chaiyasit, D., Champakaew, D., Tuetun, B. and Pitasawat B. 2006. *Journal of Vector Ecology: Journal of the Society for Vector Ecology* 31: 138–144.

19. Choochote, W., Chaithong, U., Kamsuk, K., Rattanachanpichai, E., Jitpakdi, A., Tippawangkosol, P., Chaiyasit, D., Champakaew, D., Tuetun, B. and Pitasawat, B. 2006. *Revista do Instituto de Medicina Tropical de São Paulo* 48: 33–37.
20. Chanwitheesuk, A., Teerawutgulrag, A. and Rakariyatham, N. 2005. *Food Chemistry* 92: 491–497.

Plantago major L.

1. Hsuan, K. 1992. Orders and Families of Malayan Seed Plants. Singapore University Press: Singapore.
2. eFloras.org 2008 http://www.efloras.org/florataxon.aspx?flora_id=5&taxon_id=200022050 Date Accessed: 30ᵗʰ March 2008
3. Wee, Y. C. and Hsuan, K. 1990. An illustrated Dictionary of Chinese Medicinal Herbs. Times Edition and Eu Yan Seng Holdings Ltd: Singapore.
4. Ringbom, T., Segura, L., Noreen, Y., Perera, P. and Bohlin, L. 1998. *Journal of Natural Products* 61: 1212–1215.
5. Foster, S. and Hobbs, C. 2002. A Field Guide to Western Medicinal Plants and Herbs. Houghton Mifflin Company: New York.
6. Franca, F., Lago, E. L. and Marsden, P. D. 1996. *Revista da Sociedade Brasileira de Medicina Tropical* 29: 229–232.
7. Yesilada, E., Sezik, E., Fujita, T., Tanaka, S. and Tabata, M. 1993. *Phytotherapy Research* 7: 263–265.
8. Blumenthal, M., Goldberg, A. and Gruenwald, J. (Eds.). 2000. Herbal Medicine. Expanded Commission E Monographs. Integrative Medicine Communications: Austin, Texas.
9. Atta, A. H. and El-Sooud, K. A. 2004. *Journal of Ethnopharmacology* 95: 235–238.
10. Hetland, G., Samuelsen, A. B., Lovik, M., Paulsen, B. S., Aaberge, I. S., Groeng, E. C. and Michaelsen, T. E. 2000. *Scandinavian Journal of Immunology* 52: 348–355.
11. Velasco-Lezama, R., Tapia-Aguilar, R., Román-Ramos, R., Vega-Avila, E., Pérez-Gutiérrez, Ma. S. 2006. *Journal of Ethnopharmacology* 103: 36–42.

12. Atta, A. H. and Mouneir, S. M. 2005. *Phytotherapy Research* 19: 481–485.
13. Murai, M., Tamayama, Y. and Nishibe, S. 1995. *Planta Medica* 61: 479–480.
14. Shipochliev, T., Dimitrov, A. and Aleksandrova, E. 1981. *Veterinarno-meditsinski nauki* 18: 87–94.
15. Ikawati, Z., Wahyuono, S. and Maeyama, K. 2001. *Journal of Ethnopharmacology* 75: 249–256.
16. Galvez, M., Martin-Cordero, C., Lopez-Lazaro, M., Cortes, F. and Ayuso, M. J. 2003. *Journal of Ethnopharmacology* 88: 125–130.
17. Lin, L. T., Liu, L. T., Chiang, L. C. and Lin, C. C. 2002. *Phytotherapy Research* 16: 440–444.
18. Ozaslan, M., Didem Karagöz, I., Kalender, M. E., Kilic, I. H., Sari, I. and Karagöz, A. 2007. *American Journal of Chinese Medicine* 35: 841–851.
19. Campos, A. M. and Lissi, E. A. 1995. *Boletin de la Sociedad Chilena de Quimica* 40: 375–381.
20. Ponce-Macotela, M., Navarro-Alegria, I., Martinez-Gordillo, M. N. and Alvarez-Chacon, R. 1994. *Revista de Investigación Clínica* 46: 343–347.
21. Weenen, H., Nkunya, M. H. H., Bray, D. H., Mwasumbi, L. B., Kinabo, L. S., Kilimali, V. A. E. B. 1990. *Planta Medica* 56: 368–370.
22. Chiang, L. C., Chiang, W., Chang, M. Y., Ng, L. T., Lin, C. C. 2002. *Antiviral Research* 55: 53–62.
23. Basaran, A. A., Ceritoglu, I., Undeger, U. and Basaran, N. 1997. *Phytotherapy Research* 11: 609–611.
24. Dorhoi, A., Dobrean, V., Zăhan, M. and Virag, P. 2006. Modulatory effects of several herbal extracts on avian peripheral blood cell immune responses. *Phytotherapy Research* 20: 352–358.
25. Chiang, L. C., Ng, L. T., Chiang, W., Chang, M. Y. and Lin, C. C. 2003. *Planta Medica* 69: 600–604.
26. Gomez-Flores, R., Calderon, C. L., Scheibel, L. W., Tamez-Guerra, P., Rodriguez-Padilla, C., Tamez-Guerra, R. and Weber, R. J. 2000. *Phytotherapy Research* 14: 617–622.
27. Lim-Sylianco, C. Y. and Shier, W. T. 1985. *Journal of Toxicology-Toxin Reviews* 4: 71–105.

28. Basaran, A. A., Yu, T. W., Plewa, M. J. and Anderson, D. 1996. *Teratogenesis, Carcinogenesis and Mutagenesis* 16: 125–138.
29. Shipochliev, T. 1981. *Veterinarno-meditsinski nauki* 18: 94–98.
30. Krasnov, M. S., Margasiuk, D. V., Iamskov, I. A. and Iamskova, V. P. 2003. *Radiatsionnaia biologiia, radioecologiia* 43: 269–272.
31. Nguyen, V. D. and Doan, T. N. 1989. Medicinal Plants in Vietnam. World Health Organisation, Regional Office for the Western Pacific, Manila, Institute of Materia Medica, Hanoi.
32. Skidmore-Roth L. 2001. Mosby's handbook of herbs and natural supplements. Mosby, Inc.: St. Louis.
33. Schmeda-Hirschmann, G., Loyola, J. I., Retamal, S. R. and Rodriguez, J. 1992. *Phytotherapy Research* 6: 184–188.
34. Angelov, A., Lambev, I., Markov, M., Yakimova, K., Leseva, M. and Yakimov, A. 1980. *Medical Archives* 18: 47–52.

Punica granatum L.

1. Hsuan, K. 1992. Orders and Families of Malayan Seed Plants. Singapore University Press: Singapore.
2. USDA, ARS GRIN Taxonomy for Plants 2008 http://www.ars-grin.gov/cgi-bin/npgs/html/taxon.pl?30372
Date Accessed: 30th June 2008
3. Duke, J. A. and Ayensu, E. S. 1985. Medicinal Plants of China, volume one and two. Reference Publications, Inc.: United States of America.
4. Goh, S. H., Chuah, C. H., Mok, J. S. L. and Soepadmo, E. 1995. Malaysian Medicinal Plants for the Treatment of Cardiovascular Diseases. Pelanduk Publications (M) Sdn Bhd: Malaysia.
5. Fleming, T. (Ed.). 2000. PDR for Herbal Medicines, 2nd edition. Medical Economics Company: New Jersey.
6. Wang, R., Wang, W., Wang, L., Liu, R., Ding, Y. and Du, L. 2006. *Fitoterapia* 77: 534–537.
7. Wang, R. F., Xie, W. D., Zhang, Z., Xing, D. M., Ding, Y., Wang, W., Ma, C. and Du, L. J. 2004. *Journal of Natural Products* 67: 2096–2098.
8. Wee, Y. C. and Hsuan, K. 1990. An illustrated Dictionary of Chinese Medicinal Herbs. Times Edition and Eu Yan Seng Holdings Ltd: Singapore.
9. Wren, R. C. 1988. Potter's new cyclopaedia of botanical drugs and preparations. The C.W. Daniel Company Ltd.: Saffron Walden, Essex.
10. Raj, R. K. 1975. *Indian Journal of Physiology and Pharmacology* 19, Issue 1.
11. Voravuthikunchai, S., Lortheeranuwat, A., Jeeju, W., Sririrak, T., Phongpaichit, S. and Supawita, T. 2004. *Journal of Ethnopharmacology* 94: 49–54.
12. Braga, L. C., Shupp, J. W., Cummings, C., Jett, M., Takahashi, J. A., Carmo, L. S., Chartone-Souza, E. and Nascimento, A. M. 2005. *Journal of Ethnopharmacology* 96: 335–339.
13. Guevara, J. M., Chumpitaz, J. and Valencia, E. 1994. *Revista de gastroenterología del Perú* 14: 27–31.
14. Aqil, F. and Ahmad, I. 2007. *Methods and Findings in Experimental and Clinical Pharmacology* 29: 79–92.
15. Aqil, F., Khan, M. S., Owais, M. and Ahmad, I. 2005. *Journal of Basic Microbiology* 45: 106–114.
16. Braga, L. C., Leite, A. A., Xavier, K. G., Takahashi, J. A., Bemquerer, M. P., Chartone-Souza, E. and Nascimento, A. M. 2005. *Canadian Journal of Microbiology* 51: 541–547.
17. Voravuthikunchai, S. P. and Limsuwan, S. 2006. *Journal of Food Protection* 69: 2336–2341.
18. Voravuthikunchai, S. P. and Kitpipit, L. 2005. *Clinical Microbiology and Infection* 11: 510–512.
19. Naz, S., Siddiqi, R., Ahmad, S., Rasool, S. A. and Sayeed, S. A. 2007. *Journal of Food Science* 72: M341–345.
20. Mathabe, M. C., Nikolova, R. V., Lall, N. and Nyazema, N. Z. 2006. *Journal of Ethnopharmacology* 105: 286–293.
21. Alanís, A. D., Calzada, F., Cervantes, J. A., Torres, J. and Ceballos, G. M. 2005. *Journal of Ethnopharmacology* 100: 153–157.
22. Mehta, R. and Lansky, E. P. 2004. *European Journal of Cancer Prevention* 13: 345–348.
23. Kim, N. D., Mehta, R., Yu, W., Neeman, I., Livney, T., Amichay, A., Poirier, D., Nicholls, P.,

Kirby, A., Jiang, W., Mansel, R., Ramachandran, C., Rabi, T., Kaplan, B. and Lansky, E. 2002. *Breast Cancer Research and Treatment* 71: 203–217.

24. Kawaii, S. and Lansky, E. P. 2004. *Journal of Medicinal Food* 7: 13–18.

25. Kohno, H., Suzuki, R., Yasui, Y., Hosokawa, M., Miyashita, K. and Tanaka, T. 2004. *Cancer Science* 95: 481–486.

26. Lansky, E. P., Jiang, W., Mo, H., Bravo, L., Froom, P., Yu, W., Harris, N. M., Neeman, I. and Campbell, M. J. 2005. *Investigational New Drugs* 23: 11–20.

27. Afaq, F., Saleem, M., Krueger, C. G., Reed, J. D. and Mukhtar, H. 2005. *International Journal of Cancer* 113: 423–433.

28. Lansky, E. P., Harrison, G., Froom, P. and Jiang, W. G. 2005. *Investigational New Drug* 23: 121–122; Erratum in: *Investigational New Drugs* 23: 379.

29. Lansky, E. P. and Newman, R. A. 2006. *Journal of Ethnopharmacology* 109: 177–206.

30. Malik, A. and Mukhtar, H. 2006. *Cell Cycle* 5: 371–373.

31. Malik, A., Afaq, F., Sarfaraz, S., Adhami, V. M., Syed, D. N. and Mukhtar, H. 2005. *Proceedings of the National Academy of Sciences of the USA* 102: 14813–14818.

32. Khan, N., Afaq, F., Kweon, M. H., Kim, K. and Mukhtar, H. 2007. *Cancer Research* 67: 3475–3482.

33. Khan, N., Hadi, N., Afaq, F., Syed, D. N., Kweon, M. H., Mukhtar, H. 2007. *Carcinogenesis* 28: 163–173.

34. Seeram, N. P., Adams, L. S., Henning, S. M., Niu, Y., Zhang, Y., Nair, M. G. and Heber, D. 2005. *Journal of Nutritional Biochemistry* 16: 360–367.

35. Seeram, N. P., Aronson, W. J., Zhang, Y., Henning, S. M., Moro, A., Lee, R. P., Sartippour, M., Harris, D. M., Rettig, M., Suchard, M. A., Pantuck, A. J., Belldegrun, A. and Heber, D. 2007. *Journal of Agriculture and Food Chemistry* 55: 7732–7737.

36. Pantuck, A. J., Leppert, J. T., Zomorodian, N., Aronson, W., Hong, J., Barnard, R. J., Seeram, N., Liker, H., Wang, H., Elashoff, R., Heber, D., Aviram, M., Ignarro, L. and Belldegrun, A. 2006. *Clinical Cancer Research* 12: 4018–4026.

37. Das, A. K., Mandal, S. C., Banerjee, S. K., Sinha, S., Saha, B. P. and Pal, M. 2001. *Phytotherapy Research* 15: 628–629.

38. Jafri, M. A., Aslam, M., Javed, K. and Singh, S. 2000. *Journal of Ethnopharmacology* 70: 309–314.

39. Li, Y., Qi, Y., Huang, T. H., Yamahara, J. and Roufogalis, B. D. 2008. *Diabetes, Obesity and Metabolism* 10: 10–17.

40. Katz, S. R., Newman, R. A. and Lansky, E. P. 2007. *Journal of Medicinal Food* 10: 213–217.

41. Huang, T. H., Peng, G., Kota, B. P., Li, G. Q., Yamahara, J., Roufogalis, B. D. and Li, Y. 2005. *Toxicology and Applied Pharmacology* 207: 160–169.

42. Das, A. K., Mandal, S. C., Banerjee, S. K., Sinha, S., Das, J., Saha, B. P. and Pal, M. 1999. *Journal of Ethnopharmacology* 68: 205–208.

43. Prakash, A. O., Saxena, V., Shukla, S., Tewari, R. K., Mathur, S., Gupta, A., Sharma, S. and Mathur, R. 1985. *Acta Europaea Fertilitatis* 16: 441–448.

44. Vasconcelos, L. C., Sampaio, M. C., Sampaio, F. C. and Higino, J. S. 2003. *Mycoses* 46: 192–196.

45. Holetz, F. B., Pessini, G. L., Sanches, N. R., Cortez, D. A., Nakamura, C. V. and Filho, B. P. 2002. *Memórias do Instituto Oswaldo Cruz* 97: 1027–1031.

46. Jung, K. H., Kim, M. J., Ha, E., Uhm, Y. K., Kim, H. K., Chung, J. H. and Yim, S. V. 2006. *Biological and Pharmaceutical Bulletin* 29: 1258–1261.

47. Adams, L. S., Seeram, N. P., Aggarwal, B. B., Takada, Y., Sand, D. and Heber, D. 2006. *Journal of Agriculture and Food Chemistry* 54: 980–985.

48. Reddy, M. K., Gupta, S. K., Jacob, M. R., Khan, S. I. and Ferreira, D. 2007. *Planta Medica* 73: 461–467.

49. Noda, Y., Kaneyuki, T., Mori, A. and Packer, L. 2002. *Journal of Agriculture and Food Chemistry* 50: 166–171.

50. Singh, R. P., Chidambara Murthy, K. N. and Jayaprakasha, G. K. 2002. *Journal of Agriculture and Food Chemistry* 50: 81–86.

51. Sezer, E. D., Akçay, Y. D., Ilanbey, B., Yildirim, H. K. and Sözmen, E. Y. 2007. *Journal of Medicinal Food* 10: 371–374.

52. Zaid, M. A., Afaq, F., Syed, D. N., Dreher, M. and Mukhtar, H. 2007. *Photochemistry and Photobiology* 83: 882–888.
53. Zhang, Q., Jia, D. and Yao, K. 2007. *Natural Product Research* 21: 211–216.
54. Kulkarni, A. P., Mahal, H. S., Kapoor, S. and Aradhya, S. M. 2007. *Journal of Agriculture and Food Chemistry* 55: 1491–1500.
55. Rout, S. and Banerjee, R. 2007. *Bioresource Technology* 98: 3159–3163.
56. Ricci, D., Giamperi, L., Bucchini, A. and Fraternale, D. 2006. *Fitoterapia* 77: 310–312.
57. Afaq, F. and Mukhtar, H. 2006. *Experimental Dermatology* 15: 678–684.
58. Kaur, G., Jabbar, Z., Athar, M. and Alam, M. S. 2006. *Food and Chemical Toxicology* 44: 984–993.
59. Rosenblat, M., Hayek, T. and Aviram, M. 2006. *Atherosclerosis* 187: 363–371.
60. Sudheesh, S. and Vijayalakshmi, N. R. 2005. *Fitoterapia* 76: 181–186.
61. Faria, A., Monteiro, R., Mateus, N., Azevedo, I. and Calhau, C. 2007. *European Journal of Nutrition* 46: 271–278.
62. Sestili, P., Martinelli, C., Ricci, D., Fraternale, D., Bucchini, A., Giamperi, L., Curcio, R., Piccoli, G. and Stocchi, V. 2007. *Pharmacology Research* 56: 18–26.
63. Neurath, A. R., Strick, N., Li, Y. Y. and Debnath, A. K. 2004. *BMC Infectious Diseases* 4: 41.
64. Zhang, J., Zhan, B., Yao, X., Gao, Y. and Shong, J. 1995. *Zhongguo Zhong Yao Za Zhi* 20: 556–568.
65. Li, Y., Ooi, L. S., Wang, H., But, P. P. and Ooi, V. E. 2004. *Phytotherapy Research* 18: 718–722.
66. Gharzouli, K., Khennouf, S., Amira, S. and Gharzouli, A. 1999. *Phytotherapy Research* 13: 42–45.
67. Ajaikumar, K. B., Asheef, M., Babu, B. H. and Padikkala, J. 2005. *Journal of Ethnopharmacology* 96: 171–176.
68. Toklu, H. Z., Dumlu, M. U., Sehirli, O., Ercan, F., Gedik, N., Gökmen, V. and Sener, G. 2007. *Journal of Pharmacy and Pharmacology* 59: 1287–1295.
69. Esmaillzadeh, A., Tahbaz, F., Gaieni, I., Alavi-Majd, H. and Azadbakht, L. 2006. *International Journal for Vitamin and Nutrition Research* 76: 147–151.
70. Gracious Ross, R., Selvasubramanian, S. and Jayasundar, S. 2001. *Journal of Ethnopharmacology* 78: 85–87.
71. Yamasaki, M., Kitagawa, T., Koyanagi, N., Chujo, H., Maeda, H., Kohno-Murase, J., Imamura, J., Tachibana, H. and Yamada, K. 2006. *Nutrition* 2006 22: 54–59.
72. West, T., Atzeva, M. and Holtzman, D. M. 2007. *Developmental Neuroscience* 29: 363–372.
73. Kwak, H. M., Jeon, S. Y., Sohng, B. H., Kim, J. G., Lee, J. M., Lee, K. B., Jeong, H. H., Hur, J. M., Kang, Y. H. and Song, K. S. 2005. *Archives of Pharmacal Research* 28: 1328–1332.
74. Hartman, R. E., Shah, A., Fagan, A. M., Schwetye, K. E., Parsadanian, M., Schulman, R. N., Finn, M. B. and Holtzman, D. M. 2006. *Neurobiology of Disease* 24: 506–515.
75. Loren, D. J., Seeram, N. P., Schulman, R. N. and Holtzman, D. M. 2005. *Pediatric Research* 57: 858–864; Erratum in: *Pediatric Research* 2007 62: 363.
76. de Nigris, F., Williams-Ignarro, S., Sica, V., Lerman, L. O., D'Armiento, F. P., Byrns, R. E., Casamassimi, A., Carpentiero, D., Schiano, C., Sumi, D., Fiorito, C., Ignarro, L. J. and Napoli, C. 2007. *Cardiovascular Research* 73: 414–423.
77. de Nigris, F., Williams-Ignarro, S., Lerman, L. O., Crimi, E., Botti, C., Mansueto, G., D'Armiento, F. P., De Rosa, G., Sica, V., Ignarro, L. J. and Napoli, C. 2005. *Proceedings of the National Academy of Sciences of the USA* 102: 4896–4901.
78. Fuhrman, B., Volkova, N. and Aviram, M. 2005. *Journal of Nutritional Biochemistry* 16: 570–576.
79. Sastravaha, G., Yotnuengnit, P., Booncong, P. and Sangtherapitikul, P. 2003. *Journal of the International Academy of Periodontology* 5: 106–115.
80. Murthy, K. N., Reddy, V. K., Veigas, J. M. and Murthy, U. D. 2004. *Journal of Medicinal Food* 7: 256–259.
81. Morsy, T. A., Mazyad, S. A. and el-Sharkawy, I. M. 1998. *Journal of Egyptian Society of Parasitology* 28: 699–709.
82. Tripathi, S. M. and Singh, D. K. 2000. *Brazilian Journal of Medical and Biological Research* 33: 1351–1355.

83. Tripathi, S. M., Singh, V. K., Singh, S. and Singh, D. K. 2004. *Phytotherapy Research* 18: 501–506.
84. Kapoor, L. D. 1990. Handbook of Ayurvedic Medicinal Plants. CRC Press: Boca Raton, FL.
85. Morton, J. F. 1981. Atlas of Medicinal Plants of Middle America: Bahamas to Yucatan. C.C. Thomas Publishers Co.: Springfield, IL.
86. Nguyen, V. D. and Doan, T. N. 1989. Medicinal Plants in Vietnam. World Health Organisation, Regional Office for the Western Pacific, Manila, Institute of Materia Medica, Hanoi.
87. Enrique, E., Utz, M., De Mateo, J. A., Castelló, J. V., Malek, T. and Pineda, F. 2006. *Annals of Allergy, Asthma and Immunology* 96: 122–123.
88. McGuffin, M., Hobbs, C., Upton, R. and Goldberg, A. (Eds). 1997. *American Herbal Products Association's Botanical Safety Handbook*. CRC Press: Boca Raton, FL.
89. Gaig, P., Botey, J., Gutierrez, V., Pena, M., Eseverri, J. L. and Marin, A. 1992. *Journal of Investigational Allergology and Clinical Immunology* 2: 216–218.
90. Fetrow, C. W. and Avila, J. R. 2000. The complete guide to herbal medicines. Springhouse Corp: Springhouse, PA.
91. Skidmore-Roth, L. 2001. Mosby's handbook of herbs and natural supplements. Mosby, Inc.: St. Louis.
92. Duke, J. A., Bogenschutz-Godwin, M. J., duCellier, J. and Duke, P. A. K. 2002. Handbook of medicinal herbs. CRC Press: Boca Raton, FL.
93. Summers, K. M. 2006. Potential drug-food interactions with pomegranate juice. *Annals of Pharmacotherapy* 40: 1472–1473.
94. Usia, T., Iwata, H., Hiratsuka, A., Watabe, T., Kadota, S. and Tezuka, Y. 2006. *Phytomedicine* 13: 67–73.
95. Hidaka, M., Okumura, M., Fujita, K., Ogikubo, T., Yamasaki, K., Iwakiri, T., Setoguchi, N. and Arimori, K. 2005. *Drug Metabolism and Disposition* 33: 644–648.
96. Nagata, M., Hidaka, M., Sekiya, H., Kawano, Y., Yamasaki, K., Okumura, M. and Arimori, K. 2007. *Drug Metabolism and Disposition* 35: 302–305.

Rhodomyrtus tomentosa (Ait.) Hassk.

1. Wee, Y. C. 1992. A Guide to Medicinal Plants. Singapore Science Centre Publication: Singapore.
2. Hsuan, K. 1990. The Concise Flora of Singapore, Gymnosperms and Dicotyledons. Singapore University Press: National University of Singapore.
3. Wee, Y. C. and Corlett, R. 1986. The City and the Forest: Plant Life in Urban Singapore. Singapore University Press: National University of Singapore.
4. USDA, ARS GRIN Taxonomy for Plants 2008 http://www.ars-grin.gov/cgi-bin/npgs/html/taxon.pl?31664
   Date Accessed: 30th June 2008
5. Duke, J. A. and Ayensu, E. S. 1985. Medicinal Plants of China, volume one and two. Reference Publications, Inc.: United States of America.
6. Dachriyanus, S., Sargent, M. V., Skelton, B. W., Soediro, I., Sutisna, M., White, A. H. and Yulinah, E. 2002. *Australian Journal of Chemistry* 55: 229–232.
7. Dachriyanus, Fahmi, R., Sargent, M. V., Skelton, B. W. and White, A. H. 2004. *Acta Crystallographica Section E-Structure Reports Online* 60: O86–O88.
8. Liu, Y. Z., Hou, A. J., Ji, C. R. and Wu, Y. J. 1997. *Chinese Chemical Letters* 8: 39–40.
9. Hou, A. J., Wu, Y. J. and Liu, Y. Z. 1999. *Zhongcaoyao* 30: 645–648.
10. Hui, W. H., Li, M. M. and Luk, K. 1975. *Phytochemistry* 14: 833–834.
11. Hui, W. H. and Li, M. M. 1976. *Phytochemistry* 15: 1741–1743.
12. Nguyen, V. D. and Doan, T. N. 1989. Medicinal Plants in Vietnam. World Health Organisation, Regional Office for the Western Pacific, Manila, Institute of Materia Medica, Hanoi.

Rhoeo spathacea (Sw.) Stearn

1. Hsuan, K., Chin, S. C. and Tan, H. T. W. 1998. The Concise Flora of Singapore volume II: Monocotyledons. Singapore University Press: Singapore.

2. eFloras.org 2008
   http://www.efloras.org/florataxon.aspx?flora_
   id=5&taxon_id=242356469
   Date Accessed: 30th March 2008
3. Wee, Y. C. and Hsuan, K. 1990. An illustrated
   Dictionary of Chinese Medicinal Herbs.
   Times Edition and Eu Yan Seng Holdings Ltd:
   Singapore.
4. Perez, G. R. M. 1996. *Phytomedicine* 3:
   163–167.
5. Garcia, M., Miyares, C., Menendez, E. and
   Sainz, F. 1971. *Canadian Journal of Physiology
   and Pharmacology* 49: 1106–1110.
6. Weniger, B., Haag-Berrurier, M. and
   Anton, R. 1982. *Journal of Ethnopharmacology* 6: 67–84.

Ricinus communis L.

1. Hsuan, K. 1990. The Concise Flora of
   Singapore, Gymnosperms and Dicotyledons.
   Singapore University Press: National
   University of Singapore.
2. USDA, ARS GRIN Taxonomy for Plants 2008
   http://www.ars-grin.gov/cgi-bin/npgs/html/
   taxon.pl?31896
   Date Accessed: 30th June 2008
3. Fleming, T. (Ed.). 2000. PDR for Herbal
   Medicines, 2nd edition. Medical Economics
   Company: New Jersey.
4. Wee, Y. C. and Hsuan, K. 1990. An illustrated
   Dictionary of Chinese Medicinal Herbs.
   Times Edition and Eu Yan Seng Holdings Ltd:
   Singapore.
5. Duke, J. A. and Ayensu, E. S. 1985. Medicinal
   Plants of China, volume one and two.
   Reference Publications, Inc.: United States of
   America.
6. Selvaraj, O. S. and Bhalla, V. 1988. Proceedings:
   Malaysian Traditional Medicine, Kuala Lumpur,
   pp. 221–225. Institute of Advanced Studies,
   University of Malaya, Kuala Lumpur.
7. Chakravartula, S. V. and Guttarla, N. 2007.
   Natural Product Research 21: 1073–1077.
8. Zhang, X., Han, F., Gao, P., Yu, D. and Liu, S.
   2007. *Natural Product Research* 2007 21:
   982–989.
9. Foster, S. and Hobbs, C. 2002. A Field Guide
   to Western Medicinal Plants and Herbs.
   Houghton Mifflin Company: New York.
10. Ajose, F. O. 2007. *International Journal of
    Dermatology* 46: 48–55.
11. McGuffin, M., Hobbs, C., Upton, R. and
    Goldberg, A. (Eds). 1997. American Herbal
    Products Association's Botanical Safety
    Handbook. CRC Press: Boca Raton, FL.
12. Isichei, C. O., Das, S. C., Ogunkeye, O. O.,
    Okwuasaba, F. K., Uguru, V. E., Onoruvwe,
    O., Olayinka, A. O., Dafur, S. J., Ekwere, E.
    O. and Parry, O. 2000. *Phytotherapy Research*
    14: 40–42.
13. Makonnen, E., Zerihun, L., Assefa, G. and
    Rostom, A. A. 1999. *East African Medical
    Journal* 76: 335–337.
14. Okwuasaba, F. K., Osunkwo, U. A., Ekwenchi,
    M. M., Ekpenyong, K. I., Onwukeme, K. E.,
    Olayinka, A. O., Uguru, M. O. and Das, S. C.
    1991. *Journal of Ethnopharmacology* 34:
    141–145.
15. Sandhyakumary, K., Bobby, R. G. and Indira,
    M. 2003. *Phytotherapy Research* 17: 508–
    511.
16. Okwusasaba, F. K., Das, S. C., Isichei, C. O.,
    Ekwenchi, M. M., Onoryvwe, O., Olayinka,
    A. O., Uguru, V. E., Dafur, S. J., Ekwere, E.
    O. and Parry, O. 1997. *Phytotherapy Research*
    11: 97–100.
17. Salhab, A. S., Shomaf, M. S., Gharaibeh, M.
    N. and Amer, N. A. 1999. *Contraception* 59:
    395–399.
18. Raji, Y., Oloyo, A. K. and Morakinyo, A. O.
    2006. *Asian Journal of Andrology* 8: 115–
    121.
19. Ilavarasan, R., Mallika, M. and Venkataraman,
    S. 2006. *Journal of Ethnopharmacology* 103:
    478–480.
20. Ferraz, A. C., Angelucci, M. E., Da Costa, M.
    L., Batista, I. R., De Oliveira, B. H. and Da
    Cunha, C. 1999. *Pharmacology, Biochemistry
    and Behaviour* 63: 367–375.
21. Wang, H. X. and Ng, T. B. 2001. *Planta
    Medica* 67: 669–672.
22. Shukla, B., Visen, P. K. S., Patnaik, G. K.,
    Kapoor, N. K. and Dhawan, B. N. 1992. *Drug
    Development Research* 26: 183–193.
23. Nisha, M., Kalyanasundaram, M., Paily, K. P.,
    Abidha, Vanamail, P. and Balaraman, K.
    2007. *Parasitology Research* 100: 575–579.
24. Upasani, S. M., Kotkar, H. M., Mendki, P. S.
    and Maheshwari, V. L. 2003. *Pest Management
    Science* 59: 1349–1354.

25. Nguyen, V. D. and Doan, T. N. 1989. Medicinal Plants in Vietnam. World Health Organisation, Regional Office for the Western Pacific, Manila, Institute of Materia Medica, Hanoi.
26. Fetrow, C. W. and Avila, J. R. 2000. The complete guide to herbal medicines. Springhouse Corp: Springhouse, PA.
27. Duke, J. A. 1985. CRC Handbook of Medicinal Herbs. CRC Press: Boca Raton, FL.
28. Balint, G. A. 1993. *Experimental Toxicology and Pathology* 45: 303–304.
29. Bradberry, S. M., Dickers, K. J., Rice, P., Griffiths, G. D. and Vale, J. A. 2003. *Toxicological Reviews* 22: 65–70.
30. Kumar, O., Sugendran, K. and Vijayaraghavan, R. 2003. *Toxicon* 41: 333–338.
31. Kumar, O., Lakshmana Rao, P. V., Pradhan, S., Jayaraj, R., Bhaskar, A. S., Nashikkar, A. B. and Vijayaraghavan, R. 2007. *Cellular and Molecular Biology* 53: 92–102.
32. Kanerva, L., Estlander, T. and Jolanki, R. 1990. *Journal of the American Academy of Dermatology* 23: 351–355.
33. Griffiths, G. D., Phillips, G. J. and Holley, J. 2007. *Inhalation Toxicology* 19: 873–887.
34. Kaur, C. and Ling, E. A. 1993. *Journal für Hirnforschung* 34: 493–501.
35. Aslani, M. R., Maleki, M., Mohri, M., Sharifi, K., Najjar-Nezhad, V. and Afshari, E. 2007. *Toxicon* 49: 400–406.
36. Wren, R. C. 1988. Potter's new cyclopaedia of botanical drugs and preparations. The C.W. Daniel Company Ltd.: Saffron Walden, Essex.
37. Skidmore-Roth L. 2001. Mosby's handbook of herbs and natural supplements. Mosby, Inc.: St. Louis.

Ruta graveolens L.

1. Tutin, T. G., Heywood, V. H., Burges, N. A., Moore, D. M., Valentine, D. H., Walters, S. M. and Webb, D. A. (Eds.). 1968. Flora Europaea 2, Rosaceae to Umbelliferae. Cambridge University Press: London.
2. USDA, ARS GRIN Taxonomy for Plants 2008 http://www.ars-grin.gov/cgi-bin/npgs/html/taxon.pl?32578 Date Accessed: 30th June 2008

3. Duke, J. A. and Ayensu, E. S. 1985. Medicinal Plants of China, volume one and two. Reference Publications, Inc.: United States of America.
4. Li, T. S. C. 2000. Medicinal Plants. Culture, Utilisation and Phytopharmacology. Technomic Publishing Co., Inc.: Lancaster.
5. Raghav, S. K., Gupta, B., Shrivastava, A. and Das, H. R. 2007. *European Journal of Pharmacology* 560: 69–80.
6. Guarrera, P. M. 1999. *Journal of Ethnopharmacology* 68: 183–192.
7. Gravot, A., Larbat, R., Hehn, A., Lievre, K., Gontier, E., Goergen, J. L. and Bourgaud, F. 2004. *Archieves of Biochemistry and Biophysics* 422: 71–80.
8. Fleming, T. (Ed.). 2000. PDR for Herbal Medicines, 2nd edition. Medical Economics Company: New Jersey.
9. McGuffin, M., Hobbs, C., Upton, R. and Goldberg, A. (Eds). 1997. American Herbal Products Association's Botanical Safety Handbook. CRC Press: Boca Raton, FL.
10. Waldstein. A. 2006. *Journal of Ethnopharmacology* 108: 299–310.
11. Atta, A. H. and Alkofahi, A. 1998. *Journal of Ethnopharmacology* 60: 117–124.
12. Alzoreky, N. S. and Nakahara, K. 2003. *International Journal of Food Microbiology* 80: 223–230.
13. Trovato, A., Monforte, M. T., Rossitto, A. and Forestieri, A. M. 1996. *Bollettino Chimico Farmaceutico* 135: 263–266.
14. Pathak, S., Multani, A. S., Banerji, P. and Banerji, P. 2003. *International Journal of Oncology* 23: 975–982.
15. Es, S., Kuttan, G., Kc, P. and Kuttan, R. 2007. *Asian Pacific Journal of Cancer Prevention* 8: 390–394.
16. Preethi, K. C., Kuttan, G. and Kuttan, R. 2006. *Asian Pacific Journal of Cancer Prevention* 7: 439–443.
17. Réthy, B., Zupkó, I., Minorics, R., Hohmann, J., Ocsovszki, I. and Falkay, G. 2007. *Planta Medica* 73: 41–48.
18 Kong, Y. C., Lau, C. P., Wat, K. H., Ng, K. H., But, P. P., Cheng, K. F. and Waterman, P. G. 1989. *Planta Medica* 55: 176–178.
19. Prakash, A. O., Saxena, V., Shukla, S., Tewari, R. K., Mathur, S., Gupta, A., Sharma, S. and Mathur, R. 1985. *Acta Europaea Fertilitatis* 16: 441–448.

20. de Freitas, T. G., Augusto, P. M. and Montanari, T. 2005. *Contraception* 71: 74–77.
21. Raghav, S. K., Gupta, B., Agrawal, C., Goswami, K. and Das, H. R. 2006. *Journal of Ethnopharmacology* 104: 234–239.
22. Oliva, A., Meepagala, K. M., Wedge, D. E., Harries, D., Hale, A. L., Aliotta, G. and Duke, S. O. 2003. *Journal of Agriculture and Food Chemistry* 51: 890–896.
23. Meepagala, K. M., Schrader, K. K., Wedge, D. E. and Duke, S. O. 2005. *Phytochemistry* 66: 2689–2695.
24. Ivanova, A., Mikhova, B., Najdenski, H., Tsvetkova, I. and Kostova, I. 2005. *Fitoterapia* 76: 344–347.
25. Al-Heali, F. M. and Rahemo, Z. 2006. *Turkiye Parazitolojii Dergisi* 30: 272–274.
26. Skidmore-Roth L. 2001. Mosby's handbook of herbs and natural supplements. Mosby, Inc.: St. Louis.
27. Blumenthal, M., Busse, W. R., Goldberg, A., Gruenwald, J., Hall, T., Klein, S., Riggins, C. W. and Rister, R. S. (Eds.). 1998. The Complete German Commission E monographs, Therapeutic guide to herbal medicines. American Botanical Council, Boston: Integrative Medicine Communications: Austin, Texas.
28. Schempp, C. M., Schopf, E. and Simon, J. C. 1999. *Hautarzt* 50: 432–434.
29. Heskel, N. S., Amon, R. B., Storrs, F. J. and White, C. R. Jr. 1983. *Contact Dermatitis* 9: 278–280.
30. Sumner J. 2000. The Natural History of Medicinal Plants. Timber Press, Portland.
31. Furniss, D. and Adams, T. 2007. *Journal of Burn Care and Research* 28: 767–769.
32. Eickhorst, K., DeLeo, V. and Csaposs, J. 2007. *Dermatitis* 18: 52–55.
33. Seak, C. J. and Lin, C. C. 2007. *Clinical Toxicology (Philadelphia)* 45: 173–175.
34. El Agraa, S. E., el Badwi, S. M. and Adam, S. E. 2002. *Tropical Animal Health and Production* 34: 271–281.

Saccharum officinarum L.

1. Hsuan, K., Chin, S. C. and Tan, H. T. W. 1998. The Concise Flora of Singapore volume II: Monocotyledons. Singapore University Press: Singapore.
2. eFloras.org 2008 http://www.efloras.org/florataxon.aspx?flora_id=2&taxon_id=200026232 Date Accessed: 30th June 2008
3. Ledon, N., Casaco, A., Rodriguez, V., Cruz, J., Gonzalez, R., Tolon, Z., Cano, M. and Rojas, E. 2003. *Planta Medica* 69: 367–369.
4. Noa, M., Herrera, M., Magraner, J. and Mas, R. 1994. *Journal of Pharmacy and Pharmacology* 46: 282–285.
5. Noa, M., Mendoza, S., Mas, R., Mendoza, N. and Leon, F. 2004. *Drugs in R&D* 5: 281–290.
6. Duarte-Almeida, J. M., Negri, G., Salatino, A., de Carvalho, J. E. and Lajolo, F. M. 2007. *Phytochemistry* 68: 1165–1171.
7. Colombo, R., Yariwake, J. H., Queiroz, E. F., Ndjoko, K. and Hostettmann, K. 2006. *Phytochemical Analysis* 17: 337–343.
8. Colombo, R., Yariwake, J. H., Queiroz, E. F., Ndjoko, K. and Hostettmann, K. 2005. *Journal of Chromatography A* 1082: 51–59.
9. Ledón, N., Romay, Ch., Rodríguez, V., Cruz, J., Rodríguez, S., Ancheta, O., González, A., González, R., Tolón, Z., Cano, M., Rojas, E., Capote, A. and Valdes, T. 2005. *Planta Medica* 71: 126–129.
10. Balick, M. J., Kronenberg, F., Ososki, A. L., Reiff, M., Fugh-Berman, A., O'Connor, B., Roble, M., Lohr, P. and Atha, D. 2000. *Economic Botany* 54: 344–357.
11. Wee, Y. C. and Hsuan, K. 1990. An illustrated Dictionary of Chinese Medicinal Herbs. Times Edition and Eu Yan Seng Holdings Ltd: Singapore.
12. Ledón, N., Casacó, A., Remirez, D., González, A., Cruz, J., González, R., Capote, A., Tolón, Z., Rojas, E., Rodríguez, V. J., Merino, N., Rodríguez, S., Ancheta, O. and Cano, M. C. 2007. *Phytomedicine* 14: 690–695.
13. Noa, M., Mendoza, S., Mas, R. and Mendoza, N. 2002. *Drugs under Experimental and Clinical Research* 28: 177–183.
14. Mendoza, S., Noa, M., Mas, R. and Mendoza, N. 2003. *International Journal of Tissue Reactions* 25: 91–98.
15. Gamez, R., Mas, R., Noa, M., Menendez, R., Garcia, H., Gonzalez, J., Perez, Y. and Goicochea, E. 2004. *Drugs under Experimental and Clinical Research* 30: 75–88.

16. Arruzazabala, M. L., Carbajal, D., Mas, R., Garcia, M. and Fraga, V. 1993. *Thrombosis Research* 69: 321–327.
17. Molina, V., Carbajal, D., Arruzazabala, L. and Más, R. 2005. *Journal of Medicinal Food* 8: 232–236.
18. Carbajal, D., Arruzazabala, M. L., Mas, R., Molina, V. and Valdes, S. 1994. *Prostaglandins, Leukotrienes, and Essential Fatty Acids* 50: 249–251.
19. Mas, R., Castano, G., Fernandez, J., Gamez, R., Illnait, J., Fernandez, L., Lopez, E., Mesa, M., Alvarez, E. and Mendoza, S. 2004. *Asia Pacific Journal of Clinical Nutrition* 13: S102.
20. Mas, R., Castano, G., Fernandez, J., Gamez, R. R., Illnait, J., Fernandez, L., Lopez, E., Mesa, M., Alvarez, E. and Mendoza, S. 2004. *Asia Pacific Journal of Clinical Nutrition* 13: S101.
21. Menendez, R., Amor, A. M., Rodeiro, I., Gonzalez, R. M., Gonzalez, P. C., Alfonso, J. L. and Mas, R. 2001. *Archieves of Medical Research* 32: 8–12.
22. Torres, O., Agramonte, A. J., Illnait, J., Mas Ferreiro, R., Fernandez, L. and Fernandez, J. C. 1995. *Diabetes Care* 18: 393–397.
23. Aneiros, E., Más, R., Calderón, B., Illnait, J., Fernández, L., Castaño, G. and Fernández, J. C. 1995. *Current Therapeutic Research* 56: 176–182.
24. Crespo, N., Alvarez, R., Más, R., Illnait, J., Fernández, L. and Fernández, J. C. 1997. *Current Therapeutic Research* 58: 44–51.
25. Hikosaka, K., Koyama, Y., Motobu, M., Yamada, M., Nakamura, K., Koge, K., Shimura, K., Isobe, T., Tsuji, N., Kang, C. B., Hayashidani, H., Wang, P. C., Matsumura, M. and Hirota, Y. 2006. *Bioscience, Biotechnology and Biochemistry* 70: 2853–2858.
26. Hikosaka, K., El-Abasy, M., Koyama, Y., Motobu, M., Koge, K., Isobe, T., Kang, C. B., Hayashidani, H., Onodera, T., Wang, P. C., Matsumura, M. and Hirota, Y. 2007. *Phytotherapy Research* 21: 120–125.
27. Amer, S., Na, K. J., El-Abasy, M., Motobu, M., Koyama, Y., Koge, K. and Hirota, Y. 2004. *International Immunopharmacology* 4: 71–77.
28. Noa, M., Mendoza, S. and Más, R. 2005. *Journal of Medicinal Food* 8: 237–241.
29. Boral, D., Chatterjee, S. and Bhattacharya, K. 2004. *Annals of Agriculture and Environmental Medicine* 11: 45–52.
30. Gamez, R., Aleman, C. L., Mas, R., Noa, M., Rodeiro, I., Garcia, H., Hernandez, C., Menendez, R. and Aguilar, C. 2001. *Journal of Medicinal Food* 4: 57–65.
31. Gamez, R., Mas, R., Noa, M., Menendez, R., Aleman, C., Acosta, P., Garcia, H., Hernandez, C., Amor, A., Perez, J. and Goicochea, 2000. *Toxicology Letters* 118: 31–41.
32. Aleman, C. L., Puig, M. N., Elias, E. C., Ortega, C. H., Guerra, I. R., Ferreiro, R. M. and Brinis, F. 1995. *Food and Chemical Toxicology* 33: 573–578.
33. Rodriguez, M. D. and Garcia, H. 1994. *Teratogen, Carcinogen and Mutagen* 14: 107–113.
34. Molina Cuevas, V., Arruzazabala, M. L., Carbajal Quintana, D., Mas Ferreiro, R. and Valdes Garcia, S. 1998. *Archieves of Medical Research* 29: 21–24.
35. Arruzazabala, M. L., Carbajal, D., Mas, R., Valdes, S. and Molina, V. 2001. *Journal of Medicinal Food* 4: 67–70.
36. Carbajal, D., Arruzazabala, M. L., Valdes, S. and Mas, R. 1998. *Pharmacology Research* 38: 89–91.

Sauropus androgynus (L.) Merr.

1. eFloras.org 2008 http://www.efloras.org/florataxon.aspx?flora_id=2&taxon_id=200012609 Date Accessed: 30th March 2008
2. Hsuan, K. 1990. The Concise Flora of Singapore, Gymnosperms and Dicotyledons. Singapore University Press: National University of Singapore.
3. USDA, ARS GRIN Taxonomy for Plants 2008 http://www.ars-grin.gov/cgi-bin/npgs/html/taxon.pl?33189 Date Accessed: 30th June 2008
4. Nguyen, V. D. and Doan, T. N. 1989. Medicinal Plants in Vietnam. World Health Organisation, Regional Office for the Western Pacific, Manila, Institute of Materia Medica, Hanoi.
5. Kanchanapoom, T., Chumsri, P., Kasai, R., Otsuka, H. and Yamasaki, K. 2003. *Phytochemistry* 63: 985–988.

6. Ching, L. S. and Mohamed, S. 2001. *Journal of Agriculture and Food Chemistry* 49: 3101–3105.

7. Lin, C. Q., Lin, W. B., Pan, W. D. and Li, Y. J. 1999. *Redai Yaredai Zhiwu Xuebao* 7: 255–256.

8. Wang, P. H. and Lee, S. S. 1997. *Journal of the Chinese Chemical Society (Taipei)* 44: 145–149.

9. Liao, X. K. and Li, Y. H. 1996. *Redai Yaredai Zhiwu Xuebao* 4: 70–71.

10. Yu, S. F., Shun, C. T., Chen, T. M. and Chen, Y. H. 2006. *Biological and Pharmaceutical Bulletin* 29: 2510–2513.

11. Jain, S. K. and DeFilipps, R. A. 1991. Medicinal Plants of India, volume one and two. Reference Publications Inc.: United States of America.

12. Muhammad, M. M., Abdul, M. A., Mohd, A. A., Rozita, M. N., Nashriyah, B. M., Abdul, R. R. and Kazuyoshi, K. 1997. *Pesticide Science* 51: 165–170.

13. Padmavathi, P. and Rao, M. P. 1990. *Plant Foods for Human Nutrition (Dordrecht, Netherlands)* 40: 107–113.

14. Ger, L. P., Chiang, A. A., Lai, R. S., Chen, S. M. and Tseng, C. J. 1997. *American Journal of Epidemiology* 145: 842–849.

15. Oonakahara, K., Matsuyama, W., Higashimoto, I., Machida, K., Kawabata, M., Arimura, K., Osame, M., Hayashi, M., Ogura, T., Imaizumi, K. and Hasegawa, Y. 2005. *Respiration* 72: 221.

16. Hayashi, M., Tagawa, A., Ogura, T., Kozawa, S., Nakamura, M., Watanuki, Y. and Takahashi, H. 2007. *Nihon Kokyuki Gakkai Zasshi* 45: 81–86.

17. Chang, Y. L., Yao, Y. T., Wang, N. S. and Lee, Y. C. 1998. *American Journal of Respiratory and Critical Care Medicine* 157: 594–598.

18. Hsiue, T. R., Guo, Y. L., Chen, K. W., Chen, C. W., Lee, C. H. and Chang, H. Y. 1998. *Chest* 113: 71–76.

19. Wu, C. L., Hsu, W. H. and Chiang, C. D. 1998. *Zhonghua yi xue za zhi (Chinese medical journal, Free China Ed)* 61: 34–38.

20. Lin, T. J., Lu, C. C., Chen, K. W., Yueh, W. C. and Deng, J. F. 1997. *Toxicon* 35: 502.

21. Kao, C. H., Hom Y. J., Wu, C. L. and ChangLai, S. P. 1999. *Respiration* 66: 46–51.

22. Yu, S. F., Chen, T. M. and Chen, Y. H. 2007. *Journal of the Formosan Medical Association* 106: 537–547.

Sesbania grandiflora Pers.

1. Hsuan, K. 1990. The Concise Flora of Singapore, Gymnosperms and Dicotyledons. Singapore University Press: National University of Singapore.

2. USDA, ARS GRIN Taxonomy for Plants 2008 http://www.ars-grin.gov/cgi-bin/npgs/html/taxon.pl?33770 Date Accessed: 30th June 2008

3. Tiwari, R. D., Bajpai, R. K. and Khana, S. S. 1964. *Archiv der Pharmazie* 297: 310–312.

4. Tiwari, R. D. and Bajpai, R. K. 1964. *Proceedings of the National Academy of Sciences, India, Section A: Physical Sciences* 34: 239–244.

5. Saxena, V. K. and Mishra, L. N. 1999. *Research Journal of Chemistry and Environment* 3: 69–70.

6. Saxena, V. K. and Mishra, L. N. 1999. *Journal of the Institution of Chemists (India)* 71: 234–236.

7. Das, C. and Tripathi, A. K. 1999. *Oriental Journal of Chemistry* 15: 561–562.

8. Andal, K. R. and Sulochana, N. 1986. *Fitoterapia* 57: 293–294.

9. Lakshminarayana, R., Raju, M., Krishnakantha, T. P. and Baskaran, V. 2005. *Journal of Agriculture and Food Chemistry* 53: 2838–2842.

10. Goh, S. H., Chuah, C. H., Mok, J. S. L. and Soepadmo, E. 1995. Malaysian Medicinal Plants for the Treatment of Cardiovascular Diseases. Pelanduk Publications (M) Sdn Bhd: Malaysia.

11. Pari, L. and Uma, A. 2003. *Therapie* 58: 439–443.

12. Nadkarni, K. M. 1982. Indian Materia Medica, Vol I. Popular Prakashan: Mumbai.

13. Solis, C. S. 1969. *Acta Manilana, Series A: Natural and Applied Sciences* 4: 52–109.

14. Kasture, V. S., Deshmukh, V. K. and Chopde, C. T. 2002. *Phytotherapy Research* 16: 455–460.

15. Fojas, F. R., Barrientos, C. M., Capal, T. V., Cruzada, S. F., Sison, F. M., Co, Y. C.,

Chua, N. G. and Gavina, T. L. 1982. *Philippine Journal of Science* 111: 157–181.

16. Boonmee, A., Reynolds, C. D. and Sangvanich, P. 2007. *Planta Medica* 73: 1197–1201.

17. Kalyanagurunathan, P., Sulochana, N. and Murugesh, N. 1985. *Fitoterapia* 56: 188–189.

18. Kumar, V. R., Murugesh, N., Vembar, S. and Damodaran, C. 1982. *Toxicology Letters* 10: 157–161.

Solanum nigrum L.

1. Hsuan, K. 1990. The Concise Flora of Singapore, Gymnosperms and Dicotyledons. Singapore University Press: National University of Singapore.

2. eFloras.org 2008 http://www.efloras.org/florataxon.aspx?flora_id=2&taxon_id=200020597 Date Accessed: 17th April 2008

3. Duke, J. A. and Ayensu, E. S. 1985. Medicinal Plants of China, volume one and two. Reference Publications, Inc.: United States of America.

4. Fleming, T. (Ed.). 2000. PDR for Herbal Medicines, 2nd edition Medical Economics Company: New Jersey.

5. Goh, S. H. 1988. Proceedings: Malaysian Traditional Medicine, Kuala Lumpur, pp. 7–26. Institute of Advanced Studies, University of Malaya, Kuala Lumpur.

6. Goh, S. H., Chuah, C. H., Mok, J. S. L. and Soepadmo, E. 1995. Malaysian Medicinal Plants for the Treatment of Cardiovascular Diseases. Pelanduk Publications (M) Sdn Bhd: Malaysia.

7. Wang, L. Y., Wang, N. L. and Yao, X. S. 2007. *Zhong Yao Cai* 30: 792–794.

8. Zhou, X. L., He, X. J., Wang, G. H., Gao, H., Zhou, G. X., Ye, W. C. and Yao, X. S. 2006. *Journal of Natural Products* 69: 1158–1163.

9. Zhou, X. L., He, X. J., Zhou, G. X., Ye, W. C. and Yao, X. S. 2007. *Journal of Asian Natural Products Research* 9: 517–523.

10. Sumner, J. 2000. The Natural History of Medicinal Plants. Timber Press, Portland.

11. Wee, Y. C. and Hsuan, K. 1990. An illustrated Dictionary of Chinese Medicinal Herbs.

Times Edition and Eu Yan Seng Holdings Ltd: Singapore.

12. Perry, L. M. 1980. Medicinal plants of East and Southeast Asia: Attributed properties and uses. The MIT Press: Cambridge, Massachusetts and London.

13. Rani, P. and Khullar, N. 2004. *Phytotherapy Research* 18: 670–673.

14. Prashanth Kumar, V., Shashidhara, S., Kumar, M. M. and Sridhara, B. Y. 2001. *Fitoterapia* 72: 481–486.

15. Zakaria, Z. A., Gopalan, H. K., Zainal, H., Mohd Pojan, N. H., Morsid, N. A, Aris, A. and Sulaiman, M. R. 2006. *Yakugaku Zasshi* 126: 1171–1178.

16. Heo, K. S., Lee, S. J. and Lim, K. T. 2004. *Environmental Toxicology and Pharmacology* 17: 45–54.

17. Lee, S. J., Oh, P. S., Ko, J. H., Lim, K. and Lim, K. T. 2004. *Cancer Chemotherapy and Pharmacology* 54: 562–572.

18. Son, Y. O., Kim, J., Lim, J. C., Chung, Y., Chung, G. H. and Lee, J. C. 2003. *Food and Chemical Toxicology* 41: 1421–1428.

19 Hu, K., Kobayashi, H., Dong, A., Jing, Y., Iwasaki, S. and Yao X. 1999. *Planta Medica* 65: 35–38.

20. Saijo, R., Murakami, K., Nohara, T., Tomimatsu, T., Sato, A. and Matsuoka, K. 1982. *Yakugaku Zasshi* 102: 300–305.

21. Ji, Y. B., Gao, S. Y., Ji, C. F. and Zou, X. 2008. *Journal of Ethnopharmacology* 115: 194–202.

22. Jeong, J. B., Jeong, H. J., Park, J. H., Lee, S. H., Lee, J. R., Lee, H. K., Chung, G. Y., Choi, J. D. and de Lumen, B. O. 2007. *Journal of Agricultural and Food Chemistry* 55: 10707–10713.

23. Li, J., Li, Q., Feng, T., Zhang, T., Li, K., Zhao, R., Han, Z. and Gao, D. 2007. *Phytotherapy Research* 21: 832–840.

24. Lin, H. M., Tseng, H. C., Wang, C. J., Chyau, C. C., Liao, K. K., Peng, P. L. and Chou, F. P. 2007. *Journal of Agricultural and Food Chemistry* 55: 3620–3628.

25. Lee, S. J. and Lim, K. T. 2008. *Journal of Nutritional Biochemistry* 19: 166–174.

26. Oh, P. S. and Lim, K. T. 2007. *Journal of Biomedical Science* 14: 223–232.

27. Lee, S. J. and Lim, K. T. 2006. *Cancer Chemotherapy and Pharmacology* 57: 507–516.

28. Lee, S. J., Ko, J. H. and Lim, K. T. 2005. *Oncology Reports* 14: 789–796.
29. Lim, K. T. 2005. *Journal of Medicinal Food* 8: 215–226.
30. Heo, K. S. and Lim, K. T. 2005. *Journal of Medicinal Food* 8: 69–77.
31. An, L., Tang, J. T., Liu, X. M. and Gao, N. N. 2006. *Zhongguo Zhong Yao Za Zhi* 31: 1225–1226.
32. An, H. J., Kwon, K. B., Cho, H. I., Seo, E. A., Ryu, D. G., Hwang, W. J., Yoo, S. J., Kim, Y. K., Hong, S. H. and Kim, H. M. 2005. *European Journal of Cancer Prevention* 14: 345–350.
33. Wang, W. and Lu, D. P. 2005. *Beijing Da Xue Xue Bao* 37: 240–244.
34. Heo, K. S. and Lim, K. T. 2004. *Journal of Medicinal Food* 7: 349–357.
35. Al-Fatimi, M., Wurster, M., Schröder, G. and Lindequist, U. 2007. *Journal of Ethnopharmacology* 111: 657–166.
36. Akhtar, M. S. and Munir, M. 1989. *Journal of Ethnopharmacology* 27: 163–176.
37. Jainu, M. and Devi, C. S. 2006. *Journal of Ethnopharmacology* 104: 156–163.
38. Ikeda, T., Ando, J., Miyazono, A., Zhu, X. H., Tsumagari, H., Nohara, T., Yokomizo, K. and Uyeda, M. 2000. *Biological & Pharmaceutical Bulletin* 23: 363–364.
39. Perez, R. M., Perez, J. A., Garcia, L. M. and Sossa, H. 1998. *Journal of Ethnopharmacology* 62: 43–48.
40. Raju, K., Anbuganapathi, G., Gokulakrishnan, V., Rajkapoor, B., Jayakar, B. and Manian, S. 2003. *Biological & Pharmaceutical Bulletin* 26: 1618–1619.
41. Sultana, S., Perwaiz, S., Iqbal, M. and Athar, M. 1995. *Journal of Ethnopharmacology* 45: 189–192.
42. Lin, H. M., Tseng, H. C., Wang, C. J., Lin, J. J., Lo, C. W. and Chou, F. P. 2008. *Chemico-biological Interactions* 171: 283–293.
43. Huseini, H. F., Alavian, S. M., Heshmat, R., Heydari, M. R. and Abolmaali, K. 2005. *Phytomedicine* 12: 619–624.
44. Lee, S. J., Ko, J. H., Lim, K. and Lim, K. T. 2005. *Pharmacological Research* 51: 399–408.
45. Yen, G. C., Chen, H. Y. and Peng, H. H. 2001. *Food and Chemical Toxicology* 39: 1045–1053.
46. Moundipa, P. F. and Domngang, F. M. 1991. *British Journal of Nutrition* 65: 81–91.
47. Ahmed, A. H., Kamal, I. H. and Ramzy, R. M. 2001. *Journal of the Egyptian Society of Parasitology* 31: 843–852.
48. Ahmed, A. H. and Rifaat M. M. 2004. *Journal of the Egyptian Society of Parasitology* 34: 1041–1050.
49. El-Ansary, A. K. and Al Daihan, S. K. 2007. *Journal of the Egyptian Society of Parasitology* 37: 39–50.
50. Ahmed, A. H. and Rifaat M. M. 2005. *Journal of the Egyptian Society of Parasitology* 35: 33–40.
51. Dugan, G. M. and Gumbmann, M. R. 1990. *Food and Chemical Toxicology* 28: 101–117.
52. Duke, J. A. 1985. CRC Handbook of Medicinal Herbs. CRC Press: Boca Raton, Florida.

Swietenia macrophylla King

1. Wiart, C. 2006. Medicinal Plants of the Asia-Pacific: Drugs for the future? World Scientific Publishing Co. Pte. Ltd.: Singapore.
2. Connolly, J. D., McCrindle, R., Overton, K. H. and Warnock, W. D. C. 1965. *Tetrahedron Letters* 33: 2937–2940.
3. Solomon, K. A., Malathi, R., Rajan, S. S., Narasimhan, S. and Nethaji, M. 2003. *Acta Crystallographica E* 59: O1519–O1521.
4. Connolly, J. D., Henderson, R., McCrindle, R., Overton, K. H. and Bhacca, N. S. 1964. *Tetrahedron Letters* 5: 2593–2597.
5. Taylor Anne, R. H. and Taylor David, A. H. 1983. *Phytochemistry* 22: 2870–2871.
6. Chan, K. C., Tang, T. S. and Toh, H. T. 1976. *Phytochemistry* 15: 429–430.
7. Mootoo, B. S., Ali, A., Motilal, R., Pingal, R., Ramlal, A., Khan, A., Reynolds, W. F. and McLean, S. 1999. *Journal of Natural Products* 62: 1514–1517.
8. Tang, T. S. 1976. Some chemical studies of oil palm & *Swietenia macrophylla* (Master's Thesis). University of Malaya (Malaysia), Faculty of Science.
9. Kadota, S., Marpaung, L., Kikuchi, T. and Ekimoto, H. 1990. *Chemical and Pharmaceutical Bulletin* 38: 894–901.
10. Munoz, V., Sauvain, M., Bourdy, G., Callapa, J., Rojas, I., Vargas, L., Tae, A. and Deharo, E.

2000. *Journal of Ethnopharmacology* 69: 139–155.

11. Sodiero, I., Padmawinata, K., Wattimena, J. R. and Rekita, S. 1990. *Acta Pharmaceutica Indonesia* 15: 1–13.

12. Maiti, A., Dewanjee, S. and Mandal, S. C. 2007. *Tropical Journal of Pharmaceutical Research* 6: 711–716.

13. Bourdy, G., DeWalt, S. J., Chávez de Michel, L. R., Roca, A., Deharo, E., Muñoz, V., Balderrama, L., Quenevo, C. and Gimenez, A. 2000. *Journal of Ethnopharmacology* 70: 87–109.

Terminalia catappa L.

1. Goh, S. H., Chuah, C. H., Mok, J. S. L. and Soepadmo, E. 1995. Malaysian Medicinal Plants for the Treatment of Cardiovascular Diseases. Pelanduk Publications (M) Sdn Bhd: Malaysia.

2. Wee, Y. C. 1992. A Guide to Medicinal Plants. Singapore Science Centre Publication: Singapore.

3. Wee, Y. C. and Corlett, R. 1986. The City and the Forest: Plant Life in Urban Singapore. Singapore University Press: National University of Singapore.

4. USDA, ARS GRIN Taxonomy for Plants 2008 http://www.ars-grin.gov/cgi-bin/npgs/html/taxon.pl?36334 Date Accessed: 30th June 2008

5. Fan, Y. M., Xu, L. Z., Gao, J., Wang, Y., Tang, X. H., Zhao, X. N. and Zhang, Z. X. 2004. *Fitoterapia* 75: 253–260.

6. Masuda, T., Yonemori, S., Oyama, Y., Takeda, Y., Tanaka, T., Andoh, T., Shinohara, A. and Nakata, M. 1999. *Journal of Agricultural and Food Chemistry* 47: 1749–1754.

7. Lin, C. C., Hsu, Y. F., Lin, T. C., Hsu, F. L. and Hsu, H. Y. 1998. *The Journal of Pharmacy and Pharmacology* 50: 789–794.

8. Nagappa, A. N., Thakurdesai, P. A., Venkat Rao, N. and Singh, J. 2003. *Journal of Ethnopharmacology* 88: 45–50.

9. Mau, J. L., Ko, P. T. and Chyau, C. C. 2003. *Food Research International* 36: 97–104.

10. Ko, T. F., Weng, Y. M. and Chiou, R. Y. 2002. *Journal of Agriculture and Food Chemistry* 50: 5343–5348.

11. Kinoshita, S., Inoue, Y., Nakama, S., Ichiba, T. and Aniya, Y. 2007. *Phytomedicine* 14: 755–762.

12. Lin, Y. L., Kuo, Y. H., Shiao, M. S., Chen, C. C. and Ou, J. C. 2000. *Journal of the Chinese Chemical Society* 47: 253–256.

13. Lin, T. C. and Hsu, F. L. 1999. *Journal of the Chinese Chemical Society* 46: 613–618.

14. Woodley, E. (Ed.). 1991. Medicinal plants of Papua New Guinea, part I. Veerlag Josef Margraf Scientific Books: Weikersheim.

15. Ghazanfar, S. A. 1994. Handbook of Arabian Medicinal Plants. CRC Press: Boca Raton, FL.

16. Jain, S. K. and DeFilipps, R. A. 1991. Medicinal Plants of India, volume one and two. Reference Publications Inc.: United States of America.

17. Pawar, S. P. and Pal, S. C. 2002. *Indian Journal of Medical Sciences* 56: 276–278.

18. Goun, E., Cunningham, G., Chu, D., Nguyen, C. and Miles, D. 2003. *Fitoterapia* 74: 592–596.

19. Kloucek, P., Polesny, Z., Svobodova, B., Vlkova, E. and Kokoska, L. 2005. *Journal of Ethnopharmacology* 99: 309–312.

20. Chu, S. C., Yang, S. F., Liu, S. J., Kuo, W. H., Chang, Y. Z. and Hsieh, Y. S. 2007. *Food and Chemical Toxicology* 45: 1194–1201.

21. Chen, P. S. and Li, J. H. 2006. *Toxicology Letters* 163: 44–53.

22. Lin, C. C., Hsu, Y. F. and Lin, T. C. 1999. *American Journal of Chinese Medicine* 27: 371–376.

23. Chyau, C. C., Ko, P. T. and Mau, J. L. 2006. *LWT — Food Science and Technology* 39: 1099–1108.

24. Chen, P. S., Li, J. H., Liu, T. Y. and Lin, T. C. 2000. *Cancer Letters* 152: 115–122.

25. Tang, X. H., Gao, J., Wang, Y. P., Xu, L. Z., Zhao, X. N. and Xu, Q. 2003. *Zhongguo Zhong Yao Za Zhi* 28: 1170–1174.

26. Tang, X. H., Gao, L., Gao, J., Fan, Y. M., Xu, L. Z., Zhao, X. N. and Xu, Q. 2004. *American Journal of Chinese Medicine* 32: 509–519.

27. Tang, X. H., Gao, J., Wang, Y., Fan, Y. M., Xu, L. Z., Zhao, X., Xu, Q. and Qian Z. M. 2006. *The Journal of Nutritional Biochemistry* 17: 177–182.

28. Gao, J., Tang, X., Dou, H., Fan, Y., Zhao, X. and Xu, Q. 2004. *Journal of Pharmacy and Pharmacology* 56: 1449–1455.

29. Gao, J., Chen, J., Tang, X., Pan, L., Fang, F., Xu, L., Zhao, X. and Xu, Q. 2006. *Journal of Pharmacy and Pharmacology* 58: 227–233.

30. Nonaka, G., Nishioka, I., Nishizawa, M., Yamagishi, T., Kashiwada, Y., Dutschman, G. E., Bodner, A. J., Kilkuskie, R. E., Cheng, Y. C. and Lee, K. H. 1990. *Journal of Natural Products* 53: 587–595.

31. Morioka, T., Suzui, M., Nabandith, V., Inamine, M., Aniya, Y., Nakayama, T., Ichiba, T. and Yoshimi, N. 2005. *European Journal of Cancer Prevention* 14: 101–105.

32. Ratnasooriya, W. D. and Dharmasiri, M. G. 2000. *Asian Journal of Andrology* 2: 213–219.

33. Ko, T. F., Weng, Y. M., Lin, S. B. and Chiou, R. Y. 2003. *Journal of Agriculture and Food Chemistry* 51: 3564–3567.

34. Morton, J. F. 1981. Atlas of Medicinal Plants of Middle America: Bahamas to Yucatan. C.C. Thomas Publishers Co.: Springfield, IL.

35. Oliveira, J. T. A., Vasconcelos, I. M., Bezerra, L. C. N. M., Silveira, S. B., Monteiro, A. C. O. and Moreira, R. A. 2000. *Food Chemistry* 70: 185–191.

36. Lin, C. C., Hsu, Y. F., Lin, T. C. and Hsu, H. Y. 2001. *Phytotherapy Research* 15: 206–212.

Thevetia peruviana (Pers.) K. Schum.

1. Hsuan, K. 1990. The Concise Flora of Singapore, Gymnosperms and Dicotyledons. Singapore University Press: National University of Singapore.

2. eFloras.org 2008
   http://www.efloras.org/florataxon.aspx?flora_id=2&taxon_id=200018473
   Date Accessed: 30th March 2008

3. Duke, J. A. and Ayensu, E. S. 1985. Medicinal Plants of China, volume one and two. Reference Publications, Inc.: United States of America.

4. Goh, S. H., Chuah, C. H., Mok, J. S. L. and Soepadmo, E. 1995. Malaysian Medicinal Plants for the Treatment of Cardiovascular Diseases. Pelanduk Publications (M) Sdn Bhd: Malaysia.

5. Wiart, C. 2000. Medicinal plants of Southeast Asia. Pelanduk Publications (M) Sdn Bhd: Malaysia.

6. Tewtrakul, S., Nakamura, N., Hattori, M., Fujiwara, T. and Supavita, T. 2002. *Chemical & Pharmaceutical Bulletin* 50: 630–635.

7. Morton, J. F. 1981. Atlas of Medicinal Plants of Middle America: Bahamas to Yucatan. C.C. Thomas Publishers Co.: Springfield, IL.

8. Gata-Goncalves, L., Nogueira, J. M., Matos, O. and Bruno de Sousa, R. 2003. *Journal of Photochemistry and Photobiology B* 70: 51–54.

9. Lapcharoen, P., Apiwathnasorn, C., Komalamisra, N., Dekumyoy, P., Palakul, K. and Rongsriyam, Y. 2005. *Southeast Asian Journal of Tropical Medicine and Public Health* 36: 167–175.

10. Singh, A. and Singh, S. K. 2005. *Fitoterapia* 76: 747–751.

11. Eddleston, M., Ariaratnam, C. A., Sjostrom, L., Jayalath, S., Rajakanthan, K., Rajapakse, S., Colbert, D., Meyer, W. P., Perera, G., Attapattu, S., Kularatne, S. A., Sheriff, M. R. and Warrell, D. A. 2000. *Heart* 83: 301–306.

12. Eddleston, M., Ariaratnam, C. A., Meyer, W. P., Perera, G., Kularatne, A. M., Attapattu, S., Sheriff, M. H. and Warrell, D. A. 1999. *Tropical Medicine and International Health* 4: 266–273.

13. Duke, J. A. 1985. CRC Handbook of Medicinal Herbs. CRC Press: Boca Raton, FL.

14. Samal, K. K., Sahu, H. K., Kar, M. K., Palit, S. K., Kar, B. C. and Sahu, C. S. 1989. *Journal of the Associations of Physicians of India* 37: 232–233.

15. Maringhini, G., Notaro, L., Barberi, O., Giubilato, A., Butera, R. and Di Pasquale, P. 2002. *Italian Heart Journal* 3: 137–140.

16. Camphausen, C., Haas, N. A. and Mattke, A. C. 2005. *Zeitschrift für Kardiologie* 94: 817–823.

17. Sambasivam, S., Karpagam, G., Chandran, R. and Khan, S. A. 2003. *Journal of Environmental Biology* 24: 201–204.

18. Singh, D. and Singh, A. 2005. *Chemosphere* 60: 135–140.

19. Singh, D. and Singh, A. 2002. *Chemosphere* 49: 45–49.

20. Oji, O. and Okafor, Q. E. 2000. *Phytotherapy Research* 14: 133–135.

21. Bose, T. K., Basu, R. K., Biswas, B., De, J. N., Majumdar, B. C. and Datta, S. 1999. *Journal of the Indian Medical Association* 97: 407–410.

22. Saravanapavananthan, T. 1985. *Jaffna Medical Journal* 20: 17–21.
23. Nguyen, V. D. and Doan, T. N. 1989. Medicinal Plants in Vietnam. World Health Organisation, Regional Office for the Western Pacific, Manila, Institute of Materia Medica, Hanoi.
24. Kee, C. H. 1998. The Pharmacology of Chinese Herbs, 2nd edition. CRC Press: Boca Raton, Fl.

Tinospora crispa (L.) Diels

1. Hsuan, K. 1990. The Concise Flora of Singapore, Gymnosperms and Dicotyledons. Singapore University Press: National University of Singapore.
2. Nguyen, V. D. 1993. Medicinal Plants of Vietnam, Cambodia and Laos. Nguyen Van Duong: Vietnam.
3. USDA, ARS GRIN Taxonomy for Plants 2008 http://www.ars-grin.gov/cgi-bin/npgs/html/taxon.pl?428217
   Date Accessed: 30th June 2008
4. Kongkathip, N., Dhumma-upakorn, P., Kongkathip, B., Chawananoraset, K., Sangchomkaeo, P. and Hatthakitpanichakul, S. 2002. *Journal of Ethnopharmacology* 83: 95–99.
5. Goh, S. H., Chuah, C. H., Mok, J. S. L. and Soepadmo, E. 1995. Malaysian Medicinal Plants for the Treatment of Cardiovascular Diseases. Pelanduk Publications (M) Sdn Bhd: Malaysia.
6. Kalsom, Y. U. and Noor, H. 1995. *Fitoterapia* 66: 280.
7. Cavin, A., Hostettmann, K., Dyatmyko, W. and Potterat, O. 1998. *Planta Medica* 64: 393–396.
8. Pachaly, P., Adnan, A. Z. and Will, G. 1992. *Planta Medica* 58: 184–187.
9. Marzuki, A. and Iljaz, S. 1988. Proceedings: Malaysian Traditional Medicine, Kuala Lumpur, pp. 183–191. Institute of Advanced Studies, University of Malaya, Kuala Lumpur.
10. Noor, H., Hammonds, P., Sutton, R. and Ashcroft, S. J. 1989. *Diabetologia* 32: 354–359.
11. Higashino, H., Suzuki, A., Tanaka, Y. and Pootakham, K. 1992. *Nippon Yakurigaku Zasshi* 100: 339–344.
12. Rahman, N. N. N. A., Furuta, T., Kojima, S., Takane, K. and Mustafa Ali Mohd. 1999. *Journal of Ethnopharmacology* 64: 249–254.
13. Bertani, S., Bourdy, G., Landau, I., Robinson, J. C., Esterre, P. and Deharo, E. 2005. *Journal of Ethnopharmacology* 98: 45–54.
14. Zaridah, M. Z., Idid, S. Z., Omar, A. W. and Khozirah, S. 2001. *Journal of Ethnopharmacology* 78: 79–84.
15. Noor, H. and Ashcroft, S. J. 1998. *Journal of Ethnopharmacology* 62: 7–13.
16. Noor, H. and Ashcroft, S. J. 1989. *Journal of Ethnopharmacology* 27: 149–161.
17. Sangsuwan, C., Udompanthurak, S., Vannasaeng, S. and Thamlikitkul, V. 2004. *Journal of the Medical Association of Thailand* 87: 543–546.
18. Subehan, Usia, T., Iwata, H., Kadota, S. and Tezuka, Y. 2006. *Journal of Ethnopharmacology* 105: 449–455.

Vitex rotundifolia L. f.

1. Byung, H. H., Youngbae, S. and Hyung-Joon, C. 1998. Medicinal plants in the Republic of Korea. World Health Organization, Regional Office for the Western Pacific, Manila, Natural Products Research Institute, Seoul National University.
2. USDA, ARS GRIN Taxonomy for Plants 2008 http://www.ars-grin.gov/cgi-bin/npgs/html/taxon.pl?41839
   Date Accessed: 2nd July 2008
3. Chen, Y. L., Wang, Y. S., Kao, B. C. and Chang, Y. A. 1978. *Zhongguo Nongye Huaxue Huizhi* 16: 99–102.
4. Tada, H. and Yasuda, F. 1984. *Heterocycles* 22: 2203–2205.
5. Ono, M., Ito, Y. and Nohara, T. 2001. *Chemical and Pharmaceutical Bulletin (Tokyo)* 49: 1220–1222.
6. Ono, M., Yamamoto, M., Masuoka, C., Ito, Y., Yamashita, M. and Nohara, T. 1999. *Journal of Natural Products* 62: 1532–1537.
7. Ono, M., Yanaka, T., Yamamoto, M., Ito, Y. and Nohara, T. 2002. *Journal of Natural Products* 65: 537–541.
8. Hu, Y., Xin, H. L., Zhang, Q. Y., Zheng, H. C., Rahman, K. and Qin, L. P. 2007. *Phytomedicine* 14: 668–674.

9. Yoshioka, T., Inokuchi, T., Fujioka, S. and Kimura, Y. 2004. *Zeitschrift für Naturforschung. C, Journal of biosciences* 59: 509–514.

10. de Padua, L. S., Bunyapraphatsara, N. and Lemmens, R. H. M. J. (Eds). 1999. Plant resources of South-East Asia, No. 12, Medicinal and Poisonous Plants 1. Backhuys Publishers: Leiden, the Netherlands.

11. Li, W. X., Cui, C. B., Cai, B. and Yao, X. S. 2005. *Journal of Asian Natural Product Research* 7: 95–105.

12. Okuyama, E., Fujimori, S., Yamazaki, M. and Deyama, T. 1998. *Chemical and Pharmaceutical Bulletin (Tokyo)* 46: 655–662.

13. Hossain, M. M., Paul, N., Sohrab, M. H., Rahman, E. and Rashid, M. A. 2001. *Fitoterapia* 72: 695–697.

14. Kawazoe, K., Yutani, A., Tamemoto, K., Yuasa, S., Shibata, H., Higuti, T. and Takaishi, Y. 2001. *Journal of Natural Products* 64: 588–591.

15. Hernandez, M. M., Heraso, C., Villarreal, M. L., Vargas-Arispuro, I. and Aranda, E. 1999. *Journal of Ethnopharmacology* 67: 37–44.

16. Shin, T. Y., Kim, S. H., Lim, J. P., Suh, E. S., Jeong, H. J., Kim, B. D., Park, E. J., Hwang, W. J., Rye, D. G., Baek, S. H., An, N. H. and Kim, H. M. 2000. *Journal of Ethnopharmacology* 72: 443–450.

17. You, K. M., Son, K. H., Chang, H. W., Kang, S. S. and Kim, H. P. 1998. *Planta Medica* 64: 546–550.

18. Ko, W. G., Kang, T. H., Lee, S. J., Kim, N. Y., Kim, Y. C., Sohn, D. H. and Lee, B. H. 2000. *Food and Chemical Toxicology* 38: 861–865.

19. Ko, W. G., Kang, T. H., Lee, S. J., Kim, Y. C. and Lee, B. H. 2001. *Phytotherapy Research* 15: 535–537.

20. Kobayakawa, J., Sato-Nishimori, F., Moriyasu, M. and Matsukawa, Y. 2004. *Cancer Letters* 208: 59–64.

21. Kiuchi, F., Matsuo, K., Ito, M., Qui, T. K. and Honda, G. 2004. *Chemical and Pharmaceutical Bulletin (Tokyo)* 52: 1492–1494.

22. Ko, S. T., Moon, Y. H. and Ko, O. H. 1977. *Saengyak Hakhoechi* 8: 55–60.

23. Miyazawa, M., Shimamura, H., Nakamura, S. and Kameoka, H. 1995. *Journal of Agricultural and Food Chemistry* 43: 3012–3015.

Zingiber officinale Roscoe.

1. Palaniswamy, U. R. 2003. A guide to Medicinal Plants of Asian origin and culture. CPL Press: United States of America.

2. USDA, ARS GRIN Taxonomy for Plants 2008 http://www.ars-grin.gov/cgi-bin/npgs/html/taxon.pl?42254 Date Accessed: 30th June 2008

3. Li, T. S. C. 2000. Medicinal Plants. Culture, Utilisation and Phytopharmacology. Technomic Publishing Co., Inc.: Lancaster.

4. Ma, J., Jin, X., Yang, L. and Liu, Z. L. 2004. *Phytochemistry* 65: 1137–1143.

5. Jolad, S. D., Lantz, R. C., Chen, G. J., Bates, R. B. and Timmermann, B. N. 2005. *Phytochemistry* 66: 1614–1635.

6. Tao, Q. F., Xu, Y., Lam, R. Y., Schneider, B., Dou, H., Leung, P. S., Shi, S. Y., Zhou, C. X., Yang, L. X., Zhang, R. P., Xiao, Y. C., Wu, X., Stöckigt, J., Zeng, S., Cheng, C. H. and Zhao, Y. 2008. *Journal of Natural Products* 71: 12–17.

7. Sekiwa, Y., Kubota, K. and Kobayashi, A. 2000. *Journal of Agriculture and Food Chemistry* 48: 373–377.

8. Duke, J. A. and Ayensu, E. S. 1985. Medicinal Plants of China, volume one and two. Reference Publications, Inc.: United States of America.

9. Wee, Y. C. and Hsuan, K. 1990. An illustrated Dictionary of Chinese Medicinal Herbs. Times Edition and Eu Yan Seng Holdings Ltd: Singapore.

10. Wee, Y. C. 1992. A Guide to Medicinal Plants. Singapore Science Centre Publication: Singapore.

11. Blumenthal, M., Goldberg, A. and Gruenwald, J. (Eds.). 2000. Herbal Medicine. Expanded Commission E Monographs. Integrative Medicine Communications: Austin, Texas.

12. Bradley, P. R. (Ed.). 1992. British Herbal Compendium, vol. 1. British Herbal Medicine Association: Bournemouth.

13. ESCOP. 1997. "Zingiberis rhizoma." Monographs on the Medicinal Uses of Plant Drugs. European Scientific Cooperative on Phytotherapy: Exeter, UK.

14. Ojewole, J. A. 2006. *Phytotherapy Research* 20: 764–772.

15. Young, H. Y., Luo, Y. L., Cheng, H. Y., Hsieh, W. C., Liao, J. C. and Peng, W. H. 2005. *Journal of Ethnopharmacology* 96: 207–210.

16. Cady, R. K., Schreiber, C. P., Beach, M. E. and Hart, C. C. 2005. *Medical Science Monitor* 11: PI65–69.
17. Iqbal, Z., Lateef, M., Akhtar, M. S., Ghayur, M. N. and Gilani, A. H. 2006. *Journal of Ethnopharmacology* 106: 285–287.
18. Phan, P. V., Sohrabi, A., Polotsky, A., Hungerford, D. S., Lindmark, L. and Frondoza, C. G. 2005. *Journal of Alternative and Complementary Medicine* 11: 149–154.
19. Shukla, Y., Prasad, S., Tripathi, C., Singh, M., George, J. and Kalra, N. 2007. *Molecular Nutrition and Food Research* 51: 1492–1502.
20. Shukla, Y. and Singh, M. 2007. *Food and Chemical Toxicology* 45: 683–690.
21. Sagar, S. M., Yance, D. and Wong, R. K. 2006. *Current Oncology* 13: 99–107.
22. Sagar, S. M., Yance, D. and Wong, R. K. 2006. *Current Oncology* 13: 14–26.
23. Yance, D. R. Jr. and Sagar, S. M. 2006. *Integrative Cancer Therapy* 5: 9–29.
24. Rhode, J., Fogoros, S., Zick, S., Wahl, H., Griffith, K. A., Huang, J. and Liu, J. R. 2007. *BMC Complementary and Alternative Medicine* 20: 7: 44.
25. Lee, S. H., Cekanova, M. and Baek, S. J. 2008. *Molecular Carcinogenesis* 47: 197–208.
26. Vijaya Padma, V., Arul Diana Christie, S. and Ramkuma, K. M. 2007. *Basic and Clinical Pharmacology and Toxicology* 100: 302–307.
27. Hsu, M. H., Kuo, S. C., Chen, C. J., Chung, J. G., Lai, Y. Y. and Huang, L. J. 2005. *Leukemia Research* 29: 1399–1406.
28. Hofbauer, S., Kainz, V., Golser, L., Klappacher, M., Kiesslich, T., Heidegger, W., Krammer, B., Hermann, A. and Weiger, T. M. 2006. *Forsch Komplementmed* 13: 18–22.
29. Chen, C. Y., Liu, T. Z., Liu, Y. W., Tseng, W. C., Liu, R. H., Lu, F. J., Lin, Y. S., Kuo, S. H. and Chen, C. H. 2007. *Journal of Agriculture and Food Chemistry* 55: 948–954.
30. Kim, E. C., Min, J. K., Kim, T. Y., Lee, S. J., Yang, H. O., Han, S., Kim, Y. M., Kwon, Y. G. 2005. *Biochemical and Biophysical Research Communications* 335: 300–308.
31. Ishiguro, K., Ando, T., Maeda, O., Ohmiya, N., Niwa, Y., Kadomatsu, K. and Goto, H. 2007. *Biochemical and Biophysical Research Communications* 362: 218–223.
32. Vimala, S., Norhanom, A. W. and Yadav, M. 1999. *British Journal of Cancer* 80: 110–116.
33. Surh, Y. J., Park, K. K., Chun, K. S., Lee, L. J., Lee, E. and Lee, S. S. 1999. *Journal of Environmental Pathology, Toxicology and Oncology* 18: 131–139.
34. Al-Amin, Z. M., Thomson, M., Al-Qattan, K. K., Peltonen-Shalaby, R. and Ali, M. 2006. *British Journal of Nutrition* 96: 660–666.
35. Kato, A., Higuchi, Y., Goto, H., Kizu, H., Okamoto, T., Asano, N., Hollinshead, J., Nash, R. J. and Adachi, I. 2006. *Journal of Agriculture and Food Chemistry* 54: 6640–6644.
36. Chen, J. C., Huang, L. J., Wu, S. L., Kuo, S. C., Ho, T. Y. and Hsiang, C.Y. 2007. *Journal of Agriculture and Food Chemistry* 55: 8390–8397.
37. Keating, A. and Chez, R. A. 2002. *Alternative Therapies in Health and Medicine* 8: 89–91.
38. Eberhart, L. H., Mayer, R., Betz, O., Tsolakidis, S., Hilpert, W., Morin, A. M., Geldner, G., Wulf, H. and Seeling, W. 2003. *Anesthesia and Analgesia* 96: 995–998.
39. Sharma, S. S., Kochupillai, V., Gupta, S. K., Seth, S. D. and Gupta, Y. K. 1997. *Journal of Agriculture and Food Chemistry* 57: 93–96.
40. Sharma, S. S. and Gupta, Y. K. 1998. *Journal of Ethnopharmacology* 62: 49–55.
41. Vishwakarma, S. L., Pal, S. C., Kasture, V. S. and Kasture, S. B. 2002. *Phytotherapy Research* 16: 621–626.
42. Abdel-Aziz, H., Windeck, T., Ploch, M. and Verspohl, E. J. 2006. *European Journal of Ethnopharmacology* 530: 136–143.
43. Abdel-Aziz, H., Nahrstedt, A., Petereit, F., Windeck, T., Ploch, M. and Verspohl, E. J. 2005. *Planta Medica* 71: 609–616.
44. Borrelli, F., Capasso, R., Aviello, G., Pittler, M. H. and Izzo, A. A. 2005. *Obstetrics and Gynecology* 105: 849–856.
45. Chittumma, P., Kaewkiattikun, K. and Wiriyasiriwach, B. 2007. *Journal of the Medical Association of Thailand* 90: 15–20.
46. Apariman, S., Ratchanon, S. and Wiriyasirivej, B. 2006. *Journal of the Medical Association of Thailand* 89: 2003–2009.
47. Nanthakomon, T. and Pongrojpaw, D. 2006. *Journal of the Medical Association of Thailand* 89: S130–136.

48. Pongrojpaw, D., Somprasit, C. and Chanthasenanont, A. 2007. *Journal of the Medical Association of Thailand* 90: 1703–1709.
49. Ben-Arye, E., Oren, A. and Ben-Arie, A. 2006. *Harefuah* 145: 738–742.
50. Chaiyakunapruk, N., Kitikannakorn, N., Nathisuwan, S., Leeprakobboon, K. and Leelasettagool, C. 2006. *American Journal of Obstetric and Gynecology* 194: 95–99.
51. Boone, S. A. and Shields, K. M. 2005. *Annals of Pharmacotherapy* 39: 1710–1713.
52. Thompson, H. J. and Potter, P. J. 2006. *Evidence Based Nursing* 9: 80.
53. Bryer, E. 2005. *Journal of Midwifery and Women's Health* 50: e1–3.
54. Kadnur, S. V. and Goyal, R. K. 2005. *Indian Journal of Experimental Biology* 43: 1161–1164.
55. Ghayur, M. N. and Gilani, A. H. 2005. *Journal of Cardiovascular Pharmacology* 45: 74–80.
56. Kim, S. O., Chun, K. S., Kundu, J. K. and Surh, Y. J. 2004. *Biofactors* 21: 27–31.
57. Shen, C. L., Hong, K. J. and Kim, S. W. 2003. *Journal of Medicinal Food* 6: 323–328.
58. Penna, S. C., Medeiros, M. V., Aimbire, F. S., Faria-Neto, H. C., Sertie, J. A. and Lopes-Martins, R. A. 2003. *Phytomedicine* 10: 381–385.
59. Thomson, M., Al-Qattan, K. K., Al-Sawan, S. M., Alnaqeeb, M. A., Khan, I. and Ali, M. 2002. *Prostaglandins, Leukotrienes, and Essential Fatty Acids* 67: 475–478.
60. Altman, R. D. and Marcussen, K. C. 2001. *Arthritis and Rheumatism* 44: 2531–2538.
61. Minghetti, P., Sosa, S., Cilurzo, F., Casiraghi, A., Alberti, E., Tubaro, A., Loggia, R. D. and Montanari, L. 2007. *Planta Medica* 73: 1525–1530.
62. Aimbire, F., Penna, S. C., Rodrigues, M., Rodrigues, K. C., Lopes-Martins, R. A. and Sertié, J. A. 2007. *Prostaglandins, Leukotrienes, and Essential Fatty Acids* 77: 129–138.
63. Levy, A. S., Simon, O., Shelly, J. and Gardener, M. 2006. *BMC Pharmacology* 6: 12.
64. Aktan, F., Henness, S., Tran, V. H., Duke, C. C., Roufogalis, B. D. and Ammit, A. J. 2006. *Planta Medica* 72: 727–734.
65. Lantz, R. C., Chen, G. J., Sarihan, M., Sólyom, A. M., Jolad, S. D. and Timmermann, B. N. 2007. *Phytomedicine* 14: 123–128.
66. Grzanna, R., Lindmark, L. and Frondoza, C. G. 2005. *Journal of Medicinal Food* 8: 125–132.
67. Mahady, G. B., Pendland, S. L., Yun, G. S., Lu, Z. Z. and Stoia, A. 2003. *Anticancer Research* 23: 3699–3702.
68. Akoachere, J. F., Ndip, R. N., Chenwi, E. B., Ndip, L. M., Njock, T. E. and Anong, D. N. 2002. *East African Medical Journal* 79: 588–592.
69. Agarwal, M., Walia, S., Dhingra, S. and Khambay, B. P. 2001. *Pest Management Science* 57: 289–300.
70. Ficker, C. E., Arnason, J. T., Vindas, P. S., Alvarez, L. P., Akpagana, K., Gbeassor, M., De Souza, C. and Smith, M. L. 2003. *Mycoses* 46: 29–37.
71. Betoni, J. E., Mantovani, R. P., Barbosa, L. N., Di Stasi, L. C. and Fernandes Junior, A. 2006. *Memórias do Instituto Oswaldo Cruz* 101: 387–390.
72. Nagoshi, C., Shiota, S., Kuroda, T., Hatano, T., Yoshida, T., Kariyama, R. and Tsuchiya, T. 2006. *Biological and Pharmaceutical Bulletin* 29: 443–447.
73. Gupta, S. and Ravishankar, S. 2005. *Foodborne Pathogens and Disease* 2: 330–340.
74. López, P., Sánchez, C., Batlle, R. and Nerín, C. 2005. *Journal of Agriculture and Food Chemistry* 53: 6939–6946.
75. Norajit, K., Laohakunjit, N. and Kerdchoechuen, O. 2007. *Molecules* 12: 2047–2060.
76. Mahady, G. B., Pendland, S. L., Stoia, A., Hamill, F. A., Fabricant, D., Dietz, B. M. and Chadwick, L. R. 2005. *Phytotherapy Research* 19: 988–991.
77. Wang, H. and Ng, T. B. 2005. *Biochemical and Biophysical Research Communions* 336: 100–104.
78. Samy, R. P. 2005. *Fitoterapia* 76: 697–699.
79. Ansari, M. N., Bhandari, U. and Pillai, K. K. 2006. *Indian Journal of Experimental Biology* 44: 892–897.
80. Natarajan, K. S., Narasimhan, M., Shanmugasundaram, K. R. and Shanmugasundaram, E. R. 2006. *Journal of Ethnopharmacology* 105: 76–83.
81. Ippoushi, K., Ito, H., Horie, H. and Azuma, K. 2005. *Planta Medica* 71: 563–566.

82. Asnani, V. and Verma, R. J. 2007. *Acta Poloniae Pharmaceutica* 64: 35–37.
83. Nurtjahja-Tjendraputra, E., Ammit, A. J., Roufogalis, B. D., Tran, V. H. and Duke, C. C. 2003. *Thrombosis Research* 111: 259–265.
84. Koo, K. L., Ammit, A. J., Tran, V. H., Duke, C. C. and Roufogalis, B. D. 2001. *Thrombosis Research* 103: 387–397.
85. Guh, J. H., Ko, F. N., Jong, T. T. and Teng, C. M. 1995. *Journal of Pharmacy and Pharmacology* 47: 329–332.
86. Young, H. Y., Liao, J. C., Chang, Y. S., Luo, Y. L., Lu, M. C. and Peng, W. H. 2006. *American Journal of Chinese Medicine* 34: 545–551.
87. Jantan, I., Rafi, I. A. and Jalil, J. 2005. *Phytomedicine* 12: 88–92.
88. Riyazi, A., Hensel, A., Bauer, K., Geissler, N., Schaaf, S. and Verspohl, E. J. 2007. *Planta Medica* 73: 355–362.
89. Yoshikawa, M., Hatakeyama, S., Taniguchi, K., Matuda, H. and Yamahara, J. 1992. *Chemical and Pharmaceutical Bulletin Tokyo* 40: 2239–2241.
90. Siddaraju, M. N. and Dharmesh, S. M. 2007. *Molecular Nutrition and Food Research* 51: 324–332.
91. Koch, C., Reichling, J., Schneele, J. and Schnitzler, P. 2008. *Phytomedicine* 15: 71–78.
92. Schnitzler, P., Koch, C. and Reichling, J. 2007. *Antimicrobial Agents and Chemotherapy* 51: 1859–1862.
93. Sookkongwaree, K., Geitmann, M., Roengsumran, S., Petsom, A. and Danielson, U. H. 2006. *Pharmazie* 61: 717–721.
94. Vishwakarma, S. L., Pal, S. C., Kasture, V. S. and Kasture, S. B. 2002. *Phytotherapy Research* 16: 621–626.
95. Ajith, T. A., Hema, U. and Aswathy, M. S. 2007. *Food and Chemical Toxicology* 45: 2267–2272.
96. Yemitan, O. K. and Izegbu, M. C. 2006. *Phytotherapy Research* 20: 997–1002.
97. Nie, H., Meng, L. Z. and Zhang, H. 2006. *Zhongguo Zhong Xi Yi Jie He Za Zhi* 26: 529–532.
98. Bhandari, U., Sharma, J. N. and Zafar, R. 1998. *Journal of Ethnopharmacology* 61: 167–171.

99. Verma, S. K., Singh, M., Jain, P. and Bordia, A. 2004. *Indian Journal of Experimental Biology* 42: 736–738.
100. Akhani, S. P., Vishwakarma, S. L. and Goyal, R. K. 2004. *Journal of Pharmacy and Pharmacology* 56: 101–105.
101. Srinivasan, K. 2005. *International Journal of Food Science and Nutrition* 56: 399–414.
102. Bhandari, U., Kanojia, R. and Pillai, K. K. 2005. *Journal of Ethnopharmacology* 97: 227–230.
103. Ghayur, M. N. and Gilani, A. H. 2005. *Journal of Cardiovascular Pharmacology* 45: 74–80.
104. Chang, C. P., Chang, J. Y., Wang, F. Y. and Chang, J. G. 1995. *Journal of Ethnopharmacology* 48: 13–19.
105. Imanishi, N., Andoh, T., Mantani, N., Sakai, S., Terasawa, K., Shimada, Y., Sato, M., Katada, Y., Ueda, K. and Ochiai, H. 2006. *American Journal of Chinese Medicine* 34: 157–169.
106. Shen, C. L., Hong, K. J. and Kim, S. W. 2005. *Journal of Medicinal Food* 8: 149-153.
107. Kim, D. S., Kim, J. Y. and Han, Y. S. 2007. *Journal of Alternative and Complementary Medicine* 13: 333–340.
108. Prajapati, V., Tripathi, A. K., Aggarwal, K. K. and Khanuja, S. P. 2005. *Bioresource Technology* 96: 1749–1757.
109. Jagetia, G., Baliga, M. and Venkatesh, P. 2004. *Cancer Biotherapy and Radiopharmaceuticals* 19: 422–435.
110. Haksar, A., Sharma, A., Chawla, R., Kumar, R., Arora, R., Singh, S., Prasad, J., Gupta, M., Tripathi, R. P., Arora, M. P., Islam, F. and Sharma, R. K. 2006. *Pharmacology, Biochemistry and Behavior* 84: 179–188.
111. Sharma, A., Haksar, A., Chawla, R., Kumar, R., Arora, R., Singh, S., Prasad, J., Islam, F., Arora, M. P. and Kumar Sharma, R. 2005. *Pharmacology, Biochemistry and Behavior* 81: 864–870.
112. Bisset, N. G. 1994. Herbal drugs and phytopharmaceuticals. CRC Press: Boca Raton, FL.
113. McGuffin, M., Hobbs, C., Upton, R. and Goldberg, A. (Eds.). 1997. *American Herbal Products Association's Botanical Safety Handbook.* CRC Press: Boca Raton, FL.

114. Seetheram, K. A. and Pasricha, J. S. 1987. *Indian Journal of Dermatology, Venereology and Leprology* 53: 325–328.

115. Weidner, M. S. and Sigwart, K. 2001. *Reproductive Toxicology* 15: 75–80.

116. Iwu, M. M. 1993. Handbook of African Medicinal Plants. CRC Press: Boca Raton, FL.

117. Meyer, K., Schwartz, J., Crater, D. and Keyes, B. 1995. *Dermatology Nursing* 7: 242–244.

118. Shalansky, S., Lynd, L., Richardson, K., Ingaszewski, A. and Kerr, C. 2007. *Pharmacotherapy* 27: 1237–1247.

119. Jiang, X., Williams, K. M., Liauw, W. S., Ammit, A. J., Roufogalis, B. D., Duke, C. C., Day, R. O. and McLachlan, A. J. 2005. *British Journal of Clinical Pharmacology* 59: 425–432.

120. Chiang, H. M., Chao, P. D., Hsiu, S. L., Wen, K. C., Tsai, S. Y. and Hou, Y. C. 2006. *American Journal of Chinese Medicine* 34: 845–855.

Botanical Glossary

abaxial: away from the axis, referring to the surface of an organ that is furthest from the axis in bud

actinomorphic: having radially arranged floral segments which are more or less equal in size and shape

acuminate: tapering gradually to an extended point

acute: terminating in a distinct but not extended point, the converging edges forming an angle of less than 90 degrees

adaxial: towards the axis, referring to the surface of an organ that is closest to the axis in bud

alternate: borne singly and spaced around and along the axis, applied to leaves or other organs on an axis

annual: a plant/tree whose life cycle is only one year

articulated: consisting of segments held together by joints

axillary: situated at the angle between one part of a plant and another part, e.g., a branch and a leaf

basal: arising from or positioned at the base

bipinnate: 2-pinnate; twice pinnately divided

blade: part of the leaf above the sheath or petiole

bract: a leaf-like structure, usually different in form from the normal leaves, associated with the inflorescence

bulbous: shaped like a bulb; having an underground storage organ made up of enlarged and fleshy scales wrapped around each other from which flowers and leaves are produced

calyx: the outermost part of a flower, usually green

capsule: a dry fruit formed from two or more carpels that splits at maturity to release the seeds

carpel: the female reproductive organ of a flower

chartaceous: papery

compound: consisting of two or more anatomically or morphologically equivalent units

cordate: heart-shaped in outline

coriaceous: leathery

corolla: consists of petals or a corolla tube and corolla lobes

crenate: with obtuse or rounded teeth which either point forwards or are perpendicular to the margin

cuneate: obtriangular, i.e., wedge-shaped

cylindrical: tubular- or rod-shaped

cyme: an inflorescence in which each flower, in turn, is formed at the tip of a growing axis, further flowers being formed on branches arising below. *adj.* cymose

decussate: having paired organs with successive pairs at right angles to give four rows

dehiscent: breaking open at maturity to release the contents

dentate: with sharp, spreading, rather coarse teeth standing out from the margin

denticulate: finely dentate

discoid: resembling a disc

drupe: a fleshy or pulpy fruit with the inner portion of the pericarp hard or stony

ellipsoid: elliptic in outline and with a length:breadth ratio between 3:2 and 2:1

elliptic: widest at the middle of the blade, with curved margins and pointed apex and base

elongate: lengthened; stretched out

endocarp: the innermost layer of the fruit wall, derived from the innermost layer of the carpel wall

endospermous: possessing the nutritive tissue of a seed, consisting of carbohydrates, proteins, and lipids

entire: without any incisions or teeth

epiphyte: a plant growing on, but not parasitic on, another plant. *adj.* epiphytic

erect: upright; perpendicular

evergreen: a plant/tree that has leaves all year round

exocarp: the outermost layer of the fruit wall, derived from the outermost layer of the carpel wall. Sometimes called epicarp

exstipulate: leaves without stipules

filament: the stalk of a stamen below the point of attachment to the anther

-foliolate: used with a number prefix to denote the number of leaflets

follicle: a dry fruit, derived from a single carpel and dehiscing along one suture

frond: the leaf of a fern or cycad

fusiform: spindle-shaped, i.e., is circular in cross-section and tapering at both ends

glabrous: without hairs

glaucous: blue-green in colour, with a whitish bloom

globose: spherical or globular; circular in outline

herb: a plant which is non-woody or woody at the base only; some having medicinal properties

herbaceous: herb-like; often applied to bracts, bracteoles or floral parts that are green and soft in texture

hermaphrodite: self-pollinating plants containing both male and female reproductive organs, namely stamens and ovary respectively

imbricate: closely packed and overlapping

indehiscent: not opening or splitting to release the contents at maturity

inflorescence: the arrangement of flowers in relation to the axis and to each other

internode: the part of an axis between two successive nodes, joints or point of attachment of the leaves

introduced: not indigenous; not native to the area in which it now occurs

lanceolate: lance-shaped, much longer than wide, the widest point below the middle

leaflet: one segment of a compound leaf

leathery: possess the feel or texture of leather

linear: long and narrow, with essentially parallel margins

lobe: a usually rounded or pointed projecting part, usually one of two or more, each separated by a gap

margin: the edge of the leaf blade

mesocarp: the middle layer of the fruit wall derived from the middle layer of the carpel wall

mucilaginous: soft, moist, viscous and slimy

native: a plant indigenous to the locality

oblanceolate: reverse lanceolate, widest above the middle of the blade

oblong: rectangular, with nearly parallel margins, about two times as long as wide

obovate: reverse ovate, with the broadest part above

obtuse: blunt or rounded at the apex, the converging edges separated by an angle greater than 90 degrees

opposite: describing leaves or other organs which are at the same level but on opposite sides of the stem

ovate: broader part at the end of the base

ovoid: egg-shaped; ovate in outline

palmate: describing a leaf which is divided into several lobes

panicle: a compound raceme; an indeterminate inflorescence in which the flowers are borne on branches of the main axis or on further branches of these. *adj.* paniculate

paripinnate: pinnate with an even number of leaflets and without a terminal leaflet

pedicel: the stalk of an individual flower

perennial: with a life span extending over more than two growing seasons

perianth: the outer floral whorl or whorls of a monocotyledonous flower

pericarp: the wall of a fruit developed from the ovary wall. Composed of the exocarp, mesocarp and endocarp

petal: free segment of the corolla

petiole: the stalk of a leaf

pinna: a primary segment of the blade of a compound leaf or frond. *pl.* pinnae

pinnate: with the same arrangement as a feather

plumule: the primary bud of an embryo or germinating seed

pod: a dry dehiscent fruit containing many seeds

prostrate: lying flat on the ground

raceme: an unbranched flower spike where the flowers are borne on pedicels

rachis: the axis of a pinna in a bipinnate leaf

receptacle: the region at the end of a pedicel or on an axis which bears one or more flowers

reniform: kidney-shaped in outline

rhizome: a creeping stem, usually below ground, consisting of a series of nodes and internodes with adventitious roots

rhombic: diamond-shaped, widest at the middle and with straight margins

rosette: a tuft of leaves or other organs resembling the arrangement of petals in a rose

serrate: toothed so as to resemble a saw; with regular, asymmetric teeth pointing forward

sessile: without a stalk

shrub: a woody plant usually less than 5 m high and many-branched without a distinct main stem except at ground level

simple: not divided, e.g., applied to a leaf not divided into leaflets

spathe: a large bract ensheathing an inflorescence or its peduncle

spike: an unbranched inflorescence of sessile flowers or spikelets. *adj.* spicate

stamen: one of the male organs of a flower, consisting typically of a stalk (filament) and a pollen-bearing portion (anther)

stellate: star-shaped, usually referring to hairs with radiating branches

stipe: a stalk or support such as the petiole of a frond or the stalk of an ovary or fruit

stipule: one of a pair of leaf-like, scale-like or bristle-like structures inserted at the base or on the petiole of a leaf or phyllode. *adj.* stipulate

stolon: the creeping stem of a rosetted or tufted plant, giving rise to another plant at its tip

strobilus: a cone-like structure formed from sporophylls or sporangiophores. *pl.* strobili

succulent: fleshy, juicy, soft in texture and usually thickened

tendril: a slender organ formed from a modified stem, leaf or leaflet which, by coiling around objects, supports a climbing plant

terminal: at the apex or distal end

tomentose: covered with not very long cottony hairs

trigonous: obtusely 3-angled; triangular in cross-section with plane faces

truncate: with an abruptly transverse end as if cut off

tuber: a stem, usually underground, enlarged as a storage organ and with minute scale-like leaves and buds or "eyes"

umbel: an inflorescence in which the pedicels originate from one point on top of the peduncle and are usually of equal length

undulate: with an edge or edges wavy in a vertical plane

urceolate: urn-shaped

variegate: diverse in colour or marked with irregular patches of different colours

vein: the vascular tissue of the leaf

verrucose: warty

whorl: a ring-like arrangement of similar parts arising from a common point or node

Medical Glossary

abortifacient: inducing abortion

abscesses: areas with necrotic liquid or pus accumulation

aching: causing physical pain or distress

acne: a disorder of the skin caused by inflammation of the skin glands and hair follicles

acrodynia: a disease of infancy and early childhood marked by pain and swelling in, and pink coloration of, the fingers and toes and by listlessness, irritability, failure to thrive, profuse, perspiration, and sometimes scarlet coloration of the cheeks and tip of the nose

adenitis: inflammation of a gland

AIDS: Acquired Immune Deficiency Syndrome

albuminuria: presence of serum albumin (protein) in the urine

aldosteronism: hyperaldosteronism; an abnormality of electrolyte balance caused by excessive secretion of aldosterone

alexipharmic: medicine that is intended to counter the effects of poison; an antidote to poison or infection

alexiteric: protecting against infection, venom and poison

alopecia: hair loss; baldness; absence of hair from skin areas where it is normally present

amylase: an enzyme that catalyses the hydrolysis of starch into sugars

analgesic: an agent that reduces or eliminates pain

anaphylaxis: hypersensitivity resulting from sensitisation following prior contact with the causative agent

anasarca: generalised massive oedema

angiogenic: an agent that promotes the growth of new blood vessels from pre-existing vessels

anodyne: an agent that relieves pain

anorexia: lack or loss of appetite for food

anthelmintic: an agent that destroys or expels parasitic intestinal worms

anthrax: a serious disease caused by *Bacillus anthracis*, a bacterium that forms spores

antianaphylactic: an agent that prevents anaphylaxis (exaggerated reaction of an organism to which it has previously become sensitised)

antiangiogenic: preventing the growth of new blood vessels

antianxiety: allaying anxiety

antiapoptotic: preventing genetically programmed cell death

antiarrhythmic: preventing or alleviating irregular heart beats

antiatherogenic: an agent that is capable of stopping the formation of fat deposits in blood vessels

antibacterial: an agent that destroys or stops the growth of bacteria

anticancer: reducing the frequency or rate of uncontrolled cell growth

anticarcinogenic: describing an agent that will stop the formation of a cancer

anticholinergic: opposing or blocking the physiological action of a neuro-transmitter, acetylcholine, which contract muscles and causing excitatory actions in the central nervous system

anticoagulant: acting to prevent clotting of blood

anticonvulsant: an agent that prevents or relieves seizures or epileptic attack

antidepressant: an agent that stimulates the mood of a patient

antidiabetic: an agent that controls high blood sugar

antidiarrhoeal: substances used to prevent or treat diarrhoea

antidotal: counteracting the effects of a poison

antiemetic: an agent that prevents or alleviates nausea and vomiting

antifertility: agent that is capable of reducing or adversely affecting fertility

antifilarial: an agent that counters nematode (roundworm) of the superfamily Filarioidea

antifungal: an agent that destroys or inhibits the growth of fungi

antigenotoxic: an agent that is capable of preventing damage to DNA molecules in genes, causing mutations, tumours, etc.

antihyaluronidase: directed or effective against an enzyme (hyaluronidase) that breaks down hyaluronic acid, thus facilitating the spread of fluid through tissues

antihypertensive: an agent that controls high blood pressure

anti-inflammatory: an agent that counteracts inflammation

antileishmanial: directed or effective against leishmania, a parasitic protozoan that causes leishmaniasis

antimalarial: an agent that prevents or treat the disease malaria caused by the Plasmodium parasite

antimicrobial: destroying or inhibiting the growth of microorganisms

antimitotic: pertaining to certain substances capable of stopping cell division (mitosis)

antimutagenic: directed or effective against substances that tends to increase the frequency or extent of mutation

antineoplastic: inhibiting or preventing the growth and spread of neoplasms or cancerous cells

antinociceptive: relieving pain

antiosteoporotic: directed or effective against osteoporosis (characterised by decrease in bone mass with increased porosity and brittleness)

antioxidant: inhibiting oxidation or an agent that does so

antioxidative: protect body cells from the damaging effects of oxidation

antiplasmodial: directed or effective against the malarial parasites (Plasmodium species)

antiplatelet: inhibiting or stopping the formation of platelet aggregation

antiproliferative: inhibiting cell growth

antiprotozoal: tending to destroy or inhibit the growth of protozoans

antipruritic: preventing or relieving itching

antipsychotic: any of the powerful tranquilisers used especially to treat psychosis and believed to act by blocking dopamine nervous receptors

antipyretic: reducing fever or an agent that reduces fever

antirheumatic: relieving or preventing rheumatism

antischistosomal: directed or effective against infection caused by trematode worms (schistosomes) which are parasitic in the blood of humans and other mammals

antiseptic: any substance that inhibits the growth and reproduction of microorganisms

antispasmodic: an agent that prevents or relieves muscle spasms

antithyroidal: opposing thyroid function

antitrichomonal: directed or effective against some protozoa (Trichomonas)

antituberculosis: effective in the treatment of tuberculosis, a condition due to *Mycobacterium tuberculosis*

antitumour: inhibiting the growth of tumour cells

antitussive: effective against cough; an agent that suppresses coughing

antiulcer: prevent or cure ulcers or irritation of the gastrointestinal tract

antiulcerogenic: directed or effective against development of ulcers

antivenom: an agent used in the treatment of poisoning by animal toxins

antiviral: acting, effective, or directed against viruses

anuria: lack of urine production

anxiolytic: relieving anxiety or an agent that relieves anxiety

aperient: a mild laxative or gentle purgative

aphrodisiac: an agent that stimulates sexual desire

apnoea: a condition whereby breathing is stopped

apoptotic: a term used to describe the state of programmed cell death

appendicitis: inflammation of the appendix

arrhythmia: abnormal heart beat

arthritis: inflammation of joints

ascites: effusion and accumulation of fluid in the abdominal cavity

asthma: a chronic, inflammatory lung disease characterised by recurrent breathing problems usually triggered by allergens

astringent: having the property of causing contraction of soft organic tissues for the control of bleeding or secretions

atherosclerosis: thickening, hardening and loss of elasticity of the walls of arteries

athlete's foot: fungal infections on the feet, often smelly

atrophy: a wasting away; reduction in the size of a cell, tissue or organ

bacillary: pertaining to the bacteria bacilli or to rod-like structures

bechic: pertaining to cough

beriberi: a disease due to vitamin B1 deficiency, marked by inflammation of nerves, heart problems and oedema

bilious: relating to or containing bile

blennorrhoea: any free discharge of mucus, especially from urethra or vagina

blisters: a vesicle

boils: furuncle; tender, swollen areas of infection that usually form around hair follicles

bronchial: pertaining to or affecting one or both windpipe

bronchiolitis obliterans: a disease of the lungs where the bronchioles are inflamed and plugged with granulation tissue

bronchitis: inflammation of mucous membrane of the windpipe (bronchus)

cachexia: general ill health and malnutrition

calculus: abnormal hard deposit, usually composed of mineral salts, occurring within the body

cancer: a malignant tumour of growth that expands locally by invasion and systemically by metastasis

carbuncles: swollen lumps or masses under the skin due to skin infections that often involve a group of hair follicles

cardiac: pertaining to the heart

cardioprotective: protecting the heart function

cardiotonic: an agent that increases tonicity of the heart muscles

carminative: an agent that relieves and removes gas from the digestive system

cataracts: conditions in which there is a loss or reduction of transparency of the lens of the eye causing progressive loss of clarity and detail of images

catarrh: a condition of the mucous membranes characterised by inflammation and mucous

cathartic: an active purgative, producing bowel movements and evacuation of the bowels

cephalgia: headache

chemopreventive: a chemical agent that can help prevent the development of cancer

chilblains: inflammation or blister of the hands and feet caused by exposure to cold or moisture

cholera: a form of intestinal infection that results in frequent watery stools, cramping abdominal pain and eventual collapse from dehydration

cicatrizant: a medicine or application that promotes the healing of a sore or wound, or the formation of a cicatrix, which is a scar left by the formation of new connective tissue over a healing sore or wound

coagulant: an agent promoting clotting of blood

cold sore: a recurrent, small blister in and around the mouth caused by the herpes simplex virus

colic: acute paraxysmal abdominal pain or pertaining to the colon

coma: a state of profound unconsciousness from which the patient cannot be aroused, even by powerful stimuli

condiment: something used to give a special flavour to food, as mustard, ketchup, salt, or spices

congestive heart failure: the heart fails to pump efficiently, resulting in swelling, shortness of breath, weakness, etc.

conjunctivitis: inflammation of the conjunctiva, generally associated with discharge

constipation: infrequent or difficult evacuation of faeces

contraceptive: an agent capable of preventing the formation of a foetus

contusion: injury to a part without a break in the skin; a bruise

convulsions: spasms, epilepsy

coolant: an agent that produces cooling

coryza: profuse discharge from the mucous membrane of the nose

counter-irritant: an agent producing counter-irritation, so that less pain at a particular site is experienced

cramps: a painful spasmodic muscular contraction

Crohn's disease: a chronic inflammatory disease of the digestive tract, particularly the small intestine and colon

cystitis: inflammation of the urinary bladder

cytotoxic: toxic to cells, preventing their reproduction or growth

dehydration: depletion of body fluids

delirium: mental state characterised by reduced ability to maintain attention to external stimuli and disorganised thinking

demulcent: an agent that soothes and protects the part to which it is applied

deobstruent: removes obstructions by ducts or pores of the body

depressant: an agent that diminishes any functional activity

depurative: purifying the blood or the humours

dermatitis: inflammation of the skin

dermatosis: any skin disease, especially one not characterised by inflammation

diabetes: metabolic disorder resulting in high blood sugar and discharge of large amounts of sugar in the urine.

diaphoretic: an agent that increases perspiration

diarrhoea: abnormally frequent discharge of watery stools

digitalis: the dried leaf of *Digitalis purpurea*; used as a tonic for the heart

dipsia: thirst

discutient: an agent that causes the dispersal or causes something, such as a tumour or any pathologic accumulation, to disappear

diuretic: an agent that increases the production of urine

dropsy: an abnormal accumulation of serous fluid in cellular tissues or in a body cavity

dysentery: infection of the gut caused by a bacterium called shigella, characterised by abdominal pain, diarrhoea with passage of mucus or blood

dysmenorrhoea: difficult and painful menstruation

dyspepsia: indigestion

dyspnoea: difficult or laboured breathing

dysrhythmias: an abnormality in an otherwise normal rhythmic pattern

dysuria: painful or difficult urination

ecchymoses: small haemorrhagic spots in the skin or mucous membrane, forming a non-elevated, rounded or irregular, blue or purplish patch

eczema: an inflammatory condition of the skin characterised by redness, itching, scales, crusts or scabs alone or in combination

elephantiasis: a disease caused by tumours in lymph node or filarial worms, causing excessive swelling in limbs and genitalis

embrocations: alcohol-based treatments rubbed into the skin to relieve pain or which produce reddening of the skin

embryotoxicity: any toxicity that affects an embryo

emetic: an agent that induces vomiting

emmenagogue: a substance that promotes or assists the flow of menstrual fluid

emollient: an agent that will soften, soothe and protect the part when applied locally

enteritis: inflammation of the intestine, especially the small intestine

epigastric: pertaining to the upper and middle region of the abdomen

epilepsy: convulsions

epistaxis: nosebleed

eruptions: lesions on the skin that are usually red, raised, and easily visible

erysipelas: is an acute streptococcus bacterial infection marked by deep red inflammation of the skin and mucous membranes. This disease is also known as Saint Anthony's fire

erythema: redness of the skin due to congestion of the capillaries

excrescences: abnormal outgrowth; a projection of morbid origin

expectorant: an agent that promotes the discharge or expulsion of mucus from the respiratory tract

febrifuge: an agent that reduces fever

fertility: the capacity to conceive or induce conception

fibroids: non-cancerous tumours made of muscle cells and tissues that grow in and around the wall of the uterus or womb

fibrosis: an abnormal formation of fibrous tissue as a reparative or reactive process, as opposed to the formation of fibrous tissue that is a normal constituent of an organ or tissue

filaricidal: an agent that kills filariae, nematodes that as adults are parasites in the blood or tissues of mammals and as larvae usually develop in biting insects

fistulae: abnormal connections between an organ, vessel or intestine and another structure, usually due to injury, surgery or result from infection or inflammation

flatulence: the presence of excessive gas in the intestinal tract, causing discomfort

fluxes: excessive flow or discharge

furuncle: infection of a hair follicle

furunculosis: the persistent sequential occurrence of furuncles over a period of weeks or months or the simultaneous occurrence of a number of furuncles

galactophoritis: inflammation of the milk ducts

galactorrhoea: excessive or spontaneous milk flow; persistent secretion of milk irrespective of nursing

gangrenous: characterised by the decay of body tissues, which become black and smelly

gastritis: inflammation of the stomach

gastroenteritis: inflammation of the lining membrane of the stomach and the intestinal tract

gastroprotective: protect the stomach

gavage: forced feeding, especially through a tube passed into the stomach

genotoxic: a toxic agent that damages DNA molecules in genes, causing mutations, tumours, etc.

glottal: pertaining to the vocal apparatus of the larynx, consisting of the true vocal cords and the opening between them

glucosidase: an enzyme of the hydrolase class that breaks down a glucoside

gonorrhoea: infectious sexual disease due to *Neisseria gonorrhoeae*

gout: a metabolic disease that is a form of acute arthritis, marked by inflammation of the joints and great pain

gravel: calculi occurring in small particles

haemaglutination: agglutination of red blood cells

haematemesis: the vomiting of blood

haematometra: an accumulation of blood in the uterus

haematuria: the finding of blood in the urine

haemiplegia: paralysis of one side of the body

haemolytic: lysis of red blood cells liberating haemoglobin in the plasma

haemoptysis: coughing up of blood from the airways

haemorrhage: profuse bleeding from the blood vessels

haemorrhoids: piles

haemostatic: stop bleeding

heartburn: painful, burning feeling in the chest caused by stomach acid flowing back into the esophagus

hepatitis: inflammation of the liver, usually due to viruses or toxins

hepatoprotective: protecting the liver functions

hepatotoxic: an agent that is toxic to liver cells

herpes: inflammation of the skin or mucuous membrane with clusters of deep seated vesicles; a family of viruses that infect humans: herpes simplex causes lip and genital sores; herpes zoster causes shingles

hiccups: sharp inspiratory sound with spasm of the glottis and diaphragm

HIV: Human Immunodeficiency Virus, the virus that causes AIDS

hydragogue: producing watery discharge, especially from the bowels or a cathartic that causes watery purgation

hydrocele: swelling of and fluid on the testicles

hydrothorax: a collection of serous fluid within the pleural cavity

hypercholesterolaemic: high level of cholesterol in the blood

hyperemesis: excessive vomiting

hyperkalemia: high concentration of potassium in the blood

hypertension: high blood pressure

hypocholesterolaemic: an abnormal deficiency of cholesterol in the blood

hypoglycaemic: producing a decrease in the blood sugar level

hypolipidaemic: producing or resulting from a decrease in the level of lipids in the blood

hypotension: low blood pressure

hypotensive: marked by low blood pressure

immunoglobulin: a protein of animal origin with known antibody activity

immunomodulatory: having the ability to modify the immune response or function of the immune system

immunostimulatory: the ability to stimulate the immune response

immunosuppressive: being able to reduce the immune response

impetigo: a highly contagious skin infection caused by bacteria, usually occurring around the nose and mouth; commonly occurring in children

infertility: diminution or absence of ability to produce offspring

inflammation: localised protective reaction of tissue to irritation, injury, infection, chemicals, electricity, heat, cold or microorganisms. Characterised by pain, redness, swelling and possible loss of function

influenza: the flu; an infectious viral respiratory disease characterised by chills, fever, prostration, headache, muscle aches, sore throat, and a dry cough

inotropic: affecting the force or energy of muscular contractions

insanity: a legal term for mental illness, roughly equivalent to psychosis and implying inability to be responsible for one's acts

insomnia: inability to sleep

intoxication: poisoning

jaundice: yellowish discoloration of the whites of the eyes, skin and mucous membranes caused by deposition of bile salts in these tissues

keratitis: inflammation of the cornea

lactogogue: a substance which stimulates the flow of milk

lameness: a state of being incapable of normal locomotion

larvicidal: larvae killing (especially of parasites)

laryngitis: inflammation of the lining of the larynx causing hoarseness

laxative: having a tendency to loosen or relax, specifically in relieving constipation or an agent that relieves constipation

leishmaniasis: infection with Leishmania, a parasitic protozoa

leprosy: a chronic mycobacterial disease caused *Mycobacterium leprae*, characterised by skin lesions and necrosis

lethargy: a condition of indifference or drowsiness

leucoderma: partial or total loss of skin pigmentation, often occurring in patches. Also called vitiligo.

leucodermic: an agent that can cause partial or total loss of skin pigmentation, often occurring in patches. Also called vitiligo

leucorrhoea: a gynecologic disorder resulting in abnormal, thick, whitish, non-bloody discharge from the genital tract (uterus of vagina)

leukemia: a progressive, maglinant disease of the blood-forming organs, marked by distorted proliferation and development of leukocytes and their precursors in the blood and bone marrow

lipolytic: an agent that breaks down fats

lumbago: pain in the lumbar or loin region

lymphadenopathy: enlargement of lymph nodes usually associated with inflammation or infection

malaria: a febrile disease caused by infection with Plasmodial parasites, causing periodic attacks of chills, fever and sweating

mange: a skin disease of domestic animals or pets due to mites

mastitis: inflammation of the breast

measles: acute highly infectious viral human disease caused by a virus, specifically a paramyxovirus of the genus Morbillivirus, characterised by cough, fever, small red lesions each with whitish centre, formed in the mouth in early stages of the measles

melancholic: a depressed and unhappy emotional state with abnormal inhibition of mental and bodily activity

menorrhagia: excessive or prolonged menstruation

metritis: inflammation of the uterus

metrorrhagia: uterine bleeding; usually abnormal amount, occurring at completely irregular intervals, the period of flow sometimes being prolonged

molluscicidal: effective for destroying molluscs

mumps: infectious disease caused by a paramyxovirus, marked by fever and inflammation of parotid gland and swelling of the neck and throat

mydriasis: pupil dilation

myocardial: referring to the middle and the thickest layer of the heart wall, composed of cardiac muscles

narcotic: a drug or agent which in moderate doses depresses the central nervous system, relieving pain and producing sleep but in large doses, produces unconsciousness, stupor, coma and possibly death

nausea: an unpleasant sensation vaguely referred to the epigastrium and abdomen, with a tendency to vomit

necrosis: the morphological changes indicative of cell death

neonatal: pertaining to the first four weeks after birth

nephritis: inflammation of the kidneys

nephropathy: disease of the kidneys

nephroprotective: an agent that protects or prevents damage to the kidney cells

nephrotoxic: toxic to kidney cells

neuralgia: pain occurring in the area served by a sensory nerve

neurasthenia: nervous debility dependent upon impairment in the functions of the spinal cord

neurodermatitis: a general term for a dermatosis presumed to be caused by itching due to emotional causes

neuroleptics: effects on cognition and behavior of antipsychotic drugs, which produce a state of apathy, lack of initiative and limited range of emotion

neuroprotective: an agent that protects or prevents damage to the nerves

neurotoxic: toxic to the nerves or nervous tissue

neurotoxicity: the quality of exerting a destructive or poisonous effect on nerves

nits: the eggs or youngs of a parasitic insect, such as a lice

nodal rhythm: a type of heart rhythm disorder; the cardiac rhythm that results when the heart is controlled by the atrioventricular node in which the impulse arises in the atrioventricular node, ascends to the atria, and descends to the ventricles more or less simultaneously

non-teratogenic: referring to substance which will not cause malformations of an embryo or fetus

oedema: the presence of abnormally large amounts of fluid in the intercellular tissue spaces of the body

oesophagitis: inflammation of the esophagus

oliguria: diminished urine secretion in relation to fluid intake

ophthalmia: severe inflammation of the eye

osteoclasis: surgical fracture or refracture of bones

osteodynia: ostealgia; pain in the bones

otitis: inflammation of the ear

oxytocic: an agent that promotes rapid labour by stimulating contractions of the uterus

oxyuriasis: infection with *Enterobius vermicularis* (in humans) or with other worms

pacemaker: a device that regulates the rhythm of the heart beat

palsy: paralysis

panacea: a remedy for all diseases

parasiticidal: destructive to the parasites

parturition: the act or process of giving birth to a child

periodontitis: inflammation of the tissues surrounding and supporting the teeth

pertussis: whooping cough

pharyngitis: inflammation of the pharynx (throat)

photoprotective: protecting against harmful effects of lights (UV-irradiation)

phototoxicity: toxic effect triggered by exposure to light

piles: haemorrhoids

piscicidal: a substance that kills fish

pleurisy: inflammation of the serous membrane investing the lungs and lining the walls of the thoracic cavity

pneumonia: inflammation of the lungs due to a bacterial or viral infection, which causes fever, shortness of breath, and the coughing up of phlegm

pollakiuria: abnormally frequent urination

postpartum: occurring after childbirth, with reference to the mother

prolapse: the falling down, or downward displacement, of a part or viscus

prophylactic: an agent that prevents or protects against a disease or condition

prurigo: any of several itchy skin eruptions in which the characteristic lesion is dome-shaped with a small transient vesicle on top, followed by crusting or lichenification

psoriasis: a chronic skin disease characterised by inflammation of the skin and formation of red patches

puerperal: pertaining to a woman who has just given birth to a child or to the period or state of confinement after childbirth

purgative: an agent that will cause evacuation of the intestinal contents

pustule: a small, circumscribed elevation of the skin containing pus

radioprotective: serving to protect or aiding in protecting against the harmful effects of radiation

rectocele: hernial protrusion of part of the rectum into the vagina

refrigerant: an agent that produces coolness or reduces fever, allays thirst and gives a sensation of coolness to the system.

resolvent: promoting resolution or the dissipation of a pathologic growth

retching: strong involuntary effort to vomit

Reye's syndrome: a rare disorder in children and teenagers while recovering from childhood infections, such as chicken pox, flu, and other viral infections. Reye's syndrome include nausea, severe vomiting, fever, lethargy, stupor, restlessness, and possibly delirium. Also caused by taking aspirin in children less than 16 years old

rheumatism: general term for acute and chronic conditions characterised by inflammation (arthritis, tendonitis and bursitis), soreness and stiffness of muscles, and pain in joints

rhinitis: inflammation of the nasal mucous membrane

ringworm: any of a number of contagious skin diseases caused by certain parasitic fungi and characterised by the formation of ring-shaped eruptive patches

rubefacient: an agent that produces a mild irritation, reddening of the skin, and local vasodilation increasing the blood supply to the area of application.

saluretics: agents that promotes urinary excretion of sodium and chloride ions

scabies: a contagious skin disease characterised by itching, inflammation, hair loss and secondary bacterial infection.

schistosomiasis: infection with Schistosoma (blood flukes) which causes infection in man by penetrating the skin of people coming in contact with infected waters

sclerosis: an induration or hardening, especially from inflammation and in diseases of the interstitial tissues

scrofula: primary tuberculosis of the cervical lymph nodes; the inflamed structures being subject to a cheesy degeneration

scurvy: disease characterised by spongy bleeding gums, loosening of teeth, and mucous membranes that is caused by lack of Vitamin C

sedative: tending to calm, moderate, or transquilise nervousness or excitement or an agent that does so

senility: the physical and mental deterioration associated with old age

septicaemia: blood poisoning

sialogogue: an agent that stimulates the flow of saliva

sinus bradycardia: a normal but slow heart rhythm

sinusitis: inflammation of a sinus

smallpox: an acute infectious disease due to poxvirus, marked by sustained high fever and the appearance of skin eruptions and pustules, leaving small, depressed, depigmented scars

sores: any lesion of the skin or mucous membranes

spasm: a sudden, violent, involuntary muscular contraction

spastic paraparesis: partial paralysis of the lower extremities due to muscular stiffness and spasms

spermatogenesis: formation and development of sperms

spermatorrhoea: involuntary escape of semen, without orgasm

spermatotoxic: an agent that is toxic to spermatozoa

splenitis: inflammation of the spleen

splenosis: Implantation and subsequent growth of splenic tissue within the abdomen as a result of disruption of the spleen

sprue: a chronic form of malabsorption syndrome

stammering: a disorder of speech behavious marked by involuntary pauses in speech

stasis: stoppage of flow, as of blood or other body fluid

stimulant: an agent that excites the functional activity of an organ or system

stomachic: an agent that stimulates the appetite and gastric secretion

strabismus: squint

sty: a circumscribed abscess caused by bacterial infection of the glands on the edge of the eyelid

styptic: astrigent or arresting haemorrhage by means of an astringent quality

sudorific: an agent that promotes sweating; diaphoretic

sunstroke: a condition caused by excessive exposure to the sun, marked by high skin temperature, convulsions and coma

suppository: an easily fusible medicated mass to be introduced into a body orifice

syphilis: a contagious sexual disease caused by *Treponema pallidum*, characterised by local formation of ulcerous skin eruptions and systemic infection

tachycardia: increased heart beat

tachypnea: very rapid respiration

taeniasis: infection with tapeworms

taenifuge: an agent that expels tapeworms

teratogenic: an agent that that causes physical defects in the developing embryo

tetanus: an acute, sometimes fatal, disease of the central nervous system; caused by the toxin of the tetanus bacterium

thrush: candidiasis of the oral mucous membranes, fungal infection by *Candida albicans* with formation of whitish spots

tonic: producing and restoring normal tone or characterised by continuous tension

tonsillitis: inflammation of the tonsils

trachoma: a contagious disease of the conjunctiva and cornea, producing photophobia, pain, and lacrimation, caused by *Chlamydia trachomatis*

tuberculosis: any of the infectious diseases of man and other animals due to species of Mycobacterium and marked by formation of tubercles and caseous necrosis in tissues of any organs, usually the lung

tumour: an abnormal growth of tissue resulting from uncontrolled multiplication of cells and serving no physiological function

tympanites: abnormal distention due to the presence of gas or air in the intestine or the peritoneal cavity

typhoid: acute infectious disease caused by *Salmonella typhii* and characterised by fever, severe physical and mental depression, diarrhoea and headache

tyrosinase: an oxidising enzyme, occurring in plant and animal tissues, that catalyses the aerobic oxidation of tyrosine into melanin and other pigments

ulcerative colitis: a chronic inflammatory disease of the colon or colon and rectum, characterised by abdominal pain and diarrhoea, often mixed with blood and mucus

urinary calcification: the deposit of calcium salts in the tissues of the urinary tract

urinary lithiasis: the formation of urinary stones

urodynia: pain in urination

urticaria: an allergic disorder marked by raised patches of skin causing itchy swellings of the skin

uteral: relating to uterus

uterine haemorrhage: bleeding in the uterus

uterotonic: increasing the tone of uterine muscle

vaginismus: painful spasm of the vagina, hence often resulting in unsuccessful penetration

varices: enlarged tortuous vein, artery or lymphatic vessel. *adj.* varicose

vasodilation: dilation of blood vessels

vasorelaxant: an agent that relaxes the blood vessels

venereal: due to or propagated by sexual intercourse

vermifuge: an agent that causes the expulsion of intestinal worms

verruca: a wart or one of the wart-like elevations on the endocardium in various types of endocarditis

virility: the condition or quality of being virile

vitiligo: a condition in which destruction of melanocytes in small or large circumscribed areas results in patches of depigmentation often having a hyperpigmented border, and often enlarging slowly

vulnerary: a remedy useful in healing wounds

wandering pacemaker: an atrial arrhythmia that occurs when the natural cardiac pacemaker site shifts between the sino-atrial node, the atria and/or the atrioventricular node

warts: abnormal growths on the skin due to viral infection caused by human papillomavirus

whitlow: an infection of the fingers caused by herpes simplex virus

whooping cough: a bacterial infection especially of the children caused by bacterium *Bordetella pertussis* marked by convulsive spasmodic coughs

Index

www.ingramcontent.com/pod-product-compliance
Lightning Source LLC
Chambersburg PA
CBHW060334220326
41598CB00023B/2700